W9-DIZ-483

Shaft Alignment Handbook

MECHANICAL ENGINEERING

A Series of Textbooks and Reference Books

L. L. Faulkner

Columbus Division, Battelle Memorial Institute
and Department of Mechanical Engineering
The Ohio State University
Columbus, Ohio

1. *Spring Designer's Handbook*, Harold Carlson
2. *Computer-Aided Graphics and Design*, Daniel L. Ryan
3. *Lubrication Fundamentals*, J. George Wills
4. *Solar Engineering for Domestic Buildings*, William A. Himmelman
5. *Applied Engineering Mechanics: Statics and Dynamics*, G. Boothroyd and C. Poli
6. *Centrifugal Pump Clinic*, Igor J. Karassik
7. *Computer-Aided Kinetics for Machine Design*, Daniel L. Ryan
8. *Plastics Products Design Handbook, Part A: Materials and Components; Part B: Processes and Design for Processes*, edited by Edward Miller
9. *Turbomachinery: Basic Theory and Applications*, Earl Logan, Jr.
10. *Vibrations of Shells and Plates*, Werner Soedel
11. *Flat and Corrugated Diaphragm Design Handbook*, Mario Di Giovanni
12. *Practical Stress Analysis in Engineering Design*, Alexander Blake
13. *An Introduction to the Design and Behavior of Bolted Joints*, John H. Bickford
14. *Optimal Engineering Design: Principles and Applications*, James N. Siddall

15. *Spring Manufacturing Handbook*, Harold Carlson
16. *Industrial Noise Control: Fundamentals and Applications*, edited by Lewis H. Bell
17. *Gears and Their Vibration: A Basic Approach to Understanding Gear Noise*, J. Derek Smith
18. *Chains for Power Transmission and Material Handling: Design and Applications Handbook*, American Chain Association
19. *Corrosion and Corrosion Protection Handbook*, edited by Philip A. Schweitzer
20. *Gear Drive Systems: Design and Application*, Peter Lynwander
21. *Controlling In-Plant Airborne Contaminants: Systems Design and Calculations*, John D. Constance
22. *CAD/CAM Systems Planning and Implementation*, Charles S. Knox
23. *Probabilistic Engineering Design: Principles and Applications*, James N. Siddall
24. *Traction Drives: Selection and Application*, Frederick W. Heilich III and Eugene E. Shube
25. *Finite Element Methods: An Introduction*, Ronald L. Huston and Chris E. Passerello
26. *Mechanical Fastening of Plastics: An Engineering Handbook*, Brayton Lincoln, Kenneth J. Gomes, and James F. Braden
27. *Lubrication in Practice: Second Edition*, edited by W. S. Robertson
28. *Principles of Automated Drafting*, Daniel L. Ryan
29. *Practical Seal Design*, edited by Leonard J. Martini
30. *Engineering Documentation for CAD/CAM Applications*, Charles S. Knox
31. *Design Dimensioning with Computer Graphics Applications*, Jerome C. Lange
32. *Mechanism Analysis: Simplified Graphical and Analytical Techniques*, Lyndon O. Barton

33. *CAD/CAM Systems: Justification, Implementation, Productivity Measurement*, Edward J. Preston, George W. Crawford, and Mark E. Coticchia
34. *Steam Plant Calculations Manual*, V. Ganapathy
35. *Design Assurance for Engineers and Managers*, John A. Burgess
36. *Heat Transfer Fluids and Systems for Process and Energy Applications*, Jasbir Singh
37. *Potential Flows: Computer Graphic Solutions*, Robert H. Kirchhoff
38. *Computer-Aided Graphics and Design: Second Edition*, Daniel L. Ryan
39. *Electronically Controlled Proportional Valves: Selection and Application*, Michael J. Tonyan, edited by Tobi Goldoftas
40. *Pressure Gauge Handbook*, AMETEK, U.S. Gauge Division, edited by Philip W. Harland
41. *Fabric Filtration for Combustion Sources: Fundamentals and Basic Technology*, R. P. Donovan
42. *Design of Mechanical Joints*, Alexander Blake
43. *CAD/CAM Dictionary*, Edward J. Preston, George W. Crawford, and Mark E. Coticchia
44. *Machinery Adhesives for Locking, Retaining, and Sealing*, Girard S. Haviland
45. *Couplings and Joints: Design, Selection, and Application*, Jon R. Mancuso
46. *Shaft Alignment Handbook*, John Piotrowski
47. *BASIC Programs for Steam Plant Engineers: Boilers, Combustion, Fluid Flow, and Heat Transfer*, V. Ganapathy
48. *Solving Mechanical Design Problems with Computer Graphics*, Jerome C. Lange
49. *Plastics Gearing: Selection and Application*, Clifford E. Adams
50. *Clutches and Brakes: Design and Selection*, William C. Orthwein

51. *Transducers in Mechanical and Electronic Design*, Harry L. Trietley
52. *Metallurgical Applications of Shock-Wave and High-Strain-Rate Phenomena*, edited by Lawrence E. Murr, Karl P. Staudhammer, and Marc A. Meyers
53. *Magnesium Products Design*, Robert S. Busk
54. *How to Integrate CAD/CAM Systems: Management and Technology*, William D. Engelke
55. *Cam Design and Manufacture: Second Edition; with cam design software for the IBM PC and compatibles*, disk included, Preben W. Jensen
56. *Solid-State AC Motor Controls: Selection and Application*, Sylvester Campbell
57. *Fundamentals of Robotics*, David D. Ardayfio
58. *Belt Selection and Application for Engineers*, edited by Wallace D. Erickson
59. *Developing Three-Dimensional CAD Software with the IBM PC*, C. Stan Wei
60. *Organizing Data for CIM Applications*, Charles S. Knox, with contributions by Thomas C. Boos, Ross S. Culverhouse, and Paul F. Muchnicki
61. *Computer-Aided Simulation in Railway Dynamics*, by Rao V. Dukkipati and Joseph R. Amyot
62. *Fiber-Reinforced Composites: Materials, Manufacturing, and Design*, P. K. Mallick
63. *Photoelectric Sensors and Controls: Selection and Application*, Scott M. Juds
64. *Finite Element Analysis with Personal Computers*, Edward R. Champion, Jr., and J. Michael Ensminger
65. *Ultrasonics: Fundamentals, Technology, Applications: Second Edition, Revised and Expanded*, Dale Ensminger
66. *Applied Finite Element Modeling: Practical Problem Solving for Engineers*, Jeffrey M. Steele

67. *Measurement and Instrumentation in Engineering: Principles and Basic Laboratory Experiments*, Francis S. Tse and Ivan E. Morse
68. *Centrifugal Pump Clinic: Second Edition*, Revised and Expanded, Igor J. Karassik
69. *Practical Stress Analysis in Engineering Design: Second Edition, Revised and Expanded*, Alexander Blake
70. *An Introduction to the Design and Behavior of Bolted Joints: Second Edition, Revised and Expanded*, John H. Bickford
71. *High Vacuum Technology: A Practical Guide*, Marsbed H. Hablanian
72. *Pressure Sensors: Selection and Application*, Duane Tandeske
73. *Zinc Handbook: Properties, Processing, and Use in Design*, Frank Porter
74. *Thermal Fatigue of Metals*, Andrzej Weronski and Tadeuz Hejwowski
75. *Classical and Modern Mechanisms for Engineers and Inventors*, Preben W. Jensen
76. *Handbook of Electronic Package Design*, edited by Michael Pecht
77. *Shock-Wave and High-Strain-Rate Phenomena in Materials*, edited by Marc A. Meyers, Lawrence E. Murr, and Karl P. Staudhammer
78. *Industrial Refrigeration: Principles, Design and Applications*, P. C. Koelet
79. *Applied Combustion*, Eugene L. Keating
80. *Engine Oils and Automotive Lubrication*, edited by Wilfried J. Bartz
81. *Mechanism Analysis: Simplified and Graphical Techniques, Second Edition, Revised and Expanded*, Lyndon O. Barton
82. *Fundamental Fluid Mechanics for the Practicing Engineer*, James W. Murdock
83. *Fiber-Reinforced Composites: Materials, Manufacturing, and Design, Second Edition, Revised and Expanded*, P. K. Mallick

84. *Numerical Methods for Engineering Applications*, Edward R. Champion, Jr.
85. *Turbomachinery: Basic Theory and Applications, Second Edition, Revised and Expanded*, Earl Logan, Jr.
86. *Vibrations of Shells and Plates: Second Edition, Revised and Expanded*, Werner Soedel
87. *Steam Plant Calculations Manual: Second Edition, Revised and Expanded*, V. Ganapathy
88. *Industrial Noise Control: Fundamentals and Applications, Second Edition, Revised and Expanded*, Lewis H. Bell and Douglas H. Bell
89. *Finite Elements: Their Design and Performance*, Richard H. MacNeal
90. *Mechanical Properties of Polymers and Composites: Second Edition, Revised and Expanded*, Lawrence E. Nielsen and Robert F. Landel
91. *Mechanical Wear Prediction and Prevention*, Raymond G. Bayer
92. *Mechanical Power Transmission Components*, edited by David W. South and Jon R. Mancuso
93. *Handbook of Turbomachinery*, edited by Earl Logan, Jr.
94. *Engineering Documentation Control Practices and Procedures*, Ray E. Monahan
95. *Refractory Linings: Thermomechanical Design and Applications*, Charles A. Schacht
96. *Geometric Dimensioning and Tolerancing: Applications and Techniques for Use in Design, Manufacturing, and Inspection*, James D. Meadows
97. *An Introduction to the Design and Behavior of Bolted Joints: Third Edition, Revised and Expanded*, John H. Bickford
98. *Shaft Alignment Handbook: Second Edition, Revised and Expanded*, John Piotrowski

Additional Volumes in Preparation

Computer-Aided Design of Polymer-Matrix Composite Structures, edited by S. V. Hoa

Friction Science and Technology, Peter Blau

Optimizing the Shape of Mechanical Elements and Structures, A. A. Seireg and Jorge Rodriguez

Introduction to Plastics and Composites: Mechanical Properties and Engineering Applications, Edward Miller

Mechanical Engineering Software

Spring Design with an IBM PC, Al Dietrich

Mechanical Design Failure Analysis: With Failure Analysis System Software for the IBM PC, David G. Ullman

Shaft Alignment Handbook

Second Edition, Revised and Expanded

John Piotrowski
President
Turvac, Inc.
Cincinnati, OH

MARCEL DEKKER, INC. New York • Basel • Hong Kong

Library of Congress Cataloging-in-Publication Data

Piotrowski, John [date]
Shaft alignment handbook.

(Mechanical engineering ; 98)
Includes bibliographies and index.
1. Machinery--Alignment. 2. Shafting. I. Title.
II. Series.

ISBN 0-8247-9666-7

Marcel Dekker, Inc.
270 Madison Ave. New York, NY 10016

Current printing (last digit)
10 9 8 7 6 5 4 3 2 1

PRINTED IN THE UNITED STATES OF AMERICA

This book is dedicated to my wife, Bobbie Jo, who has provided the inspiration and motivation for me to write this book and who has sacrificed thousands of hours of her time to allow me the chance to learn and record this information ... to my children, Tracy, Paula, and Peter, who actually proofread this information and provided me with an "outsider's" view of this material despite the fact that they have interests outside of engineering disciplines ... and to my parents, Joseph and Magdalena, who have taught me that any honorable path will be difficult, but with hard work, discipline, and inspiration ... anything is possible.

Preface

Many of the material conveniences taken for granted in today's society have been made possible by the numerous rotating machinery systems located in every part of the world. Virtually everything we use or consume has somehow been produced or touched in some way by rotating equipment. The multitude of industrial facilities that generate our electricity, extract and deliver our fossil fuels, manufacture our chemicals, produce the food we eat, provide our transportation, furnish the clothing we wear, and mine and refine metals, require millions of pieces of rotating machinery to make all of this happen.

It makes good sense to keep the motors, pumps, gears, turbines, fans, diesels, and compressors running for long periods of time to prevent financial losses due to decreased production and overhaul costs. Repair or replacement of this equipment is expensive and the loss of revenue while this machinery is down can spell the difference between continued prosperity or financial disaster for any company. Keeping these machines running requires a thorough understanding of their design and operating envelope, careful attention during their installation and overhauls, the faculty to prevent or predict imminent failures, and the expertise to modify and enhance existing hardware to extend its operating lifespan.

In the past twenty years, easily half of the rotating equipment problems I have experienced had something to do with misaligned shafts. Additionally, operating rotating equipment under misalignment conditions can be dangerous. I have seen a coupling burst apart on a 500 hp, 3600 rpm process pump that literally sheared a 10" pipe in half and coupling pieces landed 400 yards away. Keep in mind that rotor speeds above 100,000 rpm and drivers pushing 60,000+ hp are now commonplace.

With all of the rotating machinery in existence, you would think that shaft misalignment is well understood and that everyone who is involved with installing, maintaining, and operating this equipment is well versed in preventing this. In fact, just the opposite is true. The information contained in this book is not taught in high schools or junior colleges, is not a required course for mechanical engineers, is not discussed in business schools, and is typically not taught in trade schools for mechanics, millwrights, pipefitters, or electricians.

Over the past 100 years, hundreds of technical books and articles have been published on rotor balancing, flexible coupling design, vibration analysis, structural dynamics, and industrial productivity. The first technical article on shaft alignment, on the other hand, wasn't published until after World War II and not until the late 1960's did anyone begin paying attention to this. Only two universities have conducted any research into shaft alignment and they have had a considerable amount of trouble getting funding from government or industry for continued research.

There are several people who have made valuable contributions to this book, as indicated by the multitude of references at the end of each chapter. None of these people had to do any research in this area nor did they have to write down what they learned for the benefit of everyone else who works in this area ... but they did and I am thankful for that. It is very unsettling to be responsible for machinery that is typically the heart and soul of your operation and realize that you don't know enough about this equipment to keep it running satisfactorily. I have made many mistakes over the years trying to learn about rotating machinery and I wouldn't want you to go through what I had to, so this book is an opportunity to share with you what I have learned.

I have had the pleasure to work with many people in industry over the years and can't remember anyone not wanting to know how to align machinery properly or anyone who purposely wanted to damage machinery because it was operating in a misaligned condition. There are thousands of people who have expended a tremendous effort to get their equipment aligned correctly and very few of the people they work with have a clue what they had to go through to accomplish that goal. They often do it with no acknowledgment when it's done right and no increase in pay. In my opinion, there is no better feeling than walking by a piece of machinery that you aligned ten years ago and knowing that equipment is still running smoothly. These people do it right because it makes them proud of the quality of workmanship they perform and although they might not be able to take that to the bank, they can certainly take satisfaction in it when they reflect back on their accomplishments.

The primary reasons that machinery is misaligned are lack of proper training, improper tools to do the job, and insufficient time allowed to do it right. This book will help with the training and the tools: the rest is entirely up to you.

The first edition of the **Shaft Alignment Handbook** was published in 1986 and contained 278 pages. This edition is over twice that size and was written to include information that was omitted in the first edition and material that has been developed in the field of shaft alignment since that time. Chapters 2, 10, 12, and 13 are new chapters. Chapter 2 (The History of Shaft Alignment) evolved from the research I had done on patents and from my curiosity on how all of this came about. A considerable amount of new information has been added in Chapter 3 on cement, concrete, baseplate design / installation, and piping problems. Several new alignment measurement techniques have been included in Chapter 7 along with several new methods to measure off-line to running machinery movement as found in Chapter 9. In 1986, there was only one company manufacturing laser shaft alignment systems, now there are six so I tried to arrive at a means by which you could compare them so you can understand how they work and select the best system for your needs, assuming you want to purchase one. I have had numerous requests to discuss V-belt alignment over the past few years so Chapter 10 was added. Chapter 12 covers information on aligning complex multiple element drive trains which, to my knowledge, has never been published before. Chapter 13 includes specific information on different types of rotating machinery to assist you in understanding their behavior. A considerable amount of material has been included in Chapter 14 on the vibration behavior of rotating machinery and includes interesting data collected on observing misaligned equipment with infrared thermographic cameras.

In 1986, when I completed the first edition, I thought that nothing more could possibly be published on shaft alignment and here, nine years later I sit thinking the same thing again. You think I would know better by now.

John Piotrowski

Contents

1
The Importance of Shaft Alignment

Misaligned rotating machinery has caused, and will continue to cause, a tremendous financial loss to every industry worldwide. No one has ever really calculated how much money has been wasted on prematurely damaged machinery, lost production, and excessive energy consumption due to shaft misalignment just over the last fifty years. The monetary figure would indeed be staggering considering how many pieces of rotating machinery are operating today in every industry, transmission pipeline, oil field, ocean going ship, hospital, and office complex. To get a picture of just how large a problem this is, ask yourself how many pieces of rotating machinery are within a 100 mile radius of where you are right now? Statistically, over half of those machinery drives are excessively misaligned and will probably need to be shut down and repaired or replaced in the next 16 months. The other half will probably run successfully with little or no maintenance for the next 80 months.

Even today, obtaining accurate shaft alignment on the rotating machinery that exists in every industry has been sporadic at best. For many years, the responsibility for aligning equipment has been left up to millwrights and mechanics. Far too many of these people have not been adequately trained in proper alignment techniques, and they have been subjected to using inadequate tooling to measure the misalignment and move the machinery.

All too frequently there seems to be no interest on the part of management to allow the personnel enough time to do a quality job. Even in organizations where attention is paid to aligning the critical, higher horsepower units, much of the smaller horsepower units are ignored and poorly aligned, resulting in many premature bearing, seal, shaft, or coupling failures.

Who is responsible for alignment of rotating machinery? Shaft alignment should be a major concern to every conscientious manager, engineer, foreman, and mechanic. Each of these people in a typical industrial organization has a role to play. You can't blame dial indicators or lasers, shim stock, foot bolts, pipes, bearings, seals, shafts, couplings, or the rotating machinery itself if the alignment is unsatisfactory. The real problem with alignment usually boils down to 'people' problems.

The mechanic's / millwright's function is to perform the preliminary alignment steps, measure the position of the shafts, determine and perform the proper moves on the machinery to achieve acceptable alignment tolerances, and communicate final alignment results or problems encountered during the alignment job to their supervisor.

The foreman's task is to assign the right people to do the alignment job, insure the necessary and appropriate tools are available and in working order, provide adequate time to complete the job, answer any questions the trades personnel have, provide guidance for potential problems that might occur during the alignment job, and to coordinate and communicate these problems to engineers and managers for resolution, and keep records of what was done.

The engineers are responsible for specifying and procuring the types of measuring tools used to determine the positions of the machinery shafts, provide the technical expertise and tooling to measure off-line to running machinery movement, design piping/duct support mechanisms to minimize induced stress in machine casings or coordinate piping/ducting rework, review new methods and techniques that could be used on the rotating machinery in their plant, analyze failures of rotating machinery to determine if the root cause can be traced to misalignment, listen and respond to any and all problems that were reported to them by the trades personnel and the foremen, provide training to the trades personnel, foremen, and managers, and work side-by-side with the trades personnel, if necessary, to fully understand what actually occurs in an alignment job to determine if more efficient means can be found to improve the alignment process or accuracy.

The plant and/or engineering manager's responsibility is to provide the funds necessary to procure the tools needed to accomplish the job, to insure that the personnel have been given proper training to understand how to do alignment, and to provide due credit to the individuals who have done the alignment job properly.

The capacity to have all your rotating equipment well aligned and running smoothly is directly related to your knowledge, ability, and desire to do it properly. It seems foolish to install well designed or rebuilt machinery only to watch it be destroyed in six to sixteen months because no one wanted to spend six hours carefully attaching rotating machinery to frames and foundations, aligning the shafts, and monitoring the alignment position over time to allow the equipment to run for six to sixty years instead.

Figure 1-1 shows the three key items that must be known to attain accurate and precise alignment.

When aligning rotating machinery there are 3 things you need to know ...

1 - Where is the machinery at when the equipment is not running?
2 - What position will the machinery move to when operating?
3 - If the machinery moves from off-line to running conditions, what acceptable range of positions could the machinery shafts be in when the machinery is aligned off-line and still maintain acceptable alignment tolerances during operation?
or simply ...

- Where are they?
- Where will they go?
- Where should they be?

Misalignment Costs

... are incurred through ...

- lost production
- mechanical degredation
- energy consumption

... measured against your costs to ...

- measure the misalignment
- analyze the situation and determine alignment accuracy
- correct the existing misalignment

Figure 1-1. What you need to know when aligning machinery and what costs are incurred if you do or don't align your equipment.

Tremendous advances have been made just in the past thirty years on diagnosing machinery health using non-destructive mechanical defect detection testing techniques such as vibration analysis, infrared thermography, or oil analysis as 'tools' for pinpointing machinery problems. Being able to discern each of these maladies by analyzing vibration signatures and various other data such as bearing temperatures, and equipment performance data has been one of the fastest growing areas of technology. Machinery problem diagnostic techniques have come a long way from balancing nickels on bearing caps, vibrating light meters, and the human touch. Many companies have based their entire business on monitoring machinery and helping people locate their equipment problems using FFT analyzers, spike energy meters, and a host of other newly developed equipment aimed at understanding the dynamic response of turbomachinery. All of this new equipment and methodology is very impressive, but there is a grave difference between finding a problem with your machinery and fixing a problem with your machinery.

Despite the rapid advances made in these new machinery performance analysis tools, there is still a considerable amount we do not yet understand about the dynamic behavior of machinery that is forced to operate in a misaligned condition. Much of what is said to occur on machinery (e.g. high 1 or 2 times running speed vibration frequency components and high axial vibration) is not always true! Disappointingly, very little research has been conducted on what happens to machinery if it has to run misaligned, and most of the research that has been done was conducted by a few individuals working alone in industry with no funding for their tests and, in some cases, without the blessing or knowledge of the managers in the company where they work!

Surprisingly, 99% of the rotating machinery operating today is misaligned. This may sound like an extremely strong statement but perfect alignment is nearly impossible to achieve in the real world. In fact, a small amount of misalignment is really not that bad. Gear type couplings and shafts with universal joint drives, for example, must have some misalignment in order to keep lubrication at the points of power transmission during rotation. So whenever we deal with alignment, it is important to know when to stop moving the machinery. There comes a point where no beneficial returns can be made if alignment tolerances have adequately been met.

To further complicate the issue, rotating equipment never wants to stay put. As the equipment begins to rotate, a wide variety of factors contribute to moving the shafts somewhere other than where they were when the units were sitting at rest. Heat generated in the casings from frictional losses in bearings, heating elements in lubrication system reservoirs, movement of fluids, compression of gases, foundation movement and settling all have a tendency to move the equipment in every conceivable direction. Many people are surprised to learn that for shafts to run collinear at normal operating conditions, the equipment must be positioned correctly (ie. purposely misaligned) before the units are started to account for this movement. It is important to measure this movement where the equipment is located. Do not take the manufacturer's word on calculated 'growth' amounts or how the equipment moved on the test stand (assuming this was even measured). It is rare that these calculations are correct since your equipment may be piped or mounted differently from the ideal or test stand conditions. Of all of the people I have met who are responsible for aligning machinery, only 1 out of 5 people realize that machinery moves from off-line to running conditions and of that small group of people, only half of them realized that machinery can also move vertically and laterally (side to side). Please, for your own sake, don't pretend that you have the only machinery in existence that doesn't move around. If you suspect some of your equipment falls into that category, take the time to measure what it is and factor this into your alignment strategy.

What is the objective of accurate alignment?

Simply stated, the objective of shaft alignment is to increase the operating lifespan of rotating machinery. To achieve this goal, machinery components that are most likely to fail must operate within their design limits. Since the components that are most likely to fail are the bearings, seals, coupling, and shafts, accurately aligned machinery will achieve the following results ...

- Reduce excessive axial and radial forces on the bearings to insure longer bearing life and rotor stability under dynamic operating conditions.
- Eliminate the possibility of shaft failure from cyclic fatigue.
- Minimize the amount of wear in the coupling components.
- Minimize the amount of shaft bending from the point of power transmission in the coupling to the coupling end bearing. Maintain proper internal rotor clearances.
- Reduce power consumption (documented cases have shown savings ranging from 2 to 17%!)
- Lower vibration levels in machine casings, bearing housings, and rotors (*Note: frequently, slight amounts of misalignment may actually decrease vibration levels in machinery so be cautious about relating vibration with misalignment. Refer to Chapter 14 for more information on this subject).

What happens to rotating machinery when it's misaligned a little bit, or moderately, or even severely?

The drawing shown in figure 1-2 illustrates what happens to rotating machinery when its misaligned. Albeit, the misalignment condition shown here is quite exaggerated, the drawing illustrates that rotating machinery shafts will undergo distortion (i.e. bending) when vertical or lateral loads are transferred from

Rotor distortion caused by misalignment

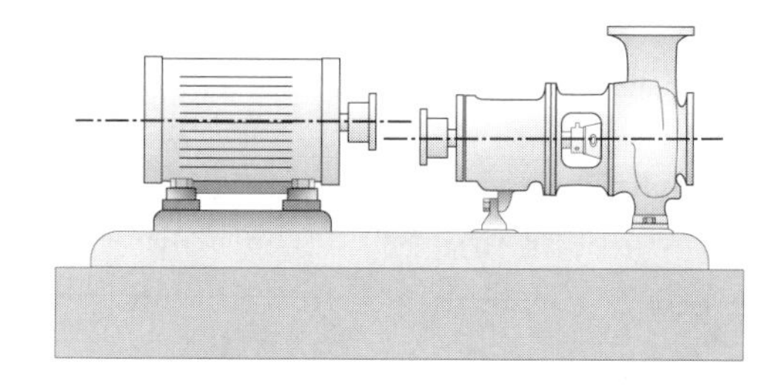

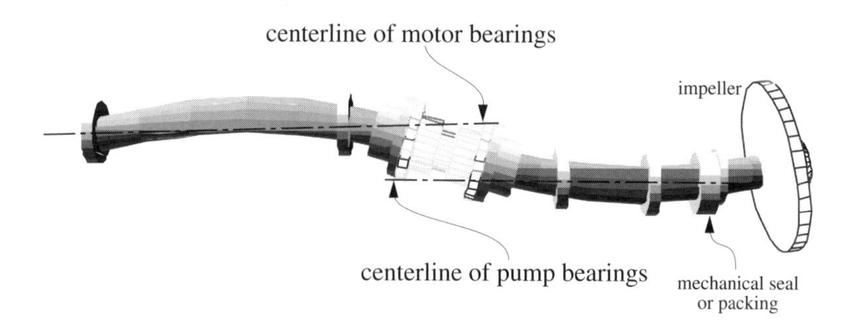

Figure 1-2. How machinery shafts elastically bend when subjected to misalignment conditions.

shaft to shaft.

Please do not misinterpret the drawing! It is fully understood that flexible couplings do just what they are designed to do ... they flex to accommodate slight misalignment. But the shafts are flexible too, and as the misalignment becomes more severe, the more the shafts also begin to flex. Keep in mind that the shafts are not permanently bent, they are just elastically bending as they undergo rotation.

Notice also that the pump shaft in this example is exerting a downward force on the inboard motor bearing as it tries to bring the motor shaft in line with its centerline of rotation. Conversely, the motor shaft is exerting an upward force on the inboard pump bearing as it tries to bring the pump shaft in line with its centerline of rotation. If the forces from shaft to shaft are great enough, the force vector on the outboard bearing of the motor may be in the upward direction and downward on the outboard bearing on the pump. Perhaps the reason why misaligned machinery may not vibrate excessively is due in part to the fact that these forces are acting in the same direction. Forces from imbalanced rotors for instance, will change their direction as the 'heavy spot' is continually moving around as the shaft rotates, thus causing vibration (i.e. motion) to occur. Shaft misalignment forces do not move around, they usually act in one direction only.

The chart shown in figure 1-3 illustrates the estimated time to failure of a typical piece of rotating equipment based on varying alignment conditions. The term 'failure' here implies a degradation of any critical component of the machine such as the seals, bearings, coupling, or rotors. The data in this graph was compiled from a large number of case histories where misalignment was found to be the root cause of the machinery failure.

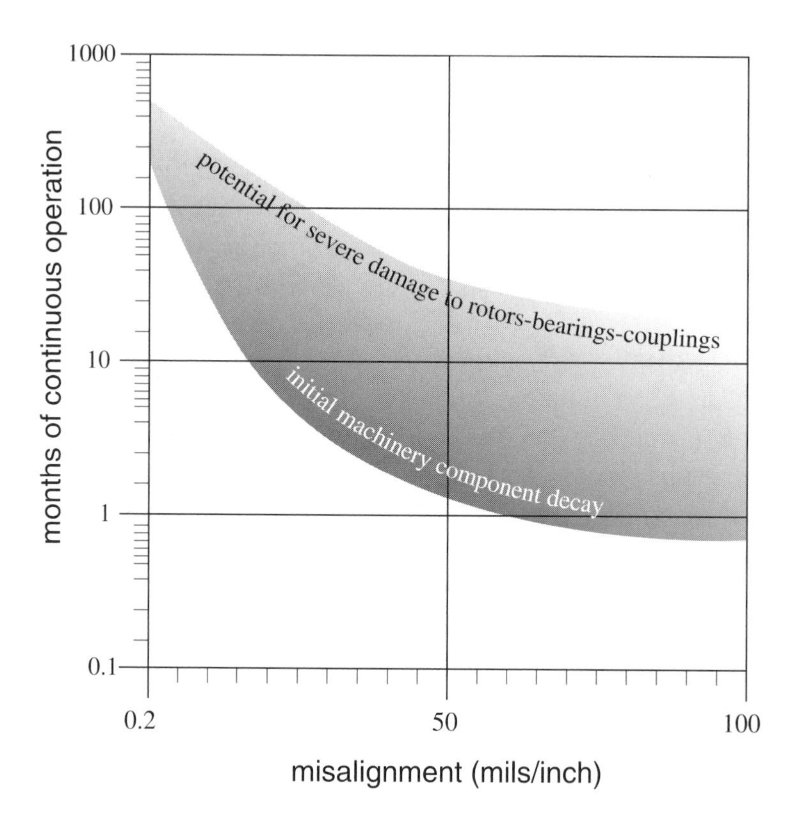

Figure 1-3. Typical operating lifespan of rotating machinery being subjected to various amounts of misalignment.

Symptoms of misalignment

Misalignment is not easy to detect on machinery that is running. The radial forces transmitted from shaft to shaft are typically static forces (i.e. uni-directional) and are difficult to measure externally. Disappointingly, there are no analyzers or sensors that you can place on the outside of a machine case to measure how much force is being applied to the bearings, shafts, or couplings. Consequently what we actually see are the secondary effects of these forces which exhibit many of the following symptoms...

- Premature bearing, seal, shaft, or coupling failures.
- Excessive radial and axial vibration. (*Note : tests have shown that different coupling designs exhibit different types of vibration behavior. It appears that the vibration is caused by the mechanical action that occurs in the coupling as it rotates. Refer to Chapter 14 for more information on this subject).
- High casing temperatures at or near the bearings or high discharge oil temperatures.
- Excessive amount of oil leakage at the bearing seals.
- Loose foundation bolts (refer to 'soft foot' in Chapter 6).
- Loose or broken coupling bolts.
- The coupling is hot while it is running and immediately after unit is shutdown (refer to Chapter 14 for more details on this subject). If it is an elastomeric type, look for rubber powder inside the coupling shroud.
- The shaft runout may have a tendency to increase after operating the equipment for some time.
- Similar pieces of equipment are vibrating less or seem to have a longer operating life.
- Unusually high number of coupling failures or they wear quickly.
- The shafts are breaking (or cracking) at or close to the inboard bearings or coupling hubs.
- Excessive amounts of grease (or oil) on the inside of the coupling guard.

Symptoms of misalignment

- premature bearing, seal, shaft, or coupling failures
- high casing temperatures at or near the bearings or high discharge oil temperatures
- excessive amount of oil leakage at the bearing seals
- coupling is hot while running or immediately after shutdown
- shafts are cracking or breaking at or near the inboard bearings or coupling hubs
- excessive amounts of grease or oil on the inside of the coupling guards
- higher than normal energy consumption
- immediate or eventual increases in vibration levels in the radial or axial directions
- loose or broken coupling bolts
- loose foot bolts, shim packs, or dowel pins

Figure 1-4. Typical symptoms that appear on misaligned rotating machinery.

Summary of an overall alignment job

There are eight basic steps involved in aligning rotating machinery.

1- Purchase or fabricate the necessary tools and measuring devices. Insure that the people involved in the alignment process have been adequately trained on: various alignment procedures and techniques, how to care for delicate measuring instruments and how to use them, what tools should be used to reposition the machinery, whether a machine is really ready to be aligned and operated or whether it should be removed and rebuilt, when a baseplate or foundation has deteriorated to the point where repairs are needed or corrections should be made, correcting problems that exist between the underside of the machine case and the points of contact on the baseplate, how to check for static and dynamic piping stress, what the desired 'off-line' machinery positions should be, how to measure off-line to running machinery movement, what the alignment tolerance is for the machine they are working on, and how to keep records of what was done during the alignment job for future reference.

2 - Obtain relevant information on the equipment being aligned. Are there any special tools needed to measure the alignment or reposition the machines? Do the machines move from off-line to running conditions? If so, how much and do you have to purposely misalign them so they move into alignment when they're running?

3 - Before you begin working on any machinery remember ... Safety First! Properly tag and lock out the equipment and inform the proper people that you're working on the machine.

4 - Insure that you perform these preliminary checks: inspect the coupling for any damage or worn components, find and correct any problems with the foundation or baseplate, perform bearing clearance or looseness checks, measure shaft and/or coupling hub runout, find and correct any 'soft foot' conditions, and eliminate excessive piping or conduit stresses on the machines.

5 - Rough align the machinery and check that all of the foot bolts are tight. Then, accurately measure the shaft positions using accurate measurement sensors (+/- 0.001" or better) such as: dial indicators, lasers, proximity probes, angular or linear resolvers, optical encoders, or CCD's. From this data, determine if the machinery is within acceptable alignment tolerances.

6 - If the machinery is not within adequate alignment tolerances ... first, determine the current positions of the centerlines of rotation of all the machinery; then, observe any movement restrictions imposed on the machines or control points; next, decide which way and how much the machinery needs to be moved; and finally, go ahead and physically reposition the machine(s) in the vertical, lateral, and axial directions. After you've made a move, be sure to re-check the alignment as described in step 5 to determine if the machines really moved the way you hoped they did. When the final desired alignment tolerance has been satisfied, record the final alignment position for future reference, the orientation of the 'soft foot' shim corrections, and the final shim packs used to adjust the height of the machinery. If lateral and axial jackscrews exist, lightly 'pinch' these screws against the sides of the machinery case (assuming excessive casing expansion doesn't occur), lock them in place, and make sure the foot bolts are secured.

7 - Install the coupling (assuming it was disassembled for inspection) and check for rotational freedom of the drive train if possible. Install the coupling guard and make any final checks on the drive train prior to removing the safety tags.

8 - Operate the unit at normal conditions checking and recording vibration levels, bearing and coupling temperatures, bearing loads, and other pertinent operating parameters.

We'll examine each of these steps in greater detail in the remainder of the book, but for now, let's look at the approximate amount of time it takes to perform each of these tasks to give you a feel for how much time this is going to take.

How long should the alignment process take?

If a mechanic performs an alignment job on a small pump for instance, once a month, and knows how to take dial indicator readings ... knows how to calculate the necessary machinery moves ... has information from his engineers on off-line to running machinery movement of the units ... has the proper tools at the job site ... does not have to fight against the pump piping if the pump has to be moved ... has a wide variety of pre-cut shim stock ... has no coupling hub or shaft runout ... with no dirt, rust, or scale buildup under the feet ... jackscrews installed on both units to lift and slide them sideways ... with shafts that rotate freely ... no coupling pieces missing ... correct shaft to shaft distance ... and no one to bother him or her ... the alignment should be completed with the coupling installed and the coupling guard in place in about three to four hours. For anyone who has never performed an alignment job, the last statement may seem quite comical, but for the people who have read this and know what I mean, there is absolutely no humor in what was written. I have never performed an alignment task and had everything fall in place, and if I ever do, I've definitely overlooked something.

There is a lot of time spent preparing for an alignment job. Cleaning baseplates and the underside of the equipment feet, purchasing alignment measuring tools, determining bracket sag, inspecting the coupling, finding and correcting soft foot problems, measuring the thickness of shim packs that are already installed, re-tapping foundation bolt holes, gathering all the tools together, and spending some time and energy training personnel to do the job right are just a few of the things that have to be done before you start. Calculating the proper moves needed to bring the shafts into alignment with computer or graphical alignment calculators can drastically reduce the amount of time spent moving the machinery around compared to trial and error methods. Measuring off-line to running machinery movement alone can take weeks on a multi-unit drive train.

Achieving what I call 'one-shot' alignment, that is, going straight from a rough alignment move to near perfect alignment happens about once in every four tries for the experienced person. One relatively large move and one 'trim' move both sideways and vertically will usually achieve the desired results. Also, the heavier the equipment is, the longer it takes to lift it or move it sideways. If four or five tries are needed to get the job done, something is wrong.

The graph in figure 1-5 shows the average amount of time taken to do steps 3 through 8. Steps 1 and 2 are not on the graph since procuring all the necessary tools, training everybody who's involved, and gathering information can take a considerable amount of time to complete. As you can see, the two most time consuming tasks in the alignment process are: performing the mechanical integrity checks and moving the machinery to align the shafts. It is not uncommon for accurate alignment to take from 3 to 8 hours, assuming everything goes just right!

Estimated time to perform the different alignment steps

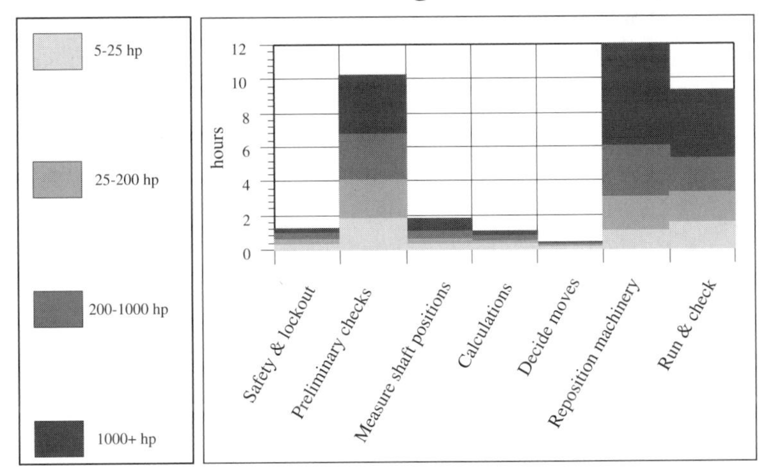

For example, it should take two people about 5 hours to align a 50 hp motor and pump unit.

Figure 1-5. Average time taken to perform primary steps in aligning machinery.

Keeping records

It is helpful to keep data on the complete alignment process for each drive train in your plant. These records should be kept in the maintenance file folder as a handy reference for future work done on each drive train. The Machinery Data Card in the Appendix can be used as a guide in planning your own record sheet.

How can I reduce the amount of time it takes to do alignment properly?

First, the people who are aligning the machinery have to know what they are doing and they have to have a goal to shoot for. They also require access to all of the tools needed to do the job and the tools must work properly. If your company purchased an expensive alignment measurement system that stays locked up in a cabinet, it's not doing anyone any good if they can't use it when they need it.

It is also important to have an alignment system that can provide you with alternative movement solutions when repositioning the machinery. The key to successfully aligning machinery comes from having the ability to arrive at a solution that is possible to perform and minimizes the required movement at the feet. Accurately calculating required movement at the machinery feet is useless if you can't move the machine the amount your alignment system is telling you.

How often should alignment be checked?

As previously mentioned, rotating machinery can move around immediately after it has been started. This is fairly rapid movement and the shafts eventually take a somewhat permanent position after the thermal and process conditions have stabilized (anywhere from 2 hours to a week in some cases). However there are slower, more subtle changes that occur over longer periods of time. Machinery will slowly change its position for the same reason your driveway buckles, or your building foundation cracks. Settling of base soils underneath the machinery will cause entire foundations to shift. As the foundations slowly move, attached piping now begins to pull and tug on the machinery cases causing the equipment to go out of alignment. Seasonal temperature changes also cause concrete, baseplates, piping, and conduits to expand and contract.

It is recommended that newly installed equipment be checked for any alignment changes anywhere from 3 to 6 months after operation has begun. Based on what you find during the first or second alignment 'checkup', tailor your alignment surveys to best suit the individual drive trains. On the average, shaft alignment on all equipment should be checked on an annual basis. Don't feel too embarrassed as you read this because you're definitely not the only person who hasn't checked your machinery since its been installed.

How much money should I be spending on tools and training?

I guess a good rule of thumb is to invest 1% of the total replacement cost of all your rotating machinery on alignment tools and training on an annual basis. For example, if you have 20 drive trains in your facility valued at $5000.00 each (total $100,000.00) then you should invest $1000.00 on alignment every year. This expenditure should only cover tools and training and should not encompass the time and materials required to do alignment jobs.

How do I know if the contractors I hired to install my machinery are doing the alignment properly?

Include some clause in your contract that requires them to provide you with the initial alignment data, 'soft foot' conditions and the corrections made, shaft and coupling hub runout information, the final alignment data, the moves made on the machinery, and the final alignment tolerance. Don't be satisfied with an answer like: "We used dial indicators and lasers." Dial indicators and laser don't move machinery ... people do!

In summary ...

The intent of this book is to explain how to obtain accurate shaft alignment to the contractor or maintenance mechanic, foreman, and engineer whose work is in the field on a day to day basis. Best of luck to each of you as you try many of these alignment methods!

Figure 1-6. Worn teeth on gear coupling running at 1.8 mils per inch misalignment for 12,000 hours.

2
The History of Machinery Alignment

The historical path of shaft alignment encompasses many interwoven engineering disciplines. To begin understanding how shaft alignment progressed to its current state of technology will require an exploration of a wide variety of topics. Since necessity is the mother of invention, the necessity of aligning rotating machinery shafts is directly linked to the development of rotating machinery and it is therefore important for us to look at the progression of this equipment throughout history. The subject of shaft alignment is also concerned with mechanical measurement, mathematics, metallurgy, vibration analysis, statics and dynamics, optics, and electronics ... each contributing to the current use of the tools and techniques described in this book.

Our story will start 450,000 years after man first began to gather in groups and use fire in a controlled fashion during the mid-Pleistocene era. The crux of engineering triumphs during the Neolithic era (11,000 to 6,000 B.C.) were the emergence of permanent dwellings, the beginnings of agriculture, and to a very basic extent metallurgy. Even through 4000 years of inhabitation from 8000 B.C. to 4000 B.C. in the oldest cities of Jericho, 'Ain Ghazal, and Catal Hüyük (located in Israel, Lebanon, and Turkey) very little developed in terms of tools, measurement, architecture, and science.

From this point forward begins the emergence of 'modern' man's eventual development of shaft alignment displayed to you in snapshots of time events.

-4000 to -3500 Smelting of gold and silver known. Copper alloys used by Egyptians and Sumerians.

-3000 to -2500 Cheops' pyramid in Egypt built to extremely precise dimensions showing knowledge of geometry and measurement. First attempts to establish a standard of measurement by Egyptians was the cubit (changed from 18 inches to 20.63 inches around 3000 B.C.). Iron objects first appear.

-2000 to -1500 Babylonia uses highly developed geometry for astronomical measurements. Assyrians and Babylonians establish the units of measurement as : the cubit (now 20.5-20.6 inches), the span (10.5 inches), and the digit (0.653 inches). Egyptians use knotted rope to construct right angles illustrating the 'Pythagorean' theorem which is also known in China during this period. Water level believed to be used in Mayan culture for construction of irrigation systems (see -600 to -500).

-1000 to -900 Chinese textbook of mathematics shows principles of plainemetry, proportions, arithmetic, root multiplication, geometry, equations with unknown quantities, and theory of motion.

-900 to -800 Iron and steel production in Indo-Caucasian culture.

-600 to -500 Theodorus of Samos, a sculptor, credited with inventing ore smelting and casting, water level?, carpenter's square, and lathe.

-384 Aristotle born. Credited with much of the initial discoveries in physics, biology, and psychology contained in his book **Historia Animalum**. Aristotle or his student Straton publishes **Mechanika** discussing the lever and gearing.

-323 Euclid writes the first book on geometry called **Elements**.

-300 to -200 Ctesibius of Alexandria invents the force pump (figure 2-1) and Archimedes of Syracuse invents the screw pump. The appearance of gears leads

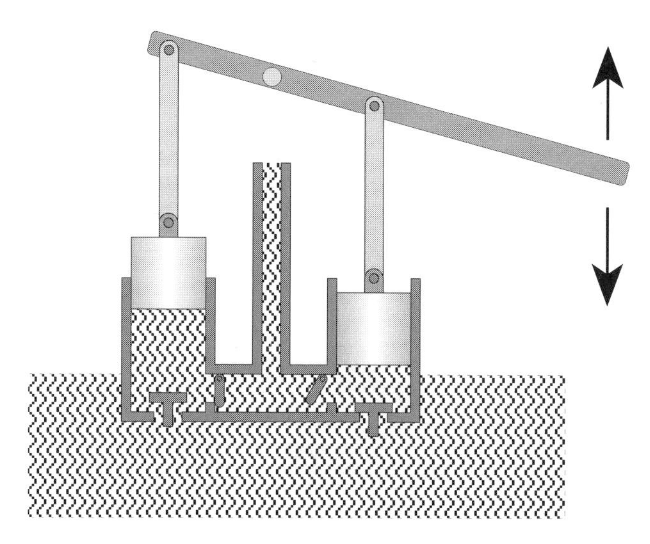

Figure 2-1. Ctesbius pump.

to the development of the ox-driven water wheel for irrigation. Universal joint used by Greeks (see 1550).

100 Hero of Alexandria describes the principles of an aeolipile (figure 2-2), a simple reaction steam turbine, in **Pneumatica** ; describes levers, gears, motion on an inclined plane, velocity, and the effects of friction in his book **Mechanics**. Hero's book **On the Dioptra** describes a type of theodolite and explanations of plain and solid geometrical figures, conic sections, formula for calculating the area of a triangle from the length of its sides, and a method for determining the square root of a non-square number appear in his book **Metrica**. Theodosius of Nithynia (aka Theodosius of Tripoli) authors **Sphaerica** dealing with spherical geometry.

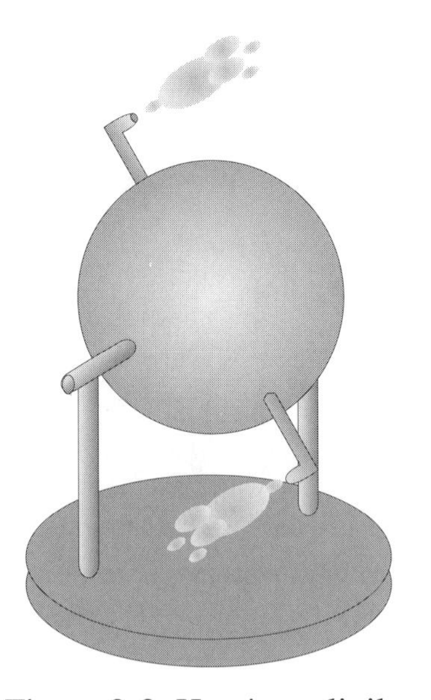

Figure 2-2. Hero's aeolipile.

250	Diophantus of Alexandria writes first book on algebra.
285	Pappus of Alexandria describes operation of current machines in use : cogwheel, lever, pulley, screw, and wedge.
700	Water wheels for mills used all over Europe.
1280	Roger Bacon discusses the basic operation of the telescope by use of "... glasses or diaphanous bodies that may be formed that the most remote objects may appear just at hand..." in his book **Epistola ad Parisiensem**. (see 1600)
1305	Edward I of England issued a decree that read: "It is ordained that three grains of barley dry and round make an inch, twelve inches make a foot, three feet make an ulna, five and a half ulna make a rod, and forty rods in length and four in breath make an acre."
1510	Leonardo da Vinci designs horizontal water wheel and proposes a 'smokejack' which uses the hot gases rising from a fire to propel a vertically oriented shaft driven by blades attached to the shaft (figure 2-3). This device was later patented by John Dumbell of England.
1540	Filde Nanez Salaciense of Portugal, while working on a system to more accurately read angles for his map making work, devises a rotary scale where a series of divisions were marked equally around the circumference of an outer scale and an inner rotary scale with a series of equally marked divisions one division less than the outer scale. Although the device didn't work very well due to the difficulty of accurately scribing equally spaced lines by hand, the idea became the basis of the vernier scale (see 1631).

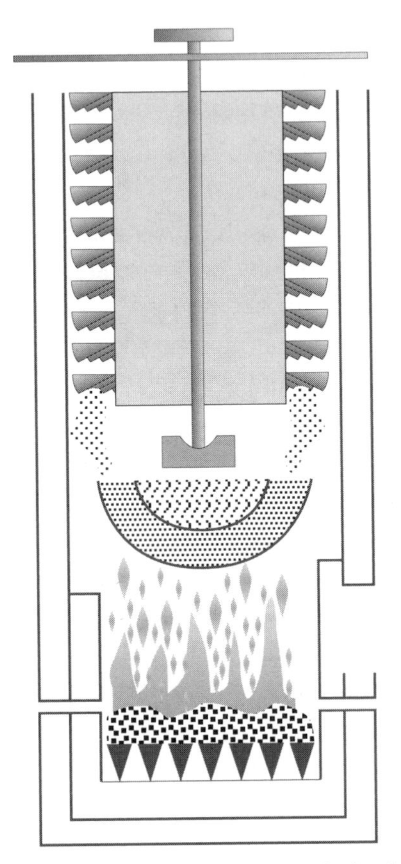

Figure 2-3. DaVinci's 'smokejack'.

1550 Jerome Cardan (Geronimo Cardano) Italian physician, mathematician, and friend of da Vinci is credited with the 'invention' of the universal joint (aka Hooke or Cardan joint). Georgius Agricola (Georg Bauer) writes **De Re Metallica**, a 10 volume series of books on chemical and mining engineering covering ores, theory of formation of mineral veins, surveying, tools, machines, pumps, hoists, water power, ore preparation, smelting, and manufacture of salt, soda, alum, vitriol, sulphur, bitumen and glass. These works were later translated to the English language by American president Herbert Hoover.

1576 François Vieté introduces use of decimal fractions.

1600 Dutch optician Johann Lippershey invents telescope which is adapted for astronomical observations a year later by Galileo Galilei who manufactured hundreds of telescopes that were in great demand by 'amateur' astronomers of his time. Galileo is also credited with: the invention of the thermometer, was the first person to coin the term 'moment' meaning: the effect of force, founded the science of strength of materials, and his fascination with the periodic swing of pendulums began his inquisition into falling objects laying the groundwork for gravity and acceleration.

1601 Giovanni Battista della Porta develops principles of condensing steam turbine.

1611 Marco de Dominus publishes scientific explanation of a rainbow (electromagnetic spectrum).

1614 Scottish mathematician, John Napier publishes **Canonis Descripto**, describing his discovery of logarithms. Napier was the first to use the decimal point to express fractions.

1621 William Oughtred devises the first 'slide rule' using Napier's logarithms.

1629 Giovanni Branca describes using a jet of steam impinging on blades projecting from a wheel to produce a rotating shaft.

1631 Pierre Vernier invents slide caliper.

1639 Prior to his death at the age of 24 in the Civil War of 1642, astronomer William Gascoigne invents micrometer from his work in attempting to determine the diameter of celestial objects. By devising a caliper whereby two fingers were moved toward or away from each other simultaneously by left hand / right hand threads, the image in the Gascoigne eyepiece could be determined by the finger distance and triangulation principles. It is not clearly known where William got the precisely cut threads for his micrometer but a French engineer Besson built a lathe capable of cutting threads in 1569 (see also 1791).

1650 Robert Hooke uses universal joint in clock. In 1655 he was employed by Robert Boyle who used his technical expertise to assist in constructing the 'air pump'. Boyle publishes a book in 1660 entitled **New Experiments Physico-Mechanical Touching the Spring of Air and Its Effects**.

1652 German scientists Otto von Guericke invents an 'air pump' and eleven years later constructs a frictional electric generating machine.

1665 Francis Grimaldi explains diffraction of light and Isaac Newton invents differential calculus.

1673 Gottfried Leibnitz invents the 'Leibnitz Wheel', the first mechanical device to perform addition, subtraction, multiplication, and division.

1690 Denis Papin devises non-condensing, single acting steam pump with piston and the steam 'safety valve'.

1698 Thomas Savery obtains patent for steam-driven, water raising engine and first coins the term

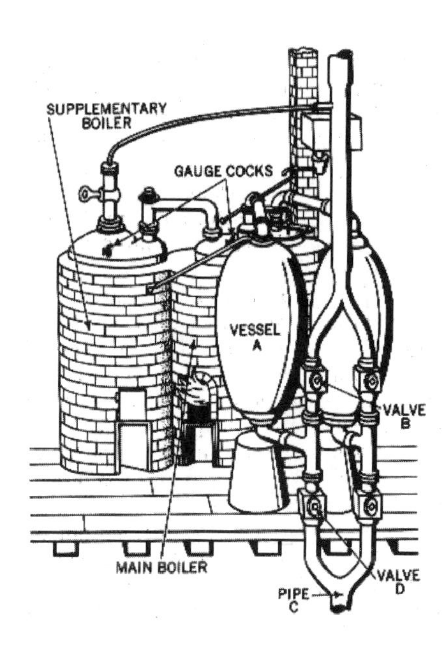

Figure 2-4. Savery's steam engine.

'horsepower' (figure 2-4). The success of the coal-fired steam engines are directly linked to the 16th century 'energy crisis' caused by the lack of wood fuel from the deforestation of England.

1702 Alain Manesson Mallet invents a telescopic sight incorporating a level bubble called the dumpy level.

1704 Isaac Newton publishes **Optics** defending the emission theory of light. From 1662 to his death in 1727 he co-invents calculus with Leibnitz, develops the generalization of the concepts of 'force', 'mass', and the principle of effect and counter-effect.

1705 Thomas Newcomen and John Cawley invent condensing steam turbine.

1714 D. G. Farenheit invents mercury thermometer followed by Réaumurs alcohol thermometer two years later, followed by Anders Celsius centigrade thermometer twenty eight years later.

1720 Theodolite made by Sisson Benjamin Cole in London was the first device to employ two spirit levels (figure 2-6).

1729 English scientist Stephen Gray discovers that some bodies are conductors and non-conductors of electricity.

1745 Ewald Jurgen von Kleist invents capacitor (Leyden jar).

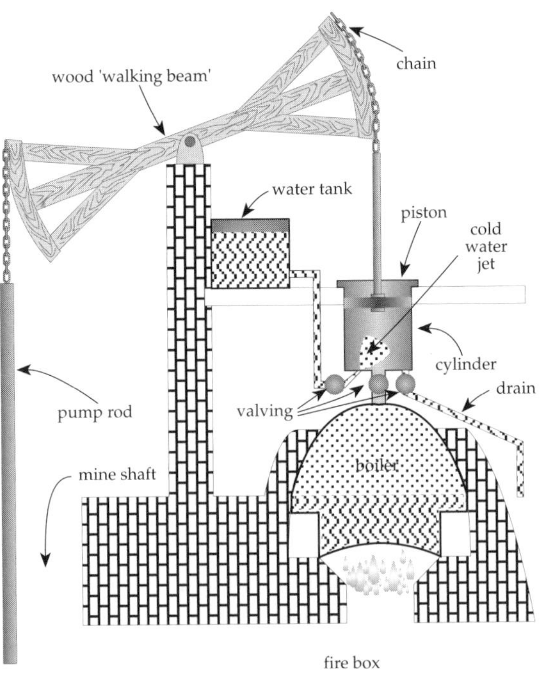

Figure 2-5. Newcomen's mine pump.

Figure 2-6. 18th century theodolite.

1750 Swiss mathematician Leonard Euler and his son Albert experiment with impulse-driven water turbines. Leonard also developed equations describing buckling of struts, the catenary curve, and formulated the laws governing the flow of fluids and the relationship of pressure to flow. (see 1770).

1754 P. Van Musschenbroek at the University of Leyden in Holland first demonstrated that when two insulated metal plates are brought in close proximity to each other without making contact, considerably more electrical charge could be stored than in a single plate (the Leyden jar) commonly known today as a capacitor. When a wire is connected to one plate of an electrically charged Leyden jar and then made to touch the other plate, an electrical discharge occurs.

1762 Cast iron first converted to malleable iron at Carron ironworks in Scotland.

1764 James Watt invents steam condenser as improvement to steam engines patenting his improvements in 1769. Files a second patent in 1781 describing sun and plant wheels and a flywheel. Files another patent in 1782 describing the use of double action whereby steam is injected on one side of a piston and vacuum on the other.

1768 French mathematician and physicist Jean Baptiste Fourier born. Fourier is credited with developing the Fourier transform used in vibration analysis (see Chapter 14).

1770 Leonard Euler publishes **Introduction to Algebra** followed by another book on mechanics, optics, acoustics, and astronomy two years later.

1774 John Wilkinson constructs boring mill to manufacture cylinders for steam engines.

1786 Galvani discovers electric current occurs when two dissimilar metals come in contact with each other. By suspending zinc and copper plates in an acid solution Galvani showed that a steady flow of current will flow (chemical battery).

1787 Ernst Chlandi experiments with sound patterns on vibrating plates.

1790 The French National Assembly committee members decided that the meter would be one ten-millionths of a quadrant of the earths meridian. In 1799 a platinum-iridium end bar was produced and became known as the 'Metre des Archives' ... the master standard of length in the world. The bar's length was based on a slightly inaccurate geodetic survey made to establish the distance of the earth's meridian. As of 1983, the meter is currently defined as the distance light travels in a vacuum after 1/299,792,458 of a second.

1791 John Barber patents first gas turbine. After working with Joseph Bramah, twenty-two year old British engineer Henry Maudslay starts his own business and develops a metal lathe (most former lathes were made from wood) capable of accurately cutting threads. Using his new machine, Maudslay cuts 50 threads per inch in a long rod that is eventually used as a micrometer he used to check his work. Maudslay is also credited with the leather 'U' seal when working with Bramah in the development of the hydraulic press.

1800 William Herschel discovers existence of infrared solar rays, Richard Trevithick constructs low pressure steam engine, and Alessandro Volta produces electricity from first zinc-copper battery.

1802 John Dalton introduces atomic theory in chemistry.

1806 Oersted discovers that a magnetic field is produced around a wire where electric current is flowing, proving up to that point the theory that electricity and magnetism are indeed interrelated.

1815 Augustin Fresnel begins research on diffraction of light.

1816 Ernst Werner von Siemens born in Hanover. Credited with the invention of the armature initially used in telegraphy and later used in the larger generators (dynamo) demonstrating the principle of dynamo-electric principle. Along

	with his brothers Karl Wilhelm (developed a type of governor for steam engines), Friedrich, and nephew Alexander, founded the Siemens Company.
1824	Joseph Aspdin of England patents Portland cement process naming it after its resemblance to portland stone, a limestone quarried at Portland, England.
1829	J. B. Nelson introduces the hot blast furnace.
1831	Michael Faraday, while taking a hiatus on his work in chemistry after discovering benzene and butylene and manufacturing the first stainless steel, began work on getting electricity from magnetism. His now infamous experiment whereby he thrust a permanent magnet into and out of a coil of wire producing a flow of current laid the ground work for much of the work about to be done in electric generation and laid the basis for operation of the vibration sensor now known as the velocity pickup (seismometer ... see Chapter 14).
1843	Jonval introduces axial flow turbine.
1848	The first screw caliper patent (i.e. micrometer) was issued to French mechanic Jean Laurent Palmer.
1849	James B. Francis builds a radial inflow water turbine wheel (172 kw / 230 hp) based on a patent by Samuel Howd achieving an efficiency of 80%. 20 year old Lester Pelton ventures to California in his failed attempt to find gold but experiments with water wheels used in gold mining to design what is now known as the Pelton wheel.
1853	Lord Kelvin discovered that current must pass

back and forth between two plates in a condenser to be able to emit electromagnetic waves introducing the concept of an oscillating electrical circuit, the key to all radio transmission.

1856 Henry Bessemer decarburized molten iron by blowing cold air through the iron, forming mild carbon steels upon cooling.

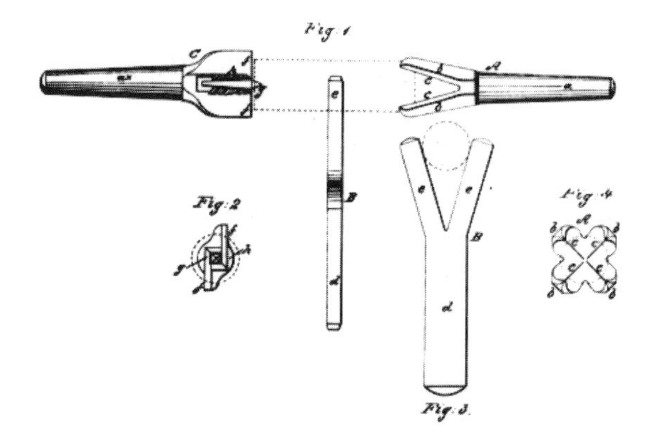

Figure 2-7. Williams' shaft centerer patent 1863.

1864 Scottish physicist James Maxwell theorizes that not all of the energy in Kelvin's oscillator is dissipated as heat but a certain percentage must radiate into space as electromagnetic radiation. Maxwell, based on the findings by Wilhelm Weber and Friedrich Kohlrausch showing that the velocity of electricity through a wire and the velocity of light in a vacuum is the same, concludes that there is a relationship between the two, giving birth to the concept of the electromagnetic theory of light.

1867 Parisian gardener Joseph Monier obtains patent for reinforced concrete. J. R. Brown and Lucian Sharpe while visiting the Paris Exposition saw a Palmer micrometer (see 1848). Taking the best features of the Palmer micrometer and another micrometer designed by S. R. Wilmot (superintendent of Bridgeport Brass) Brown & Sharpe released the first US made micrometer in 1867.

1868 R. R. Musket introduces tungsten into steel manufacturing as a self hardening metal used for cutting tools. The US Navy's Bureau of Steam Engineering adopts William Sellers' Unified Screw Thread design using 60 degrees as the thread angle. Not until after the end of World War II did American, British, and Canadian representatives finally agree on this standard thread design.

1869 W. J. Rankine publishes paper *On the Centrifugal Force of Rotating Shafts* in **Engineer** periodical.

1872 F. Stolze of Germany develops gas turbine consisting of a separately fired combustion chamber, a heat exchanger, and a multistage axial flow compressor coupled to a multistage reaction turbine.

1877 Sir Charles Parsons begins work at Armstrong Works in Elswick England. In 1884, he serves for a year on the experimental staff of Messrs. Kitson of Leeds where he patents the modern day steam turbine. Leroy S. Starrett invents combination square.

1878 Centralized generating station first proposed by St. George Lane Fox (England) and Thomas Edison (US). Carl De Laval builds a small 42,000 rpm reaction steam turbine for cream separators (figure2-8). DeLaval continues improvement to the steam turbine by utilizing hyperbolic blade designs.

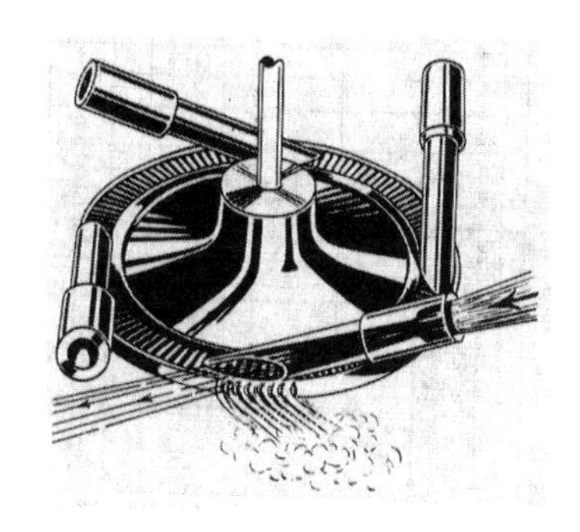

Figure 2-8. DeLaval turbine.

1880 American Society of Mechanical Engineers formed.

1881 Lucien Gulard and John Gibbs obtain English patents for 'series alternating current systems of distribution'. These patents were purchased by George Westinghouse in 1885.

1882 First electric direct current generating stations installed in London, England on January 12th and in New York City on September 4th. British electrical engineer William Ayrton invents ammeter, electrical power meter, improved voltmeters, and meters to measure self and mutual induction.

1883 John Logan of Waltham, Massachussets files a US patent for the dial indicator (figure 2-9).

1884 Nikola Tesla begins work with Thomas Edison's company patenting the induction, synchronous, and split phase electric motors and new forms of generators and transformers. In 1892, Edison General Electric and Thomson-Houston Electric companies merge to form General Electric. American Society of Electrical Engineers formed.

1886 George Westinghouse and William Stanley (credited with perfecting the transformer) first demonstrated the practicality of generating and transmitting alternating current over long distances in Great Barrington, MA. F. Hooks theorizes idea behind flexible disk coupling. British electrical engineer Sebastian Ferranti, working at Grosvenor Gallery Co. in London also proposes using high voltage alternating current for power transmission that would be utilized at discrete sites through step-down transformers.

1887 Heinrich Hertz of Liepzig experimentally confirms Maxwell's prediction by observing radiation emanating from an oscillating electric circuit. While a professor at Case School of Applied Science in Cleveland, OH, Albert Abraham Michelson devises the interferometer capable of measurements to one-millionths of an inch.

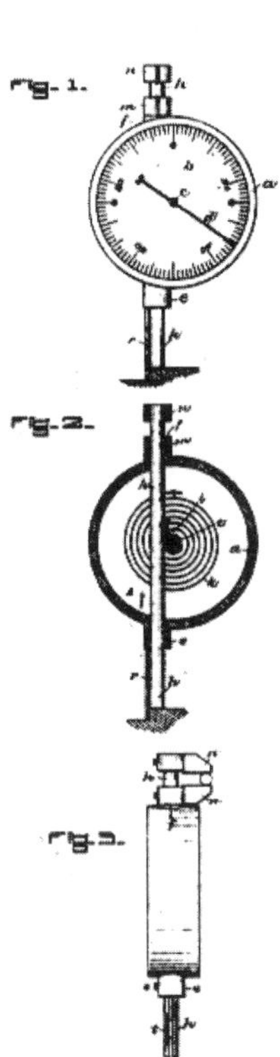

Figure 2-9. Logan's dial indicator patent (1883).

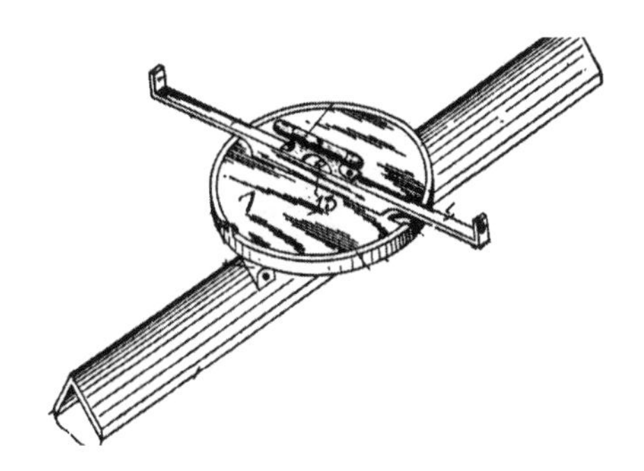

Figure 2-10. Campbell's shaft leveler patent (1893).

1889 De Laval builds a large number of steam turbines ranging in size from 5 to several hundred horsepower and in 1892 builds a 15 hp turbine for marine applications. British engineer Charles Parsons forms his own company after developing the multistage steam turbine while working at Clarke, Chapman and Co. in Gateshead England.

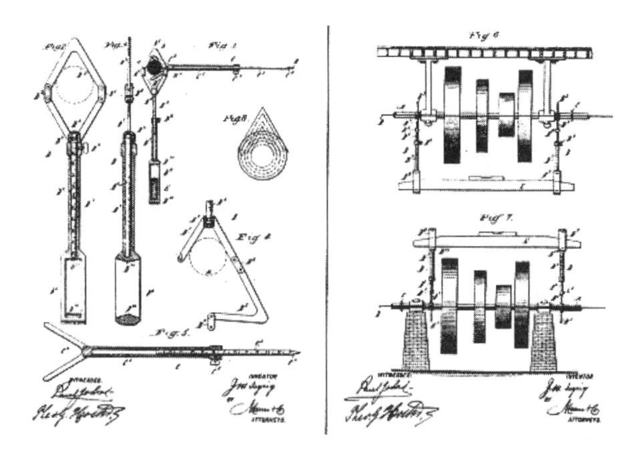

Figure 2-11. Isgrig's shaft leveler patent (1894).

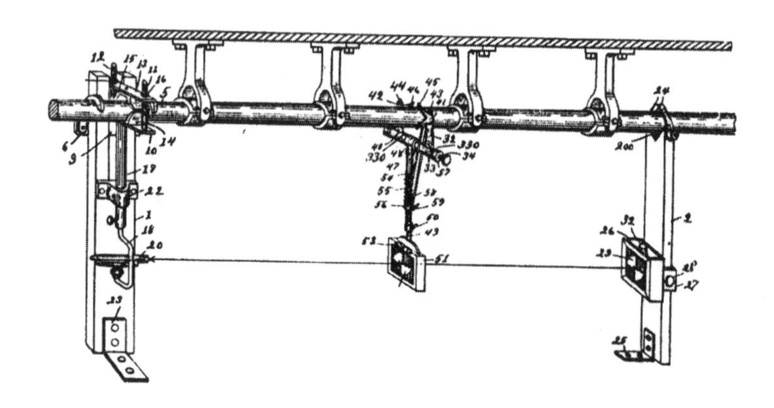

Figure 2-12. Kinkead's shaft leveler patent (1901).

1893 Rudolph Diesel invents engine named after him. Sulzer of Switzerland acquires patent rights to the diesel four years later.

1894 S. Dunkerley publishes paper *On the Whirling and Vibration of Shafts* .

1903 R. Armengaud and C. Lemale build first successful gas turbine demonstrated for Societe des Turbomoteurs in Paris achieving efficiencies of 3%.

1904 C. Chree publishes paper *The Whirling and Transverse Vibration of Rotating Shafts* . British electrical engineer, John Fleming conclu-

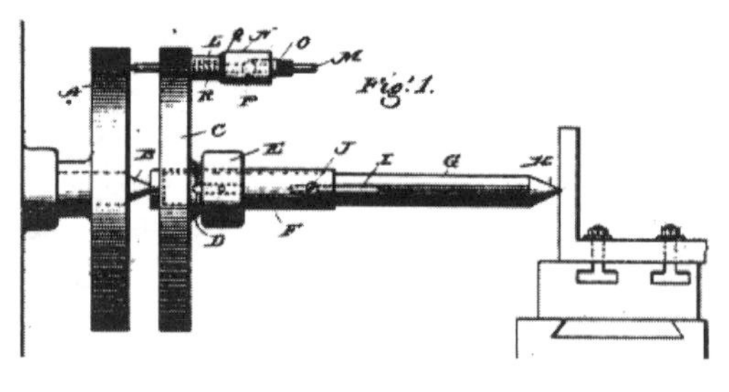

Figure 2-13. Miller's shaft centerer patent (1901).

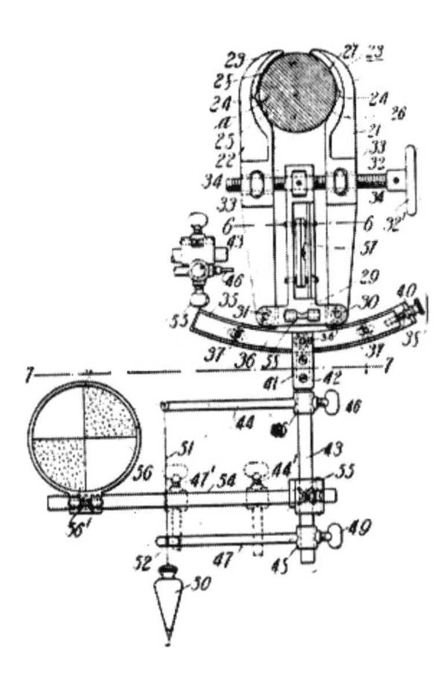

Figure 2-14. Ferris' shaft aligner patent (1908).

sively shows that the rectifying property of a thermionic valve was still operative at radio frequencies eventually leading to the 'diode valve'.

1907 US inventor Lee DeForest patents the thermionic grid-triode vacuum tube, the forerunner of the transistor.

1914 **Scientific American** describes operation of chain coupling in May issue.

1916 Albert Einstein proposes the idea of stimulated emissions of radiation.

1919 H. H. Jeffcott publishes *The Lateral Vibration of Loaded Shafts in the Neighborhood of a Whirling Speed - The Effect of Want of Balance* in **Philadelphia Magazine**.

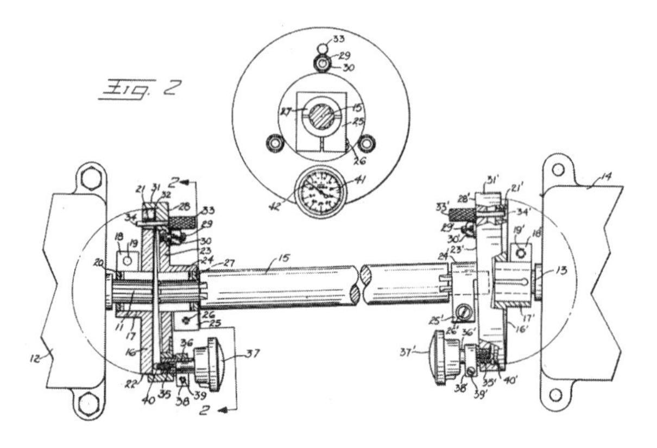

Figure 2-15. Christian's shaft alignment device (1946).

1920 Austrian engineer Viktor Kaplan patents design for adjustable rotor blades on a water driven turbine.

1924 B. L. Newkirk publishes article on *Shaft Whipping* describing the critical speeds of rotors.

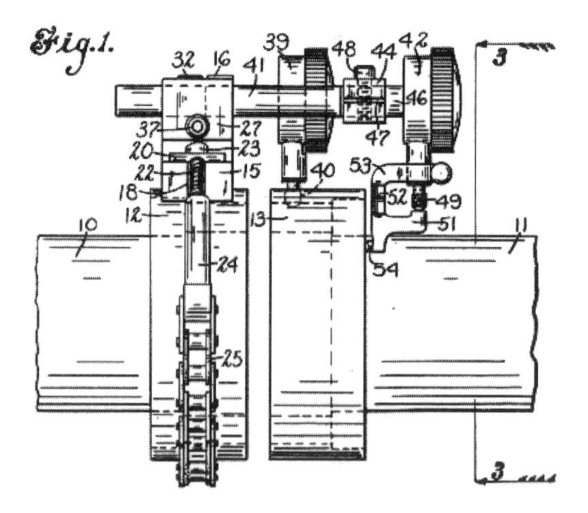

Figure 2-16. Voss' shaft alignment device (1949).

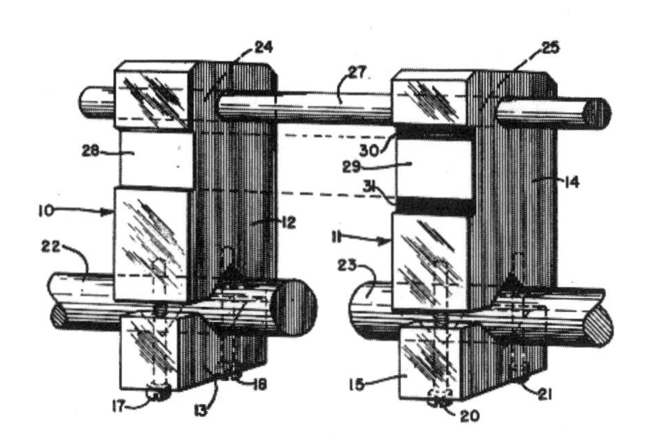

Figure 2-17. Callahan's shaft alignment device (1953).

1946 Joseph Christian files US patent for shaft alignment device utilizing the face - face measurement method (figure 2-15).

1949 Robert Voss files US patent for shaft alignment device utilizing the face - peripheral (rim) measurement method (figure 2-16).

1953 John Callahan files US patent for shaft alignment device utilizing the double radial measurement method (figure 2-17).

1954 Charles Townes at Columbia University experiments with generated radio/micro waves by stimulating the emission of energy that was stored in ammonia molecules in a container thereby naming their device the maser (acronym for microwave amplification by stimulated emission of radiation). This was the first demonstration of generating electromagnetic waves without using traditional devices such as flames, electric lamps, neon and florescent lights or natural devices such as the sun, aurora borealis, or lightning. Simultaneously, Aleksander Prohorov and Nikolai Basov in Moscow completed detailed calculations explaining the requirement for maser action to occur. All three men were awarded the Nobel Prize for their work in 1964.

1957 Gordon Gould writes proposals for visible wavelength laser (referred to as an optical maser by Townes & Schawlow) in notebooks eventually filing patents in 1977, 1979, 1987, and 1988. Gould is credited for coining the term ‘laser’.

1958 Charles Townes and Arthur Schawlow (Bell Lab researcher) file US patent for maser.

1962 The first semiconductor junction lasers were almost simultaneously demonstrated by teams at General Electric Research Laboratories, IBM Watson Research Center, and MIT Lincoln Laboratories using gallium arsenide diodes cooled to 77 degrees K and pulsed with high current.

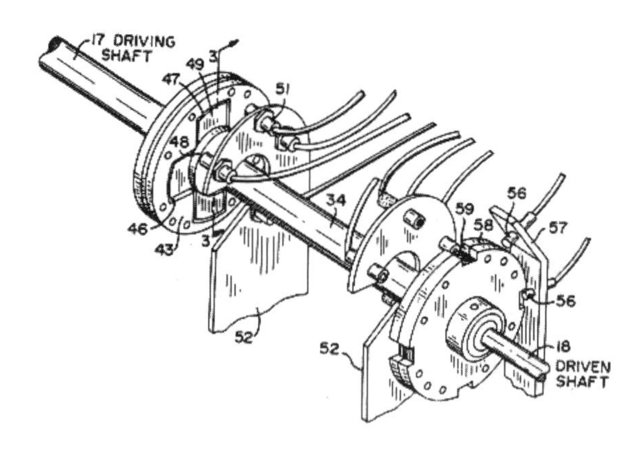

Figure 2-18. Bently's shaft alignment device (1974).

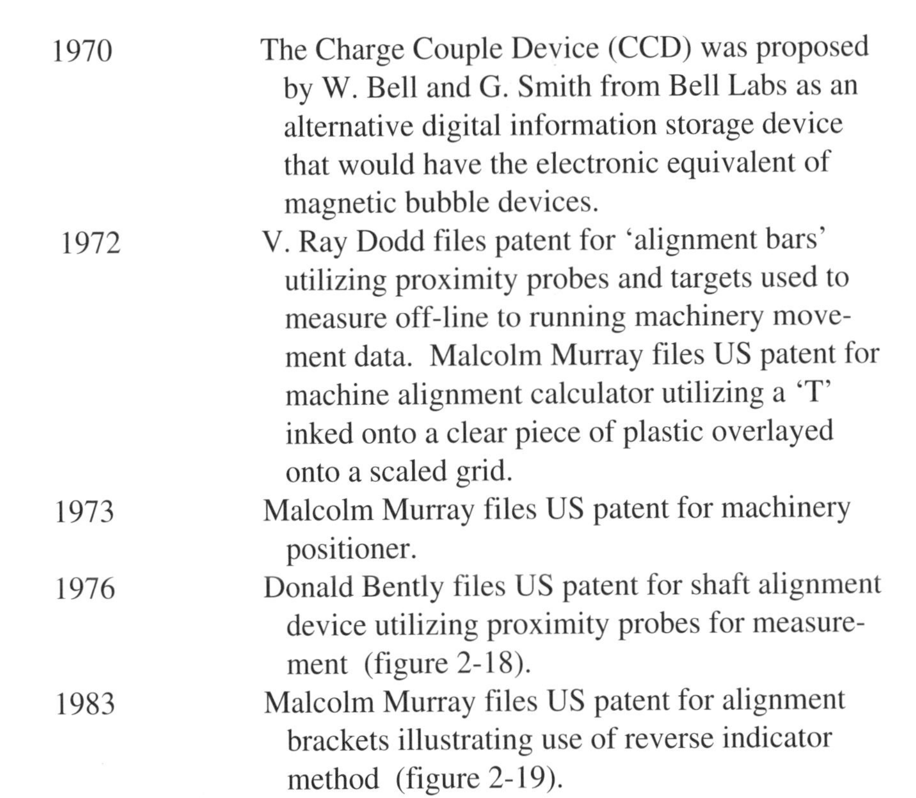

1970 The Charge Couple Device (CCD) was proposed by W. Bell and G. Smith from Bell Labs as an alternative digital information storage device that would have the electronic equivalent of magnetic bubble devices.

1972 V. Ray Dodd files patent for 'alignment bars' utilizing proximity probes and targets used to measure off-line to running machinery movement data. Malcolm Murray files US patent for machine alignment calculator utilizing a 'T' inked onto a clear piece of plastic overlayed onto a scaled grid.

1973 Malcolm Murray files US patent for machinery positioner.

1976 Donald Bently files US patent for shaft alignment device utilizing proximity probes for measurement (figure 2-18).

1983 Malcolm Murray files US patent for alignment brackets illustrating use of reverse indicator method (figure 2-19).

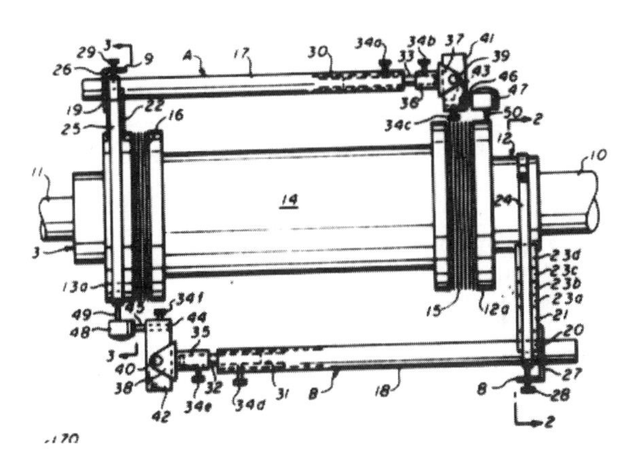

Figure 2-19. Murray's shaft alignment device (1983).

1984	John Zatazelo files US patent for electronic shaft alignment calculator.
1986	John Zatazelo files US patent for a variety of methods used to determine shaft alignment.
1987	Heinrich Lysen files US patent for laser - detector - prism shaft alignment system. Brian Morrisey files US patent for horizontal machinery positioning device.
1990	Malcolm Murray files US patent for vernier-strobe method used to measure off-line to running machinery movement data.
1991	Dieter Busch files US patent for laser - detector - prism system used to measure off-line to running machinery movement data. Paul Saunders files US patent for electronic shaft alignment system utilizing optical encoder.

There you have it. A brief run through the last 6000 years of man's existence illustrating our engineering triumphs during periods of unparalleled growth in science and technology interspersed with periods of dismal ignorance. In many cases, it is not clear who actually ‘invented’ certain methods and devices.

The vast majority of US patents for aligning and leveling machinery filed from 1860 to 1946 fell into the following categories ...

- use of spirit / dumpy levels to set the positions of shafts in line with each other
- centering devices for work pieces on lathes
- the use of a ‘tight wire’ for alignment of engine cylinder bores.

As you can tell, the technology of shaft alignment as we recognize it today, really didn’t emerge until the 1940’s and 1950’s followed by a flurry of patent activity that persists to the present day. Between 1860 and 1950, only eleven US patents were filed relating to measuring 'off-line' shaft centerline positions, the majority of which exclusively dealt with aligning and leveling line shafts typically used in the paper industry. Over sixty years had passed since John Logan patented the dial indicator before Joseph Christian actually filed a patent using this device for alignment of rotating machinery shafts.

However, this does not necessarily mean that shaft alignment was not performed on rotating machinery during the early part of the Industrial Revolution. As we have seen throughout history, several methods and devices had been in existence for long periods of time before legal patent documents were filed claiming their 'originality'. Regrettably, the true inventors of these mechanisms and procedures will probably never be known.

Bibliography

Bernard Grun, **The Timetables of History**, Simon & Schuster Inc. International Standard Book Number (ISBN) # 0-671-24987-8.

Encyclopedia Britannica, various volumes.

Jon Mancuso (1986), **Couplings and Joints**, Marcel Dekker Inc., ISBN 0-8247-7400-0.

M. Brotherton, **Masers and Lasers**, McGraw Hill Book Co., LCC# 63-23249.

John FitzMaurice Mills, **Encyclopedia of Antique Scientific Instruments**, Facts on File Inc., New York, NY, ISBN 0-87196-799-5.

Ervan G. Garrison, **A History of Engineering and Technology : Artful Methods**, CRC Press Inc., Boca Raton, FL, ISBN 0-8493-8836-8.

David Abbott, **Engineers & Inventors**, Peter Bedrick Books Inc., New York NY, ISBN 0-87226-009-7.

Jeff Hecht, **The Laser Guidebook**, McGraw Hill Inc., ISBN 0-07-027737-0.

66 Centuries of Measurement, Sheffield Corp., Dayton, OH, 1987, LCCC# 530.81.

Boyle, W. S. and G. E. Smith (1970): "Charge-coupled semiconductor devices". Bell Syst. Tech. Journal, 49, 587-593.

Amelio, G. F. et al. (1970): "Experimental verification of the charge-coupled device concept". Bell Syst. Tech. Journal, 49, 593-600.

3
Foundations, Baseplates, and Piping

Many rotating equipment alignment problems can be traced to design, installation, or deterioration problems with the foundation, base or soleplate, or the machine casings themselves. Not only is it going to be difficult to obtain accurate alignment initially, but it is going to be equally difficult to maintain satisfactory alignment over long periods of time if the machinery is sitting on unstable or improperly designed foundations and frames.

The best place to start this discussion is at the bottom of things. What are your machines and their foundations really sitting on? Imagine for a moment, the design concerns for machinery that are operating on seagoing vessels, offshore oil and gas platforms, or the 18th floor of an office complex. Not only do these support structures have to bear the weight of the machinery, but they also have to be designed to maintain a stable position if the machinery begins to vibrate.

The vast majority of rotating machinery sits on or is somehow attached to the ground. When selecting a site for rotating machinery, civil engineers must be concerned with the soil conditions and stability of the earth where the machinery is to be located. To a great extent, the earth will act as a giant shock absorber for any motion that occurs in the machinery and also act as the main support for the equipment. What is the rotating machinery you have sitting on: bedrock or sand?

All types of rotating machinery will exhibit some level of vibration during operation, so design engineers need to be concerned about how much vibration (or noise) can or will be transmitted through the structure to the surrounding environment.

On many occasions, although no one will readily admit this, I have known people who have collected vibration data on machinery that wasn't even running, but was vibrating from other machinery that was operating nearby. Much to the embarrassment of the originator, maintenance reports were then submitted stating that the 'idle' machine was out of balance or the bearings needed to be replaced.

As strange as this may sound, it is very possible for an idle machine to sustain bearing damage without even running! For example, if a stationary rotor supported in antifriction bearings is subjected to vibration for long periods of time, indentations in the balls and raceways will occur (i.e. brinnelling).

How long will rotating machinery stay accurately aligned?

It is logical to conclude that the shaft alignment will change if there is a shift in the position of the foundation. This shifting can occur very slowly as the base soils begin to compress from the weight and vibration transmitted from the machinery above. It can also occur very rapidly from radiant or conductive heat transfer from the rotating equipment itself heating the soleplate, concrete, and attached structure. There are documented case histories where drive trains were aligned well within acceptable alignment tolerances and after a 4-6 hour run, moved considerably out of alignment. Way too many people assume that when rotating equipment is aligned when it's installed or rebuilt, the alignment will stay stable forever. When was the last time you checked to see if your equipment is aligned exactly the way it was when you installed it? If you have never checked this you may be in for a big surprise!

With the advent of computer technology and better design knowledge, foundations, structures, and machine casings can be rigorously designed and checked utilizing computer aided design (CAD) and computer aided engineering (CAE) techniques before fabrication ever begins. The field of structural dynamics has provided the means to calculate structural mode shapes and system resonances of complex structures to insure that frequencies from the attached or adjacent machinery doesn't match the natural frequency of the structure itself. However, this newly applied technology can't easily remedy all the equipment already installed, and many of us are saddled with equipment sitting on poorly designed or constructed bases that are cracked or warped with piping strain that has increased from the foundation settling over a period of time or was improperly installed during the initial installation. It is unlikely that every foundation with a problem can be removed, redesigned, and installed so it is important to understand how to deal with the variety of problems that can arise.

This chapter will hopefully provide the reader with basic foundation design principles and some techniques to check equipment in the field to determine if problems exist with the foundation/ frame, or the interface between the machinery and the foundation, or piping and conduit attached to the machine itself.

Foundation Design Philosophies

There are two basic machinery foundation designs: rigid foundations and 'floating' foundations. The rigid foundation design is the most common design found in industry, whereas the 'floating' foundation design is typically found in office buildings, hospitals, etc. There are advantages and disadvantages to each design. Figure 3-1 shows a typical rigid foundation design. Figure 3-2 shows different anchor bolt designs.

Rigid Foundations

Advantages ...

- provides a stable platform to attach rotating machinery using the surrounding soil to absorb motion or vibration.
- somewhat easier to construct than 'floating' platforms.
- ability to design foundation 'inertia block' mass to effectively 'absorb' any vibration from attached machinery and isolate residual motion by segregating the foundation block

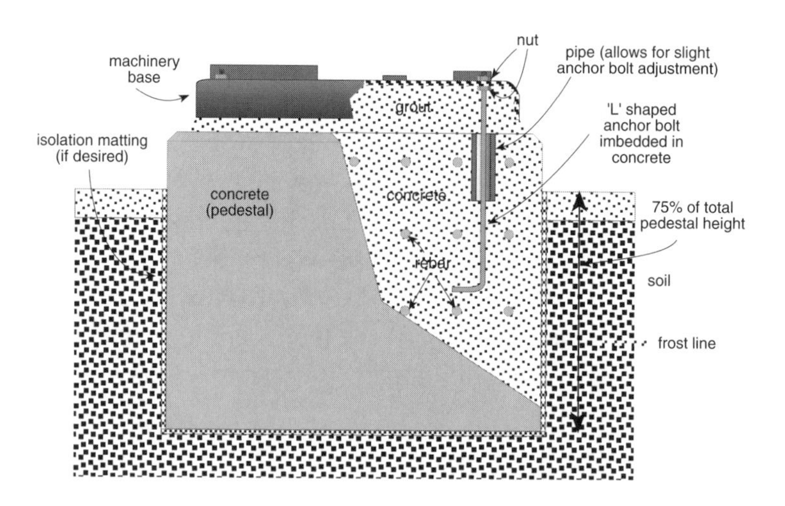

Figure 3-1. Typical rigid foundation design.

with vibration absorptive material preventing transmission of vibration to surrounding area.

Disadvantages ...

- if located outdoors, eventual degradation of foundation is imminent especially if located in geographical area where climactic conditions change radically throughout the year.
- for machinery with attached, unsupported piping or ductwork, extreme forces from improper fits can occur causing damage to machinery.

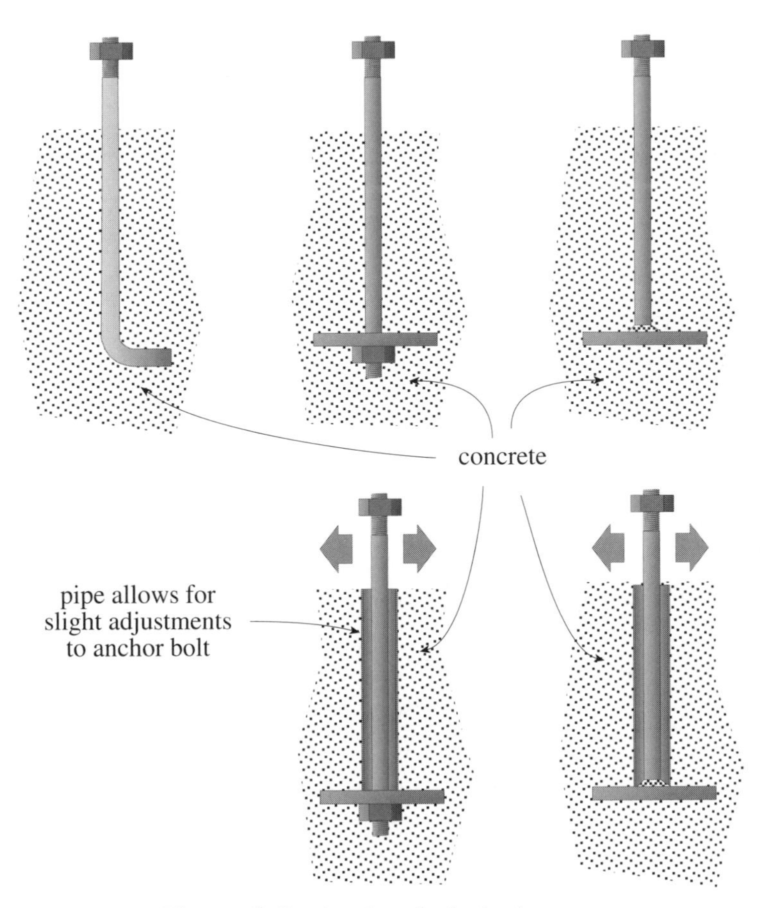

Figure 3-2. Anchor bolt designs.

- potential settling of foundation causing instability and potential transmission of forces from attached piping.
- possibility of absorbing vibration from other machinery in vicinity.

'Floating' Foundations

The 'floating' foundation design is frequently found where rotating machinery is located on upper floors of buildings whereby the machinery is attached to spring mounted concrete slabs that the machinery baseplate and the equipment is attached to. The primary purpose of this design is to isolate any vibration emanating from the machinery to the floor or building where it is attached. This design philosophy is not always used in upper floors of buildings and is sometimes found supporting machinery at ground level. Usually one drive train (e.g. motor - pump) is attached to a

Figure 3-3. Typical 'floating' foundation design.

single concrete slab. Less frequently a number of drive trains can be mounted on a single concrete slab sometimes called a 'skid' or machinery 'pallet'. Although the slab and machinery are frequently separated from the ground or the floor by coil springs, other types of designs use levelling screws and 'pads' touching the floor.

Advantages ...

- if concrete slab and baseplate act as a single 'unit' with sufficient stiffness, this design provides a stable platform to attach rotating machinery allowing the whole assembly to move in the event outside forces such as piping strain are bearing on unit.
- ability to isolate (somewhat) any vibration from attached machinery to surrounding structure and isolate transmitted vibration from other machinery in nearby area.

Disadvantages ...

- somewhat more difficult to construct and maintain than rigid foundations.
- if excessive amount of vibration exists on machinery for prolonged periods of time, potential damage may occur to the machinery or attached piping.
- potential degradation of support springs.
- frequently more difficult to align machinery and keep aligned for long periods.

Baseplates

Baseplates are typically either cast or fabricated as illustrated in figures 3-4 and 3-5. A fabricated baseplate is made by taking structural steel (e.g. I-beams, channel iron, angle iron, structural tubing, formed plate) cutting it into sections and welding the pieces together to make a base.

Figure 3-4. Typical cast baseplates.

Figure 3-5. Fabricated baseplate.

Concrete, Cement, and Grout Basics

Concrete is typically a mixture of inert materials and cement. Grout can be cement based or epoxy based. Cement based grout is typically a mixture of sand and cement. Epoxy based grouts can be pure epoxy consisting of a resin and a hardener (curing agent) or it can be mixed with inert material such as sand, steel 'shot' (small round steel balls), fly ash, etc.

The 'inert materials' in concrete are typically stone and sand but a wide variety of other material can be used. The word 'cement' is from the Latin verb 'to cut' and originally referrred to stone cuttings used in lime mortar. Lime consists of CaO (60-67%), silica SiO_2 (17-25%), alumina Al_2O_2 (3-8%), and small amounts of iron oxide, magnesia, alkali oxides, and sulfuric anhydride.

The cement, typically limestone, clay, or shale, acts as the 'glue' to bond the inert materials together by mixing water with the cement and the aggregates. When the water migrates through the mixture and eventually evaporates, the cement and aggregates chemically bond together by hydration and hydrolysis to form a continuous block. The ratio of water and cement is critical to proper curing, insuring that adequate strength is attained. Too much water will cause the paste to be too thin and will be weak when hardened. U.S. engineer Duff Abrams developed the water:cement ratio guidelines in the 1920's which are still in wide used today. The proportion of a 'typical' concrete mixture is ...

material	**low strength**	**high strength**
water	15%	20%
cement	7%	14%
aggregates	78%	66%

Table 3-1. Concrete mixing ratio.

Compressive strengths of concrete can range from 1,000 to 10,000 psi with a density of around 150 pounds per cubic foot. A compressive strength for concrete typically used in foundations for machinery is between 3000 - 4000 psi.

A 'slump test' is used to determine the consistency of concrete. A standard slump cone is filled with concrete, smoothed off at the top of the cone, and then the cone is lifted vertically clearing the top of the concrete pour allowing the concrete in the cone to 'slump' downward. The measured distance in inches from the original to the final level of the concrete mass is then observed. Concrete slump values for concrete used in machinery foundations should range from 3 to 5 inches.

Proper curing of the concrete requires that the water remain in the mixture for an acceptable period of time to insure that the chemical reaction of the cement is completed. Spraying small amounts of water on the concrete mass, and then laying wet burlap or plastic sheeting over the top of the mass will insure that the rate of water loss is minimal. Pouring concrete in extremely hot temperatures (90-120 degrees F) may cause the water to evaporate too quickly. Pouring concrete in extremely cold temperatures (below 32 degrees F) will cause the water to expand when frozen and produce a very porous mixture with diminished compressive strength. Temperatures down to 25 degrees Farenheit may be acceptable since curing of the concrete mixture is slightly exothermic but the mass should be insulated to entrain any heat during cure. The complex chemical reaction that occurs in the concrete takes place over several months of time. Concrete compressive strengths typically attain 70 - 80% of their final value 6-8 days after the initial pour. Cements may be naturally occurring (lime) or manufactured by crushing anhydrous calcium silicate-bearing rock into a powder and then heated to around 1500 degrees Fahrenheit. Manufactured cement is often called Portland cement. There are 6 basic types of cement as set forth in ASTM specification C150-61 shown in table 3-2.

Types of cement

Type	Name	Description
1	Normal	general purpose
2	Modified	low heat of hydration (curing) desired
3	High Early Strength	high strength required at an early age
4	Low Heat of Hydration	typically used in dams to reduce cracking and shrinkage
5	Sulfate Resistant	used when exposed to soils with a high alkali content
6	Air Entrained	used when severe frost action present

Table 3-2. Types of concrete.

Reinforced Concrete

Concrete is ten times stronger in compression than in tension and must therefore be reinforced by imbedding steel rods in the concrete mixture during the pour to prevent cracking while being subjected to tensile loads. The amount of reinforcement in a foundation varies and should be taken into careful consideration during the design phase. Prestressed concrete is made by placing the reinforcement rods in tension prior to pouring the concrete mixture.

Grouting

Grouting is typically used as the final binding mass between the machinery base / frame and the concrete foundation. As previously mentioned, there are two basic classes of grout - cement based and epoxy based. Proper mixing of the grout is essential to obtain the necessary strength. Be sure to carefully read the mixing instruction from the manufacturer when using any product.

Suggested Grouting Procedure

1. Planning - Prepare a materials list of all the required components (grout, wood, bracing, pump, hose, water, mixing tools, vibrators, etc.). Allow an adequate amount of time to perform the job. Instruct the personnel on the task at hand. Are there enough vent holes in the base or frame

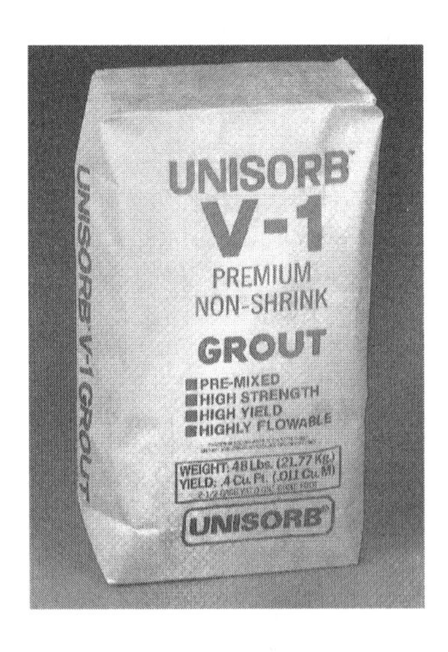

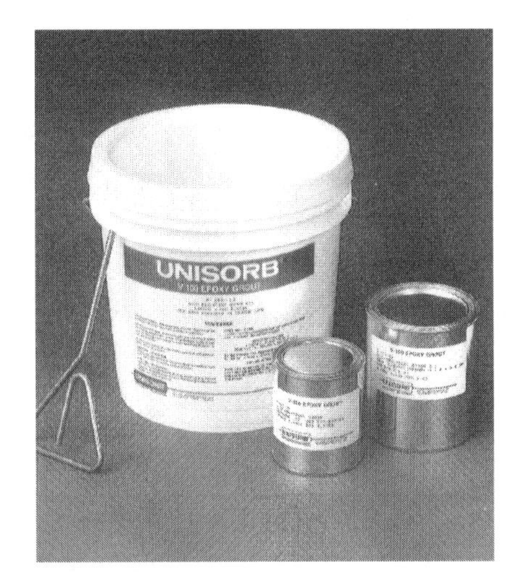

Figure 3-6. Cement-based and epoxy-based grouts. Photo courtesy of Unisorb, Jackson, MI.

Figure 3-7. Mounting isolator. Photo courtesy of Philadelphia Resins, Montgomeryville, PA.

for venting trapped air? Has the concrete foundation cured completely? Is the machinery base in the position you want it in and is it level and not warped? Will the base lift up when grout is pumped under it?

2. Machinery Base / Frame and Concrete Preparation - Insure that all contact surfaces on the undersides of the machinery base or frame are clean, rust-free, and oil-free. If possible, metal surfaces should be sand blasted and primed. The concrete surface should also be clean, dust free, and oil-free. If you are using a cement based grout, the concrete surface should be soaked with water at least 24 hours prior to grout placement to insurethat the dry concrete does not extract the water in the grout mix at an excessive rate preventing proper cure. Install any machinery positioning devices or isolators. Prior to pouring grout remove any puddles of water.
3. Building the form - Construct a form (typically wood) around the perimeter of the machinery base / frame to be grouted. Insure there is adequate clearance between the machinery base / frame and the form to allow for placement of grout and access for pumping or pushing the grout completely under the base. Build the form with a number

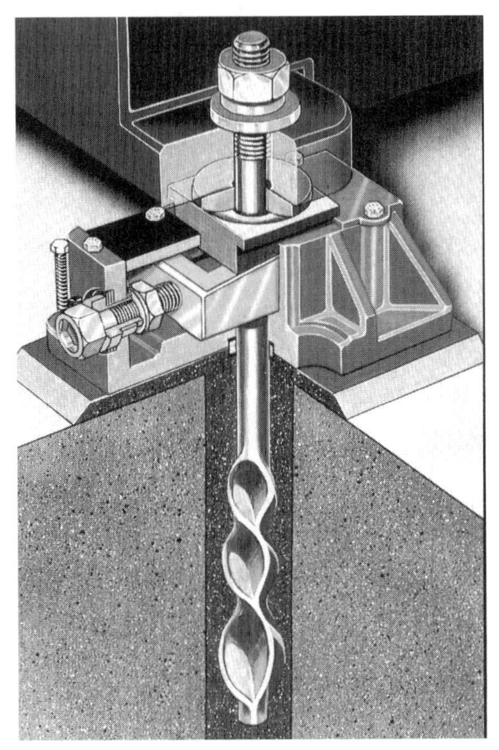

Figure 3-8. Machinery positioner. Drawing courtesy of Unisorb, Jackson, MI.

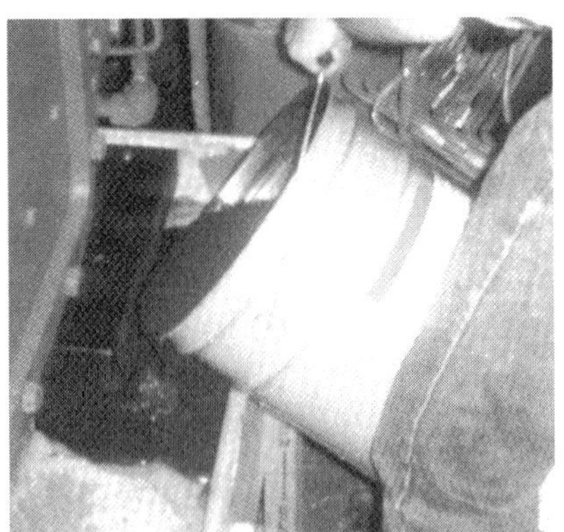

Figure 3-9. Pouring the grout. Photo courtesy of Philadelphia Resins, Montgomeryville, PA.

of pouring points around the perimeter. Insure that the height of the form is above the highest level that the grout must attain to prevent air pockets from forming. Insure there are numerous vent holes of adequate size to discharge trapped air during the pour. If you are using epoxy based grouts insure that there is a coating of wax on all of the wooden form surfaces that will be exposed to the grout so the form doesn't permanently bond to the grout. The wood form can also be wrapped in plastic sheeting as an alternative. Insure that the forms are adequately anchored.

4. Mix the Grout - Carefully follow the manufacturer's recommended mixing instructions.
5. Pour the Grout - Insure that the grout is flowing under all areas of the form, removing entrapped air at all points. Vibrators can be used with cement based grouts.
6. Cure Time - Allow adequate time for the grout to cure.
7. Remove forms.

Figure 3-10. Baseplate deteriorating from rust.

Figure 3-11. Improperly supported motor.

Figure 3-12. Loose shim pack and dowel pin.

Problems to look for ...

A complete visual inspection should be made at least once a year of all rotating equipment foundations, baseplates, piping, etc. Many of these problems are quite obvious as shown in figures 3-10 to 3-12.

Tips for designing good foundations

- Rotating equipment that will experience large amounts of thermal or dynamic movement from 'off-line to running conditions' should be spaced far enough apart to insure that the maximum allowable misalignment tolerance is not exceeded when the shafts are offset in the 'off-line to running conditions' position. Refer to Chapter 9 for more details on off-line to running machinery movement.
- Insure that the natural frequency of the foundation/structure/soil system does not match any running machinery frequencies or harmonics (such as 0.43X, 1X, 2X, 3X, 4X,

etc.) with the highest priority being placed on staying +20% away from the operating speed of the machinery sitting on the foundation being considered. Also, watch for potential problems where running speeds of any machinery nearby the proposed foundation might match the natural frequency of the system being installed. The design criterion for calculating natural frequencies of foundations and structures can be found in some of the reference literature at the end of this chapter.

- In case the calculated natural frequency of the structure does not match the actual structure when built, design in some provisions for 'tuning' the structure after erection has been completed such as : extension of the mat, 'boots' around vertical support columns, attachments to adjacent foundations, etc.
- Design the foundation and structure to provide proper clearances for piping and maintenance work to be done on the machinery, and provisions for alignment of the machine elements.
- Provide vibration joints or air gaps between the machinery foundation and the surrounding building structure to prevent transmission of vibration.
- If possible, provide centrally located, fixed anchor points at both the inboard and outboard ends on each baseplate in a drive train to allow for lateral thermal plate expansion. Insure there is sufficient clearance on the casing foot bolt holes to allow for this expansion to occur without binding against the foot bolts themselves.
- Minimize the height of the centerline of rotation from the baseplate.
- Protect the foundation from any radiant heat generated from machinery, and steam or hot process piping by insulation or heat shields where possible.

Tips on installing foundations and rotating machinery

- Select a contractor having experience in installing rotating machinery baseplates and foundations, or provide any necessary information to the contractor on: compaction of base soils, amount and design of steel reinforcement, preparing concrete joints during construction, grouting methods, etc.
- If the concrete for the entire foundation is not poured all at once, be sure to chip away the top 1/2" to 1" of concrete, remove debris, keep wet for several hours (or days if possible), allow surface to dry and immediately apply cement paste before continuing with an application of mortar (1" to 6") and then the remainder of the concrete. If not done, the existing concrete may extract the water from the freshly poured concrete too quickly and proper hydration (curing) of the new concrete will not occur.
- Use concrete vibrators to eliminate air pockets from forming during the pouring process but do not over vibrate which will cause the larger concrete particles to settle toward the bottom of the pour.
- Check for baseplate distortion. Optical or laser tooling techniques can be used for this method. Refer to Chapter 9 for more information on how to use these tools. Mounting pads should be machined flat and not exceed 2 mils difference across all pads.
- If the baseplate is distorted, relieve stress by oven baking or using vibratory shakers.
- Sandblast and coat the baseplate with inorganic zinc silicate per coating manufacturer's specifications to prevent corrosion and provide good bonding to grout.
- Try not to use wedges to level the baseplate. This requires a two step grouting procedure and may cause cracks to form in the grouting where the wedges were removed. If

the wedges are left in after the grouting has been poured, there is a great likelihood that the grout will crack and separate where the wedges are located. Instead, weld 3/4" or 1" fine threaded nuts to the outside perimeter of the baseplate to use with jackscrews for precise leveling. Machinist levels, optical or laser alignment equipment can be used to check levelness.

- Grout one bulkhead section at a time. Apply grout through a 4-6" diameter hole centrally located in each section. Provide 1" diameter vent holes near the corners of each section. Refer to API specification 610 for additional grouting instructions. Allow a minimum of 48 hours cure time before setting rotating equipment onto base.
- Install jackscrews for moving equipment in all three directions ...vertically, laterally, and axially. If jackscrews will not be used, provide sufficient clearance between baseplate and rotating equipment for insertion of hydraulic jacks for lifting equipment for shim installation. Refer to Chapter 11 on moving and repositioning the machinery.

Piping Strain

I'm almost afraid to even mention anything about this. Piping strain is a monumental problem in industry and I don't really see anyone trying to resolve this issue. The widely held design philosophy seems to be that piping should be loosely constrained so it can move and grow wherever it wants. Many people are surprised to learn that the vast majority of piping failures have occurred from cyclic fatigue, not from tension, compression or shear failures. Yet I continue to see piping being installed with feeble supports such as saddles held in place by long sections of allthread attached to an overhead beam or 'T' frames with 'bowtie' rollers. Most of the piping supports in existence were installed by pipefitters who were just supporting the pipe before all

Figure 3-13. Where are the piping supports?

the connections were made. I guess the rule of thumb for piping supports has been: the fewer, flimsier, and cheaper ... the better!

Static piping forces that result from improper fits can not be detected by simple visual inspection after the piping has been attached to a pump or compressor. Looking at a spring hanger and seeing that the spring is compressed does not indicate that the load is within acceptable limits. Also, spring hangers can only support piping loads in one direction. What if there are other forces acting in directions other than through the axis of the spring?

Even if expansion joints or flexible hose sections are included in the piping, these devices can only accept forces in one or two planes of motion. Installing flexible piping sections is just an excuse for someone to do a poor piping installation. Furthermore, flex hoses are more susceptible to failure than a rigid pipe.

Forces from the expansion or contraction of piping attached to rotating equipment carrying fluids whose temperatures are above or below the temperature of the pipe when no fluid is moving can be enormous and frequently cause drastic movement in the turbomachinery from excessive forces at the connection points.

It's this simple: the flanges and connections on pumps, compressors, fans, etc. were never meant to bear the weight or strain of piping and ductwork. They are fluid connection points. The piping must have adequate support mechanisms that bear the weight and strain of the piping in the vertical, lateral, and axial directions. A good piping design engineer should never view a pump or compressor flange as an anchor point for the piping.

I know that this has never happened at your plant, but I

Figure 3-14. Piping spring support.

have seen pipefitters attach a 20 ton chainfall around one end of a pipe and the other end around an I-beam, pull the pipe into place, install and tighten the flange bolts, and then remove the chain fall. Some of the piping in industry is so poorly installed that the pipefitter has to stand out of the way when the pipes are disconnected from machinery for fear of getting hit by the pipe when it springs away from the connection.

Figure 3-15 shows an adjustable piping anchor support. If the piping misalignment is too excessive and you are not willing to fit the pipe properly, you may want to consider using supports similar to this at the suction and discharge pipes on the pumps and compressors to prevent the pipe from 'kicking' the machinery out of alignment. If the piping strain is excessive and there are no supports to hold the poorly fit pipe in place, there is no guarantee that the equipment will stay aligned for long periods of time even if you do a great job aligning the rotating machinery.

Checking for excessive static piping forces on rotating equipment

Since a majority of rotating equipment is used to transfer liquids or gases, the connecting piping will undoubtedly have an effect on the machinery and could potentially be another source of machinery movement due to: thermal expansion of the piping, reactionary forces from the movement of the liquid in the piping itself, static weight of the piping, or from piping that has not been installed properly causing tension or compression at the piping - machine interface.

The forces that cause machinery to move from improper installation of piping can be checked by using dial indicators to monitor both the horizontal and vertical movement of the machine case. By placing indicators at each corner of the machine element, loosening all the foundation bolts, and observing the amount of movement shown on the indicators, any undesirable forces acting

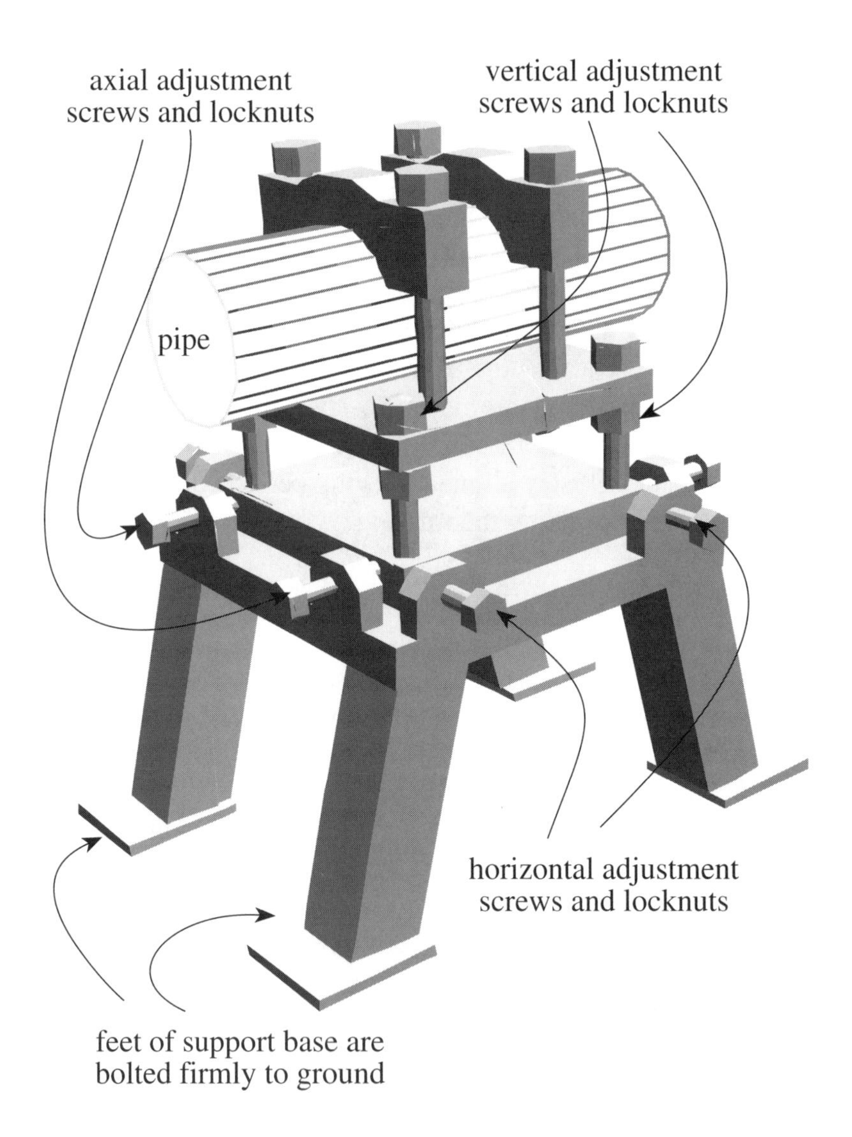

Figure 3-15. Adjustable piping anchor support.

on the machine can be determined. If more than 2 mils of movement is noticed, it may be possible to reposition the other elements in the drive train without modifying the piping to eliminate this problem.

This movement can also be checked with a shaft alignment bracket attached to one shaft with dial indicators positioned at the

How to check for excessive static piping stresses

Align the machinery and then attach brackets or clamps to one shaft and mount dial indicators in the vertical and horizontal position against the other shaft. Zero the indicators, loosen the foot bolts holding the piped machine in place and monitor the indicators for any movement. Ideally, less than 2 mils (0.002") of movement should occur on either indicator.

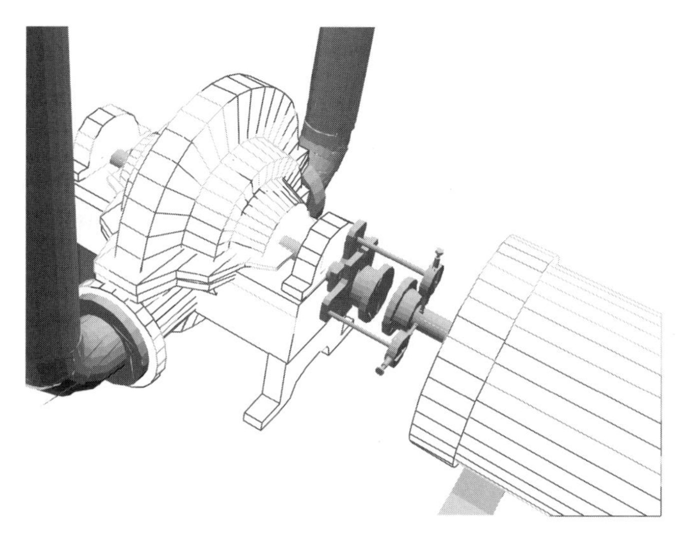

Figure 3-16. Excessive piping stress test.

12 o'clock and 3 o'clock position on the adjacent shaft as shown in figure 3-16. Movement exceeding 2 mils on either dial indicator is unacceptable after all the foundation bolts have been loosened.

Acceptable piping forces on rotating machinery

It's very easy to rant and rave about excessive piping forces but what is the bottom line here? How much force can piping exert on rotating machinery casings? If excessive piping force is exerted on rotating machinery what could or will happen? Excessive piping forces could ...

- Distort the machine case, upsetting internal clearances between moving and stationary parts of the machine.
- Cause the machine case to shift its position over a short (or long) period of time, disrupting the alignment condition.
- Cause the hold down bolts to loosen or shear along with the shim packs and dowel pins if used.

To my knowledge, no manufacturer of rotating machinery has ever conducted tests on their equipment to determine what limits of force are allowable on their piping or ductwork connections. The assumption that every rotating equipment vendor makes is that there will be no forces acting on the piping or ductwork connections when the machinery is installed. And when force is applied, everyone turns a blind eye and hopes nothing happens. The primary problem is that there are so many factors involved that it would be difficult to arrive at a formula to easily compute maximum allowable loads on fluid connections.

Visual Inspection Checklist

On an annual basis (at least), a visual inspection should be made of all rotating machinery that should include the following items ...

- properly positioned piping hangers that carry the weight of the piping.
- piping expansion joints that move freely to accept thermal or hydraulic movement.
- loose piping flange bolts.
- cracked concrete bases or support columns.
- cracks propagating at concrete joints.
- water seeping between baseplate and concrete foundation that could freeze and damage the structure.
- loose foundation bolts.
- shim packs that have worked loose.
- rusty shims.
- loose or sheared dowel pins.
- paint on shims.

Bibliography

Newcomb, W.K., "Principles of Foundation Design for Engines and Compressors", A.S.M.E. paper no 50-OGP-5, April 1951.

Witmer, F.P., "How to Cut Vibration in Big Turbine Generator Foundations", Power, Nov., 1952.

Swiger, W. F., "On the Art of Designing Compressor Foundations", A.S.M.E. paper no. 57-A-67, Nov. , 1958.

Sohre, J.S., " Foundations for High - Speed Machinery", A.S.M.E. paper no. 62-WA-250, Sept. 7, 1962.

Centrifugal Pumps for General Refinery Service, API Standard 610, American Petroleum Institute, Wash., D.C., Mar. , 1973.

Murray, M.G., " Better Pump Baseplates", Hydrocarbon Processing, Sept., 1973.

Essinger, Jack N., "A Closer Look at Turbomachinery Alignment", Hydrocarbon Processing, Sept., 1973.

Centrifugal Compressors for General Refinery Services, API Standard 617, American Petroleum Institute, Wash., D.C., Oct. 1973.

Murray, M.G., " Better Pump Grouting", Hydrocarbon Processing, Feb., 1974.

Dodd, V.R., **Total Alignment**, the Petroleum Publishing Co., Tulsa, Okla., 1975.

Massey, John R., " Installation of Large Rotating Equipment Systems-A Contractor's Comments", Proceeding - Fifth Turbomachinery Symposium, Gas Turbine Labs, Texas A&M University, College Station, Texas, Oct. 1976.

Abel, L.W., Chang, D.C., Lisnitzer, M., "The Design of Support Structures for Elevated Centrifugal Machinery", Proceedings Sixth Annual Turbomachinery Symposium , Gas Turbine Labs, Texas A&M University, Dec., 1977, pages 99-105.

Renfro, E. M., " Repair and Rehabilitation of Turbomachinery Foundations", Proceedings - Sixth Annual Turbomachinery Symposium, Gas Turbine Labs, Texas A&M University, Dec., 1977, pgs. 107-112.

Simmons, P.E., "Defining the Machine/Foundation Interface", Second International Conference, The Institution of Mechanical Engineers and A.S.M.E., Churchill College, Cambridge, England, Sept. 1-4, 1980, paper no. C252/80.

Kramer, E., "Computations of Vibration of the Coupled System Machine-Foundation", Second International Conference, The Institution of Mechanical Engineers and A.S.M.E., Churchill College, Cambridge, England, paper no. C300/80, Sept. 1-4, 1980.

Whittaker, Wayne,"Preventing Machinery Installation Problems", Manufacturing Engineering, April 1980.

Whittaker, Wayne,"Recommendations for Grouting Machinery", Plant Engineering, January 24, 1980.

Whittaker, Wayne,"Concrete : The Basics", Unisorb Technical Manual, 1980.

Renfro, E. M., "Five Years with Epoxy Grouts", Vibration Institute, Proceedings Machinery Vibration Monitoring and Analysis Meeting, New Orleans, LA, June 26-28, 1984.

Nailen, Richard L., "Installing Motors Properly", Plant Services, January 1988, pgs. 72-82.

Machinery Grouting Manual, Philadelphia Resins, Montgomeryville, PA, 1994.

4
Flexible & Rigid Couplings

One of the most important components of any drive system is the device connecting the rotating shafts together known as the coupling. Since it is nearly impossible to maintain perfectly collinear centerlines of rotation between two or more shafts, flexible couplings are designed to provide a certain degree of yielding to allow for initial or running shaft misalignment. There is a wide assortment of flexible coupling designs, each available in a variety of sizes to suit specific service conditions.

The design engineer invariably asks: why are there so many types, and is one type better than any other? Simply put, there is no 'perfect' way to connect rotating shafts (so far!). As you progress through this chapter, you will find that perhaps two or three different coupling types will fit the requirements for your drive system. One coupling being better than another is a relative term. If two or more coupling types satisfy the selection criterion and provide long, trouble free service, they are equal, not better. The ultimate challenge is for you to accurately align shafts, not to find a coupling that can accept gross amounts of misalignment to compensate for your ineptitude.

The pursuit to effectively connect two rotating shafts dates back to the beginning of the industrial era where leather straps and bushings, or lengths of rope intertwined between pins, were the medium used to compensate for shaft misalignment. Several flexible coupling designs emerged immediately after the introduction of the automobile from 1900 to 1920. As shaft speeds increased, coupling designs were continually refined to accept the new demands placed on them. As industrial competition became more severe, equipment downtime became a major concern and industry became increasingly more interested in coupling failures in an effort to prolong operating lifespan.

Patents for diaphragm couplings date back to the 1890's, but didn't become widely used until just recently as diaphragm design, material, and construction vastly improved. 'O'-ring type seals, and crowning of gear teeth in gear couplings came about during World War II. The awareness and concern for coupling and rotating machinery problems are reflected by the increase in technical information generated since the mid - 1950's. Coupling designs will continually be refined in the coming years with the ultimate goal of designing the 'perfect' coupling.

Coupling and shaft misalignment tolerances - what's the difference?

It is important for the person selecting the coupling not to be confused by the term 'allowable misalignment' in a coupling. The coupling manufacturers will often quote information on allowable misalignment for the coupling and not necessarily the equipment it is coupled to. These tolerances seem to lull the user into a sense of complacency leading one to believe that accurate shaft alignment is not necessary since " ... the coupling can take care of any misalignment" (famous last words).

It is imperative that you can differentiate between coupling tolerances and alignment tolerances. Coupling misalignment tolerances quoted by flexible coupling manufacturers typically specify the mechanical or fatigue limits of the coupling or components of the coupling. These misalignment tolerances are frequently excessive compared to the misalignment tolerances specified in Chapter 5 which deals with the rotating drive system as a whole. The misalignment tolerance guide shown in figure 5-4 is concerned with the survivability of not only the coupling, but also the shafts, seals, and bearings of the machinery over long periods of time.

The role of the flexible coupling

Exactly what is a coupling supposed to do? If a 'perfect' coupling were to exist, what would its design features include?

- allow limited amounts of parallel and angular misalignment
- transmit power
- insure no loss of lubricant in grease-packed couplings despite misalignment
- easy to install and disassemble
- accept torsional shock and dampen torsional vibration
- minimize lateral loads on bearings from misalignment
- allow for axial movement of shafts (end float) even under misaligned shaft conditions without transferring thrust loads from one machine element to another
- stay rigidly attached to the shaft without damaging or 'fretting' the shaft
- withstand temperatures from exposure to environment or

from heat generated by friction in the coupling itself
- ability to run under misaligned conditions (sometimes severe) when equipment is initially started to allow for equipment to eventually assume it's running position
- provide failure warning and overload protection to prevent coupling from bursting or flying apart
- produce minimum unbalance forces
- have a minimal effect on changing system critical speed(s).

What to consider when specifying a flexible coupling

Although some of the items listed below may not apply to your specific design criterion when specifying a flexible coupling for a rotating equipment drive system, it is a good idea to be aware of all of these items when selecting the correct coupling for the job.

- normal horsepower and speed.
- maximum horsepower/torque being transmitted at maximum speed (often expressed as hp/rpm).
- misalignment capacity: parallel, angular, and combinations of both parallel and angular.
- can the coupling accept the required amount of 'cold' offset of the shafts without failure during start-up?
- torsional flexibility.
- service factor.
- temperature range limits.
- how is the coupling attached to the shafts?
- size and number of keyways.
- type and amount of lubricant (if used).
- type and design of lubricant seals .
- actual axial end float on rotors.
- allowable axial float of shafts.
- actual axial thermal growth or shrinkage of rotors.

- type of environment coupling will be exposed to.
- will coupling be subjected to radial or axial vibration from the equipment?
- diameter of shafts and distance between shafts.
- type of shaft ends (straight-bore, tapered, threaded, etc.).
- starting and running torque requirements.
- are the running torques cyclic or steady state?
- where is a failure likely to occur and what will happen?
- noise and windage generated by the coupling.
- cost and availability of spare parts.
- lateral and axial resonances of the coupling.
- coupling guard specifications for : size, noise and windage control.
- installation procedure.
- moments of inertia
- heat generated from misalignment, windage, friction.

Types of flexible couplings

The couplings found in this chapter show some of the commonly used couplings in industry today, but in no way reflect every type, size, or manufacturer. The information presented for each coupling concerning capacity, maximum speeds, shaft bore diameters, and shaft to shaft distances are general ranges and do not reflect the maximum or minimum possible values available for each coupling design.
There are five broad categories of flexible couplings ...

- miniature
- mechanically flexible
- elastomeric
- metallic membrane / disk
- miscellaneous

Misalignment capacities will not be given for a variety of reasons: 1) manufacturers of similar couplings do not agree or

Figure 4-1. Miniature flexible couplings. Photos courtesy of (clockwise from left), Rocom Corp., Huntington Beach, CA, Metal Bellows, Sharon, MA, and Guardian Industries, Michigan City, IN.

publish identical values for angular of parallel misalignment, 2) manufacturers rarely specify if the maximum values for angular misalignment and parallel misalignment are separate or a combination of the angular and parallel values stated, 3) it is the intent of this book to provide the reader with the ability to obtain alignment accuracies well within the limits of any flexible coupling design. Coupling manufacturers assume that the user will run the coupling within their stated maximum misalignment values. If your rotating equipment or coupling has failed due to excessive misalignment, it is your fault. Good luck trying to get the coupling manufacturer to pay for the damages!

This does not infer that all couplings accept the same maximum misalignment amounts or that these allowable values should not influence the selection of a coupling. Always consult with your coupling vendor or manufacturer about your specific coupling needs. If you are not getting the satisfaction you feel you need to properly select a coupling, consult a variety of manufacturers (or end users) to comment on design selection or problem

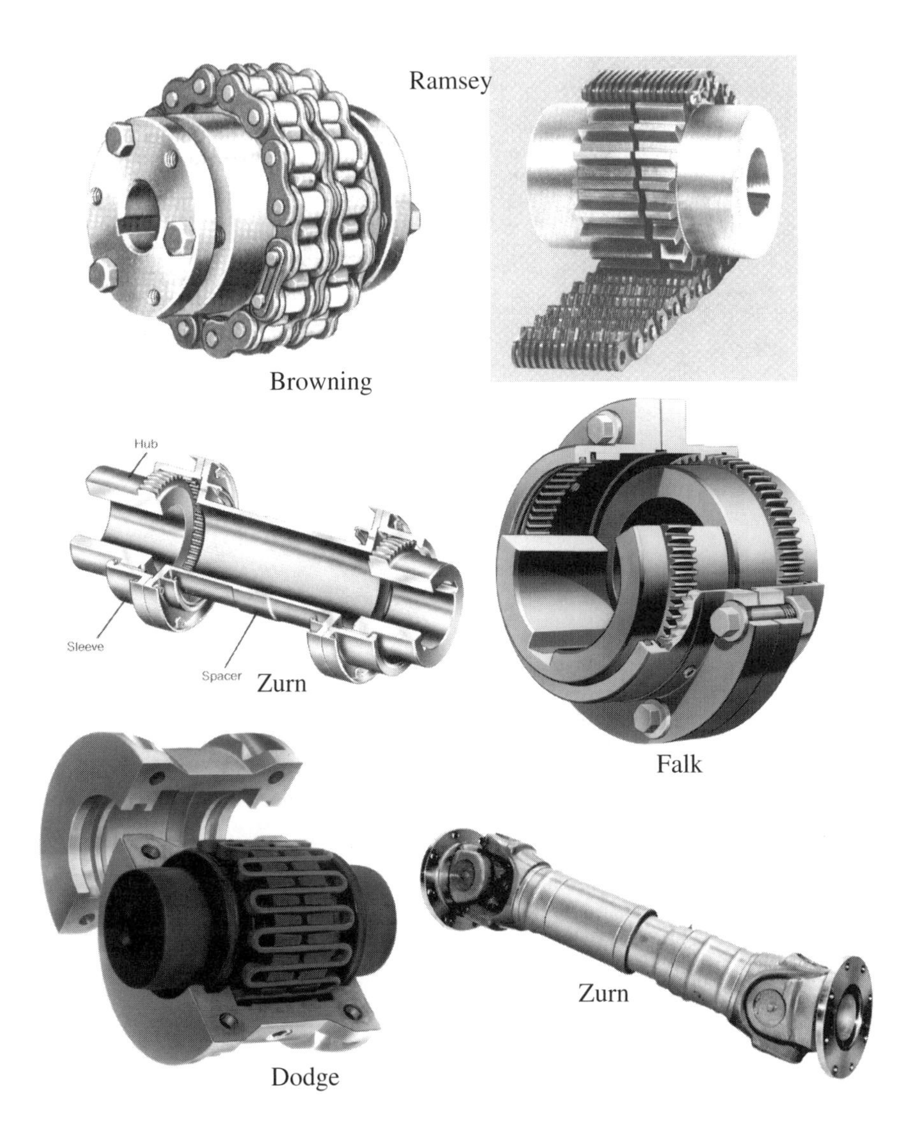

Figure 4-2. Mechanically flexible couplings. Photos courtesy of (clockwise from upper left), Browning Mfg., Maysville, KY, Ramsey Products, Charlotte, NC, Falk Corp., Milwaukee, WI, Zurn Industries, Erie, PA, Dodge-Reliance Electric, Cleveland, OH, Zurn Industries, Erie, PA.

Figure 4-3. Elastomeric flexible couplings. Photos courtesy of (clockwise from upper left), Falk Corp., Milwaukee, WI, Browning Mfg., Maysville, KY, KTR Corp., Michigan City, IN, Magnaloy Coupling Co., Alpena, MI, Lovejoy, Downers Grove, IL, T. B. Wood's and Sons, Chambersburg, PA, Dodge-Reliance Electric, Cleveland, OH.

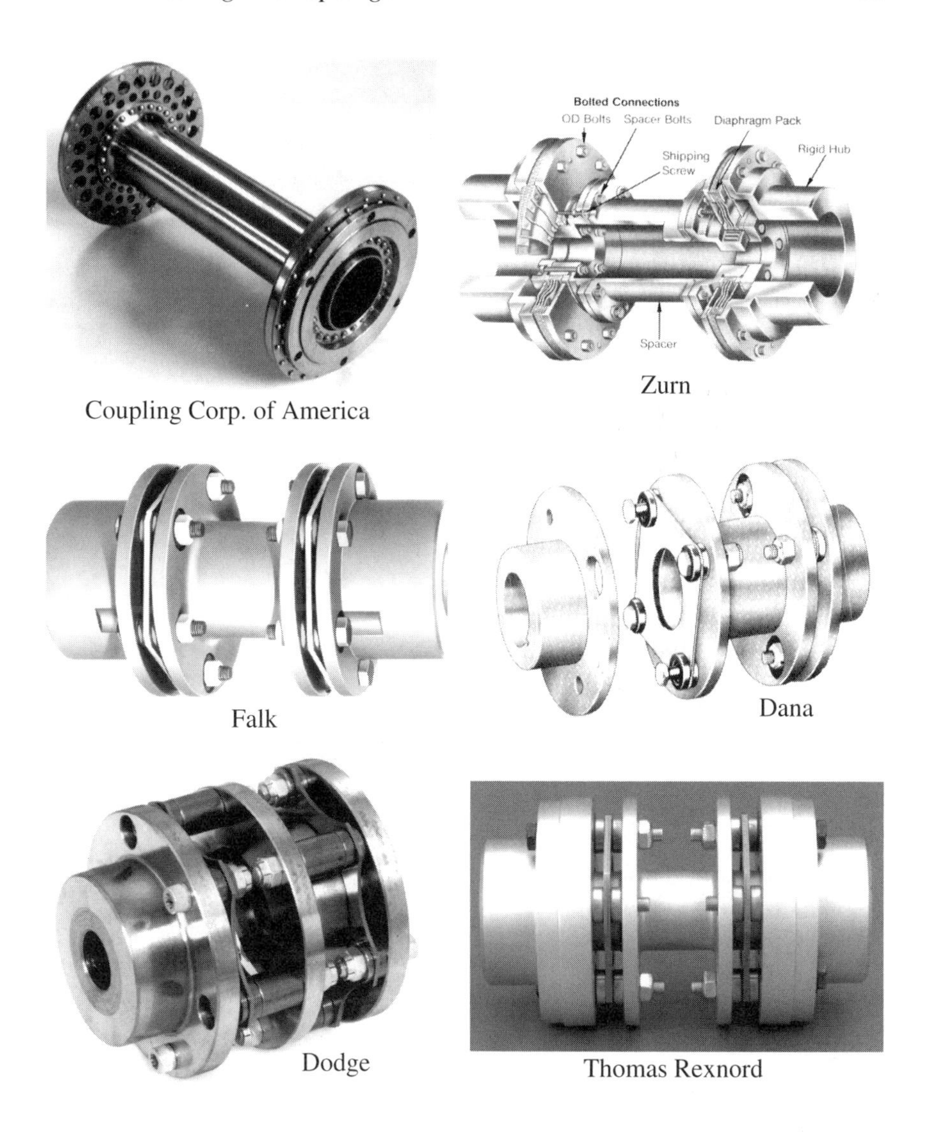

Figure 4-4. Disc / diaphragm flexible couplings. Photos courtesy of (clockwise from upper left), Coupling Corp. of America, York, PA, Zurn Industries, Erie, PA, Dana Industrial Power Transmission, San Marcos, TX, Thomas Rexnord, Warren, PA, Dodge - Reliance Electric, Cleveland, OH, Falk Corp., Milwaukee, WI.

Figure 4-5. Miscellaneous flexible couplings. Photos courtesy of (clockwise from upper left), David Brown Gear Industries, Agincourt, ONT, Browning Mfg., Maysville, KY, Schmidt Couplings, Cincinnati, OH.

identification and elimination.

Although there are a variety of coupling designs that accommodate fractional horsepower devices such as servomechanisms, this chapter will primarily show flexible couplings used on high horsepower, high speed turbomachinery. However, to give the reader an idea of the design differences between the fractional and higher horsepower couplings, Figure 4-1 illustrates a few of the fractional horsepower designs.

Mechanically flexible coupling designs

Chain couplings

The chain coupling is basically two identical gear sprockets with hardened sprocket teeth connected by a double width roller or 'silent' type chain. Packed grease lubrication is primarily used with this type of construction necessitating a sealed sprocket cover. A detachable pin or master link allows for removal of the chain. Clearances and flexing of the rollers and sprocket allow for misalignment and limited torsional flexibility.

capacity: to 1000 hp @ 1800 rpm (roller), 3000 hp @ 1800 rpm (silent)
max speed: up to 5000 rpm
shaft bores: up to 8 in.
shaft spacing: determined by chain width, generally 1/8" to 1/4"

Special designs and considerations: Wear generally occurs in sprocket teeth due to excessive misalignment or lack of lubrication. Torsional flexibility limited by yielding of chain.

Figure 4-6. Chain type flexible coupling. Photo courtesy Browning Mfg., Maysville, KY.

advantages

- easy to disassemble and reassemble
- fewer number of parts

disadvantages

- speed limited due to difficulties in maintaining balancing requirements.
- requires lubrication.
- limited allowable axial displacement.

Gear couplings

The gear coupling consists of two hubs with external gear teeth that are attached to the shafts. A hub cover or sleeve with internal gear teeth engage with the shaft hubs to provide the transmission of power. Gear tooth clearances and tooth profiles allow misalignment between shafts. Lubrication of the gear teeth is required and various designs allow for grease or oil as the lubricant.

- capacity: up to 70,000 hp
- max. speed: up to 50,000 rpm.
- shaft bores: up to 30 in.
- shaft spacing: up to 200 in.

- Special designs and considerations: A considerable amount of attention is paid to the form of the tooth itself and the tooth 'profile' has progressively evolved through the years to provide minimum wear to the mating surfaces of the internal and external gear sets.

Figure 4-7. Gear type flexible coupling. Photo courtesy of Falk Corp., Milwaukee, WI.

To provide good balance characteristics, the tip of the external gear tooth is curved and tightly fits into the mating internal gear hub cover. If the fit is too tight, the coupling will be unable to accept misalignment without damaging the coupling or the rotating equipment. If it is too loose, the excessive clearance will cause an imbalance condition. Obtaining a good fit can become very tricky when the coupling hubs have been thermally or hydraulically expanded and shrunk onto a shaft where an increase in diameter of the external gears will occur. As a rule of thumb, 1 mil per inch of external gear tooth diameter can be used as the clearance.

The amount of misalignment in a gear coupling directly affects the wear that will occur in the mating gear teeth. To better understand the motion of a gear coupling, Figure 4-8 shows the two basic positions gear teeth will take in its sleeve.

At a certain point during the rotation of the shaft, the gear tooth is in a tilted position and will completely reverse its tilt angle 180 degrees from that point relative to the coupling sleeve. 90 degrees from the tilted position, the gear tooth now assumes a pivoted position which also reverses 180 degrees. The gear tooth forces are at their maximum when in the tilted position as they supply the rotational transmission of power, and as the misalignment increases, fewer teeth will bear the load. Depending on the relative position of the two shafts, each set of gear teeth on each hub may have their maximum tilt and pivot points at different positions with respect to a fixed angular reference location.

Under excessive misalignment conditions, the load will be carried by the ends of the gear tooth flank and eventually cause gouging of the internal gear teeth and 'knife edging' of the external gear since the compressive stresses are extremely high, forcing any lubricating film out, allowing metal to metal contact to occur.

Another peculiar wear pattern evolves when the gear tooth sliding velocity falls into the 5 - 8 inches per second range and lubrication between the gear teeth diminishes. This type of wear is known as ‘worm-tracking’ where gouges occur generally from the base to the tip of the tooth flank. The formation of this type of wear pattern will occur when little or no lubrication occurs at the points of contact between teeth and the metal to metal contact fuse welds a small portion of the tooth flanks. As rotation continues, cracks begin to form at the outer edges of the weld and eventually propagate until the two welded pieces separate entirely from the mating external and internal gear teeth.

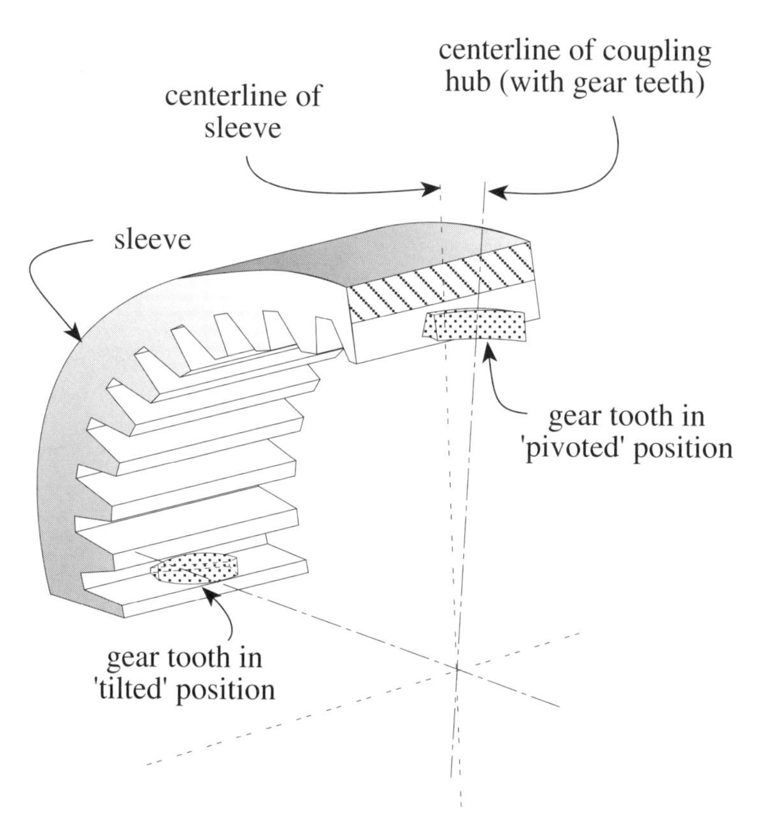

Figure 4-8. Tilted and pivoted positions of the gear teeth in its sleeve.

advantages

- allows freedom of axial movement.
- capable of high speeds.
- low overhung weight.
- good balance characteristics with proper fits and curved tooth tip profile.
- long history of successful applications.

disadvantages

- requires lubrication.
- temperature limitation due to lubricant.
- difficult to calculate reaction forces and moments of turbomachinery rotors when using these couplings since the values for the coefficient of friction between the gear teeth vary considerably.

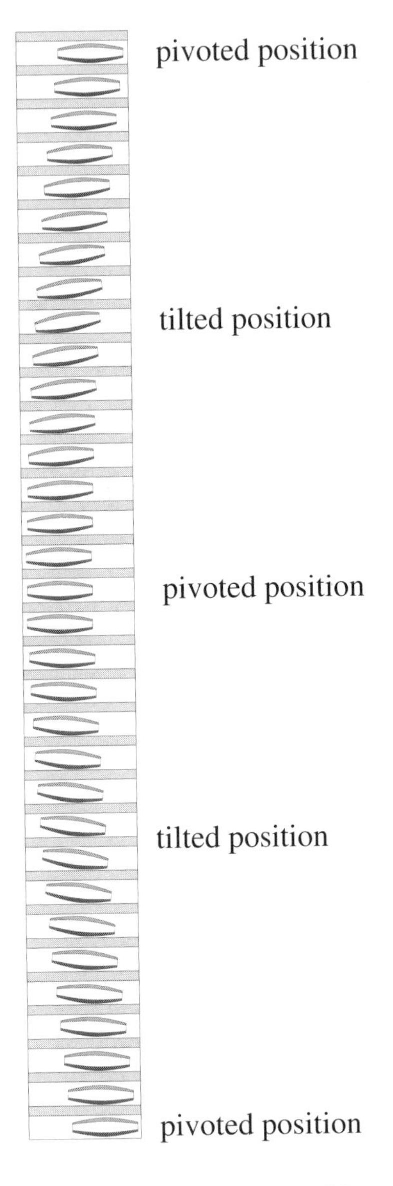

Figure 4-9. Gear tooth tracking pattern when subjected to misalignment conditions.

Metal ribbon couplings

The metal ribbon coupling was introduced around 1919 by Bibby Co. Metal ribbon couplings consist of two hubs with axial 'grooves' on the outer diameter of the hub where a continuous 'S' shaped grid meshes into the grooves. Misalignment and axial movement is achieved by flexing and sliding of the grid member in specially tapered hub 'teeth'.

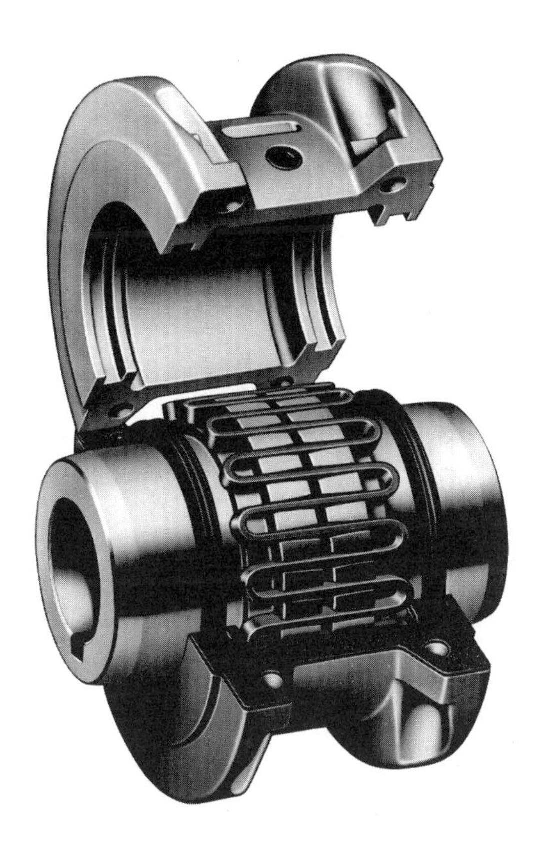

Figure 4-10. Metal ribbon type coupling. Photo courtesy of Falk Corp., Milwaukee, WI.

- capacity: up to 70,000 hp / 100 rpm.
- max. speed: to 6000 rpm.
- shaft bores: to 20 in.
- shaft spacing: to 12 in.

- Special designs and considerations: Grid fabricated from hardened, high strength steel. Close coupled hubs with removable spacer available.

advantages
- easy to assemble and disassemble.
- long history of successful applications.
- torsionally soft.

disadvantages
- requires lubrication.
- temperature limited.
- speed limited.

Universal joint couplings

Perhaps the oldest flexible coupling in existence is the universal joint coupling. This coupling is also know as the Cardan or Hooke joint (see Chapter 2). The basic design consists of 'U' shaped shaft ends with a hole drilled through each 'U' to accept a '+' shaped cross.

If one universal joint is used to connect two shafts together, then only pure angular misalignment can exist where the centerlines of rotation intersect at the center of the '+' shaped cross. As mentioned earlier, for a flexible coupling to accept both parallel and angular misalignment, there must be two flexing points. Therefore most universal joint couplings have two cross/ yolk assemblies as illustrated in figure 4-11.

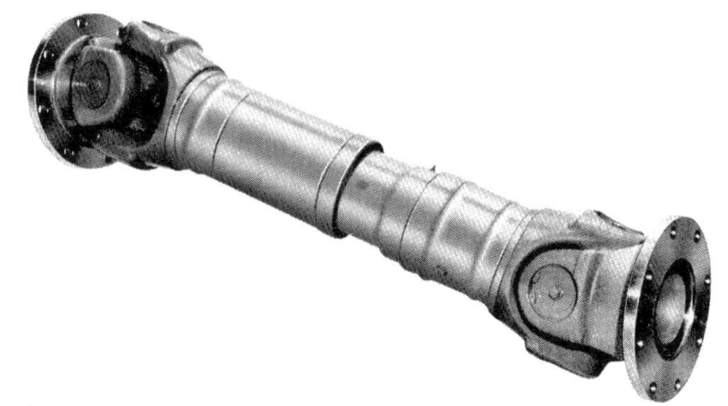

Figure 4-11. Universal joint. Photo courtesy of Zurn Industries, Erie, PA.

With a single universal joint, if the input and output shafts are not in line, a variation of the output shaft speed (ω) will result called Cardan error. When two universal joints are used where the entrance and exits angles are the same with the yokes aligned properly, the system is kinematically balanced producing synchronous shaft rotation at the input and output ends.

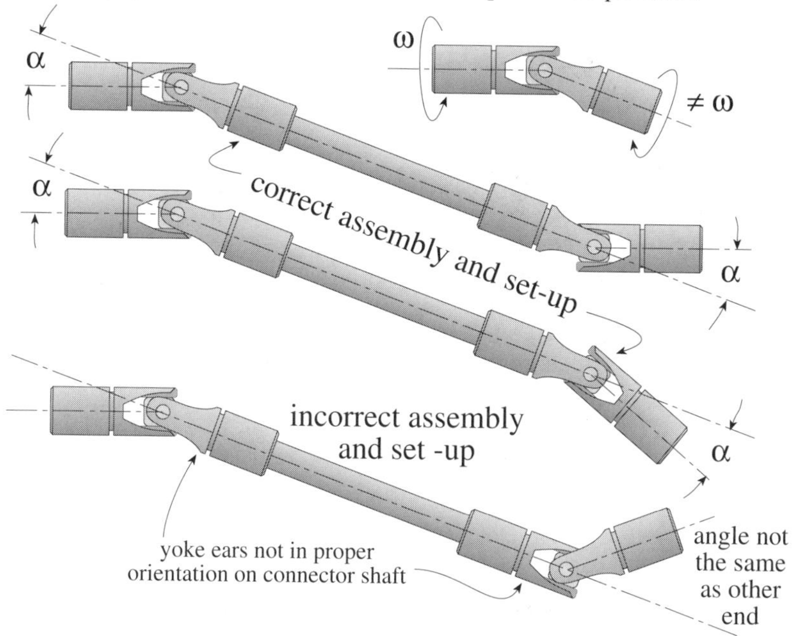

Figure 4-12. Universal joint basics.

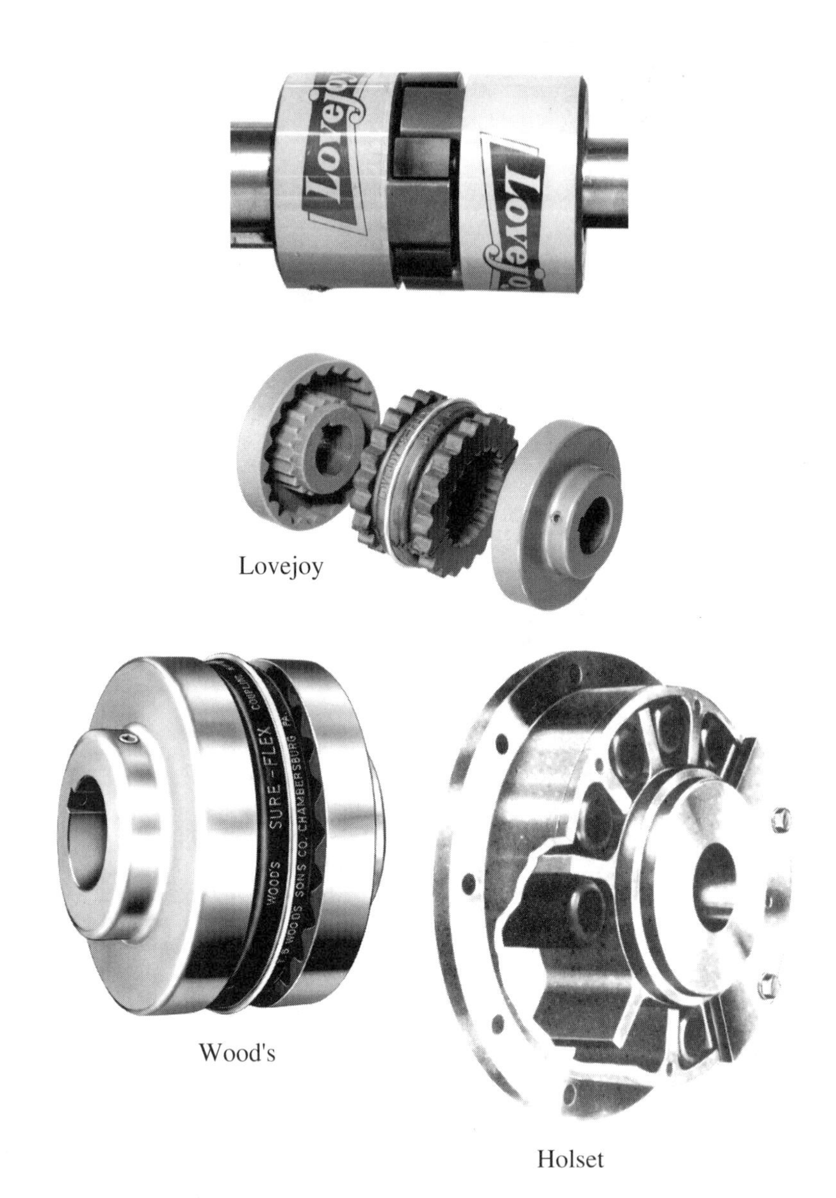

Figure 4-13. Elastomeric coupling types. Photo courtesy Lovejoy Corp., Downers Grove, IL, Holset Engineering Co., Cincinnati, OH, T. B. Wood's & Sons, Chambersburg, PA.

When one universal joint is used it is important to recognize that variations in angular velocity will occur between the two connected shafts, often referred to as the 'Cardan error'. Sinusoidal motion will occur in the axial and torsional directions producing axial vibration and torsional (i.e. 'twisting') vibration particularly if the torque being transmitted and the rotational speed is high.

When two universal joints are used it is important to recognize that sinusoidal motion will also occur in the axial and torsional directions if the 'entrance' and 'exit' angles are not the same as shown in figure 4-12. When these angles are the same in both planes, perfect kinematic balance exists across the coupling cancelling the torsional and thrust variance.

Elastomeric couplings

A wide variety of design variations employ an elastomeric medium to transmit torque and accommodate misalignment . Most of these couplings are torsionally 'soft' to absorb high starting torques or shock loads.

- capacity: up to 67,000 hp/ 100 rpm but varies widely with design.
- max. speed: approx. 5000 rpm (varies widely with design).
- shaft bores: up to 30 in.
- shaft spacing: up to 100 in. (varies widely).

- Special designs and considerations: A considerable amount of inventiveness and ingenuity has been applied to this type of coupling design through the years as evidenced by the large array of design variations. The elastomeric medium is generally natural or synthetic rubber, urethane, nylon, teflon, or oil impregnated bronze. Since the elastomer is markedly softer than the hubs and solid driving elements (wedges, pins, jaws, etc.), wear is minimal and replacement of the elastomer itself is all that is usually needed for periodic servicing.

advantages

- minimal wear in coupling.
- acts as vibration damper and isolator.
- acts as electrical shaft current insulator in some designs.
- torsionally 'soft'.
- accepts some axial movement and dampens axial vibration.
- no lubrication required.

disadvantages

- speed limited due to distortion of elastomer from high centrifugal forces causing imbalance.
- deterioration of elastomer from: temperature, oxidation of rubber, corrosive attack from undesirable environment.
- potential safety hazard if elastomeric member releases from drive elements.
- some designs may cause undesirable axial forces.
- heat generated from cyclic flexing of elastomer.

Metallic membrane / disk type coupling designs

Diaphragm couplings

Transmission of power occurs through two flexible metal diaphragms, each bolted to the outer rim of the shaft hubs and connected via a spacer tube. Misalignment and axial displacement is accomplished by flexing of the diaphragm members.

- capacity: up to 30,000 hp.
- max. speed: up to 30,000 rpm.
- shaft bores: up to 7 in.
- shaft spacing: 2 to 200 in.

- Special designs and considerations: Metal diaphragm couplings are a highly reliable drive component when operated within their rated conditions. Exceeding the maximum allowable angular and/ or parallel misalignment values or axial spacing will eventually result in disc failure. Since the diaphragm is, in effect a spring, considerations must be given to the axial spring rate and vibration characteristics to insure that the diaphragm coupling natural frequency does not match rotating speeds or harmonics in the drive system.

advantages

- excellent balance characteristics.
- no lubrication required.
- low coupling weight and bending forces on shafts when operated within alignment limits.
- accepts high temperature environment.

disadvantages

- limited axial displacement and oscillation.
- proper shaft spacing requirements are generally more stringent than other coupling types.

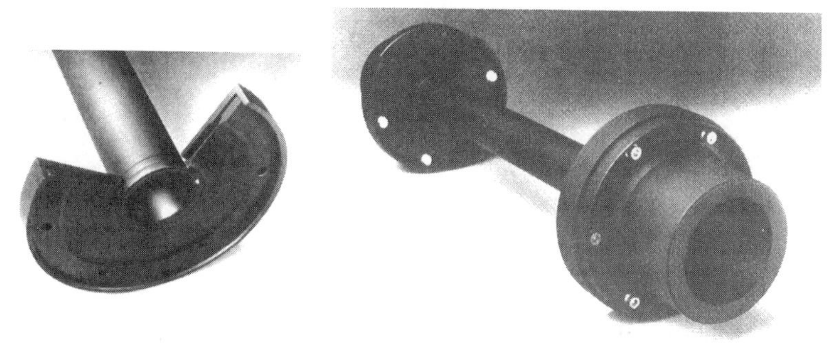

Figure 4-14. Contoured diaphragm type coupling. Photo courtesy of Kopflex Corp., Baltimore, MD.

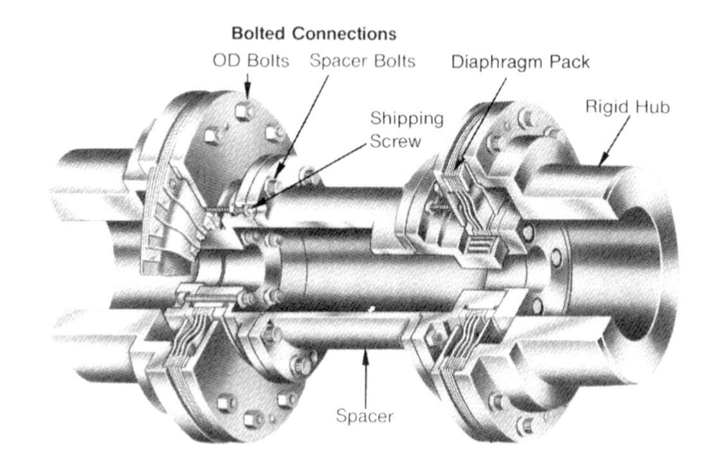

Figure 4-15. Flexible disk type couplings. Photos courtesy of Zurn Industries, Erie, PA (upper), Thomas - Rexnord, Warren, PA (lower).

- excessive misalignment will transmit high loads to shafting.

Flexible disc couplings

The flexible disc coupling is very similar in design principles to the diaphragm coupling with the exception that multiple, thinner discs or a non-circular flexing member is used as the flexing element instead of circular, contoured diaphragm elements.

- capacity: up to 65,000 hp / 100 rpm.
- max. speed: up to 30,000 rpm.
- shaft bores: to 12 in.
- shaft spacing: to 200 in.

- Special designs and considerations: It is important to note that two disc packs (or diaphragms) are needed to accommodate parallel misalignment whereas a single disc can only handle pure angular misalignment. Convolutions in the discs provide linear stiffness vs. deflection characteristics, as opposed to flat disc profiles. Once again, coupling axial resonance information must be known to prevent problems where a match may occur with machinery running speeds, higher order harmonics, or subsynchronous forcing mechanisms (oil whirl, looseness of bearing housings, clearance induced whirls, etc.).

advantages and disadvantages

- same as diaphragm couplings.

Miscellaneous coupling designs

Flexible link

The flexible link coupling utilizes a series of cross laced, metallic links with one end of each link attached to a disc mounted on the driven shaft, and the other end of each link attached to a disc mounted on the driver shaft. The links are matched in pairs so that when one is in tension, the other is in compression. Misalignment and axial displacement is accomplished by a flexing action in the series of cross links.

- capacity: up to 1100 hp / 100 rpm.
- max. speed: to 1800 rpm.
- shaft bores: up to 20 in.
- shaft spacing: close coupled or 100 mm spacer with certain designs.

Special designs and considerations: An axial 'fixation' device can be installed to prevent any axial movement if desired. Different designs can accommodate unidirectional or bi-directional rotation.

advantages
- no lubrication required.

disadvantages
- limited axial movement.
- limited misalignment capabilities (can accept pure angular misalignment only).

Figure 4-16. Flexible link type couplings. Photo courtesy of Eaton Corp., Airflex Division, Cleveland, OH, under license from Dr. Ing., Geislinger & Co., Salzburg, Austria.

Leaf spring

This coupling employs a series of radially positioned sets of leaf springs attached to an outer drive member and indexed into axial grooves in the inner drive member. The chamber around each spring set is filled with oil. When the spring pack is deflected, damping occurs as the oil flows from one side of the spring pack to the other.

- capacity: up to 15,000 hp / 100 rpm.
- max. speed: 3600 rpm.
- shaft bores: up to 12 in.
- shaft spacing: up to 40 in.

- Special designs and considerations: Designed primarily for diesel and reciprocating machines. Capable of transmitting shock

Figure 4-17. Leaf spring type coupling. Photo courtesy of Eaton Corp., Airflex Division, Cleveland, OH.

torque values substantially higher than other couplings until springs reach their maximum allowable angular movement where the radial stiffness increases substantially. Various spring stiffnesses can be installed in each size coupling to properly match the torsional requirements to the drive system.

advantages

- torsionally soft with good damping characteristics.
- freedom of axial shaft movement.

disadvantages

- requires lubricant for damping.
- temperature limitations due to lubricant.
- torsional characteristics change drastically with loss of oil.

Pin drive

A series of metal pins with leaf springs are placed near the outer diameter where they engage into a series of holes bored into both shaft hubs. Some pin designs consist of a pack of flat springs with cylindrical keepers at each end that act as the flexing element in the coupling design. The spring sets can swivel in the pin connection to allow movement across the width of the spring set.

- capacity : up to 3800 hp at 100 rpm.
- max. speed: to 4000 rpm.
- shaft bores: to 13 in.
- shaft spacing: close coupled (1/8" to 1/2").

Figure 4-18. Pin drive type coupling. Photo courtesy of David Brown Gear Industries, Agincourt, Ontario, Canada.

- Special designs and considerations: Drive pins can be fabricated to accommodate various torsional flexibility requirements and are indexed into oil impregnated bronze bushings in the coupling hubs.

advantages
- can accommodate up to 1/2" of axial displacement.
- no lubrication required.

disadvantages
- limited misalignment capability.

Rigid Coupling design

Well before flexible couplings came into existence, rigid couplings were used to connect two (or more) shafts together. Although flexible couplings are used on the vast majority of rotating machinery drive systems today, rigid couplings still have their place and are frequently used on systems where very little misalignment occurs, and in situations where high horsepowers are being transmitted from shaft to shaft, or in vertical pump applications where one of the drive motor bearings is carrying the weight (thrust) of the armature and pump rotors.

It is important to recognize that when two shafts are connected together with rigid couplings, the two separate shafts have effectively become one continuous shaft. Therefore the 'misalignment' tolerances for rigid couplings are the same tolerances that apply for acceptable runout conditions on a single shaft as discussed in Chapter 6. There are two classic rigid coupling alignment techniques - the 16 point and 20 point methods, and they are discussed in Chapter 7.

Figure 4-19. Rigid coupling. Browning Mfg., Maysville, KY.

Flexible coupling lubrication

There are basically two methods used to lubricate couplings: single charge or continuous feed. Greases are generally used in single charge lubricated couplings and the type is generally specified by the coupling manufacturer .

Problems that can occur in greased packed couplings are:

1 - loss of lubricant from leakage at : lube seals, shaft keyways, mating flange faces, or lubricant filler plugs.
2 - excessive heat generated in the coupling from an insufficient amount of lubricant, excessive misalignment, or poor heat dissipation inside the coupling shroud which reduces viscosity and accelerates oxidation.
3 - improper lubricant.
4 - the centrifugal forces generated in the coupling can be high enough to separate greases into oils and soaps.

Since soap has a higher specific gravity than oil, it will eventually collect where the force is the highest (namely where the gear teeth are located) causing a buildup of sludge.

Periodically inspect the inside of the coupling guard and directly under the coupling to see if any leakage is occurring. If so, do not continue to add more grease since the oil usually leaks out and the soaps continue to build up. Thoroughly clean the coupling, replace the seals and gaskets, and replenish with the correct kind and amount of lubricant.

Continuous feed lubrication systems generally use the same lube oil as the bearings and spray tubes are positioned to inject a directed stream of oil into the coupling as shown in Figure 4-20.

In addition to supplying lubricant to the coupling, a continuous supply of oil acts as an excellent heat transfer agent maintaining a relatively stable temperature in the coupling. However, contaminants in the oil, particularly water (which often condenses in lube oil tanks) or corrosive process gases carried over from the inboard oil seals on compressors can damage the coupling in time. Stainless steel lube oil piping, condensate and particulate matter removal with lube oil centrifuges, 5-10 micron filters, and entrained gas venting systems will alleviate many of these problems.

Coupling installation

Once a flexible coupling has been selected for a specific service, the next important step is proper installation. It is quite easy to destroy an expensive coupling assembly due to sloppy shaft fits, incorrect key dimensions, improperly measured shaft diameters, and so on. After the coupling has been uncrated the following steps should be performed before installation is even at-

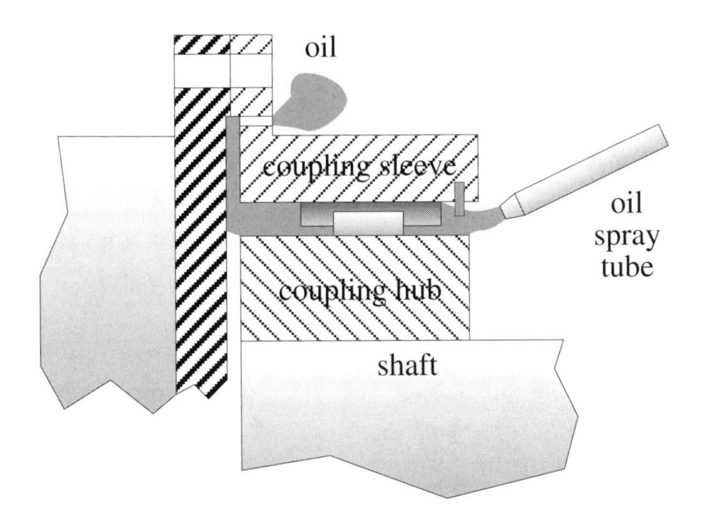

Figure 4-20. Continuous oil feed for gear type coupling.

tempted.

1) Insure that the correct type of coupling was ordered and all the parts are with it (bolts, spacer spool, hubs, cover, gaskets, etc.).

2) Physically measure all the dimensions against the coupling drawing and parts listings, paying particular attention to coupling hub bores, keyway dimensions, and spool length.

3) Measure the shafts where the coupling is going to be installed (i.e. outside diameters, tapers, keyways, etc.).

4) If possible, assemble the entire coupling before it is placed on the shaft checking for proper gear tooth clearances, elastomeric member fits, bolt hole diameter fits, and clearances.

Coupling hub attachment methods

There are a variety of methods employed to attach the coupling hubs to a shaft, each one having its advantages and disadvantages. Recommended guidelines for installing these various shaft to coupling hub fits are outlined below and should be followed to insure a proper fit to prevent slippage or unwanted shaft fretting. Shaft fretting occurs when a coupling hub is loose on its shaft and the oscillatory rocking motions of the hub cause pitting on the mating surfaces of the shaft and the coupling hub.

Classification of coupling hub to shaft fits

- straight bore, sliding clearance with keyway(s)
- straight bore, interference fit with keyway(s)
- splined shaft with end lock nut
- tapered bore, interference fit with keyway(s)
- keyless taper bore

Keys and keyways

A large percentage of shafts employ one or more keys to prevent the coupling hub from rotating on the shaft as the rotational force is applied. The American National Standards Institute (ANSI) has set up design guidelines for proper shaft sizes to key sizes and these are shown in Table 4-1.

Types of keys

There are three different classes of key fits:

- CLASS 1 - side and top clearance, relatively loose fit, only applies to parallel keys (see Table 4-2).
- CLASS 2 - minimum to possible interference, relatively

Nominal shaft diameter		Nominal Key Size			Nominal Keyseat Depth (H/2)	
			Height (H)			
over	to (inclusive)	Width (W)	Square	Rectangular	Square	Rectangular
0.31250	0.43750	0.09375	0.09375	-	0.04688	-
0.43750	0.56250	0.12500	0.12500	0.09375	0.06250	0.04688
0.56250	0.87500	0.18750	0.18750	0.12500	0.09375	0.06250
0.87500	1.25000	0.25000	0.25000	0.18750	0.12500	0.09375
1.25000	1.37500	0.31250	0.31250	0.25000	0.15625	0.12500
1.37500	1.75000	0.37500	0.37500	0.25000	0.18750	0.12500
1.75000	2.25000	0.50000	0.50000	0.37500	0.25000	0.18750
2.25000	2.75000	0.62500	0.62500	0.43750	0.31250	0.21875
2.75000	3.25000	0.75000	0.75000	0.50000	0.37500	0.25000
3.25000	3.75000	0.87500	0.87500	0.62500	0.43750	0.31250
3.75000	4.50000	1.00000	1.00000	0.75000	0.50000	0.37500
4.50000	5.50000	1.25000	1.25000	0.87500	0.62500	0.43750
5.50000	6.50000	1.50000	1.50000	1.00000	0.75000	0.50000
6.50000	7.50000	1.75000	1.75000	1.50000	0.87500	0.75000
7.50000	9.00000	2.00000	2.00000	1.50000	1.00000	0.75000
9.00000	11.00000	2.50000	2.50000	1.75000	1.25000	0.87500

All dimensions are given in inches. Square keys preferred for shaft dimensions above line ... rectangular keys below. Source - **Machinery's Handbook**, 21st Ed., Industrial Press, New York, NY, 1980.

Table 4-1. Key and Keyway Sizes for Various Shaft Diameters.

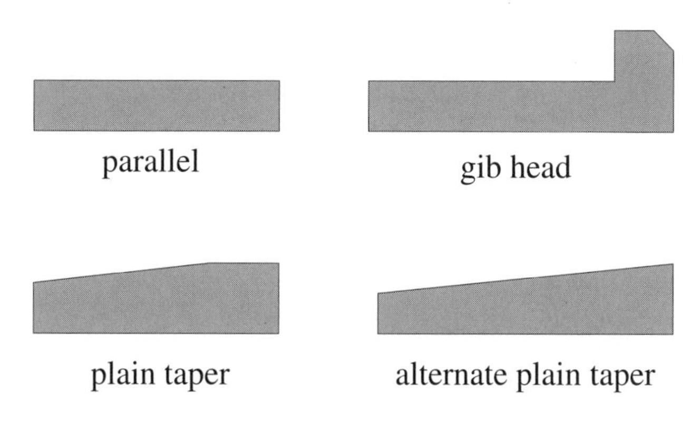

Figure 4-21. Types of keys.

Type of Key	Key Width over	Key Width ... To (inclusive)	Side Fit Width Tolerance Key	Side Fit Width Tolerance Keyseat	Side Fit Fit Range	Top and Bottom Fit Depth Tolerance Key	Top and Bottom Fit Depth Tolerance Shaft Keyseat	Top and Bottom Fit Depth Tolerance Hub Keyseat	Top and Bottom Fit Fit Range
Square	-	0.500	+0 -2	+2 -0	4 CL 0	+0 -2	+0 -15	+10 -0	32 CL 5 CL
Square	0.500	0.750	+0 -2	+3 -0	5 CL 0	+0 -2	+0 -15	+10 -0	32 CL 5 CL
Square	0.750	1.000	+0 -3	+3 -0	6 CL 0	+0 -3	+0 -15	+10 -0	33 CL 5 CL
Square	1.000	1.500	+0 -3	+4 -0	7 CL 0	+0 -3	+0 -15	+10 -0	33 CL 5 CL
Square	1.500	2.500	+0 -4	+4 -0	8 CL 0	+0 -4	+0 -15	+10 -0	34 CL 5 CL
Square	2.500	3.500	+0 -6	+4 -0	10 CL 0	+0 -6	+0 -15	+10 -0	34 CL 5 CL
Rectangular	-	0.500	+0 -3	+2 -0	5 CL 0	+0 -3	+0 -15	+10 -0	33 CL 5 CL
Rectangular	0.500	0.750	+0 -3	+3 -0	6 CL 0	+0 -3	+0 -15	+10 -0	33 CL 5 CL
Rectangular	0.750	1.000	+0 -4	+3 -0	7 CL 0	+0 -4	+0 -15	+10 -0	34 CL 5 CL
Rectangular	1.000	1.500	+0 -4	+4 -0	8 CL 0	+0 -4	+0 -15	+10 -0	34 CL 5 CL
Rectangular	1.500	3.000	+0 -5	+4 -0	9 CL 0	+0 -5	+0 -15	+10 -0	35 CL 5 CL
Rectangular	3.000	4.000	+0 -6	+4 -0	10 CL 0	+0 -6	+0 -15	+10 -0	36 CL 5 CL
Rectangular	4.000	6.000	+0 -8	+4 -0	12 CL 0	+0 -8	+0 -15	+10 -0	38 CL 5 CL
Rectangular	6.000	7.000	+0 -13	+4 -0	17 CL 0	+0 -13	+0 -15	+10 -0	43 CL 5 CL

Key dimensions are in inches. Fit dimensions are in mils (1 mil = 0.001")

Table 4-2. Key Fits.

tight fit.

- CLASS 3 - interference with degree of interference not standardized..

Shafts with two keyways can present another type of problem from improper machining of the coupling hub and/or shaft keyways as illustrated in Figure 4-22.

If the offset gap is larger than 10% of the total key width, it is recommended that the shaft and/or coupling hub be reworked for improved indexing of the keys in their mating keyways.

Straight bore - sliding clearance with keyways

This method of shaft and coupling hub fitup is used extensively in industry and provides the easiest and quickest installation of the coupling hub. However, shaft fretting is likely to occur with this sort of arrangement since there is a certain amount of clearance (0.0005" to 0.001" generally) between the coupling hub and

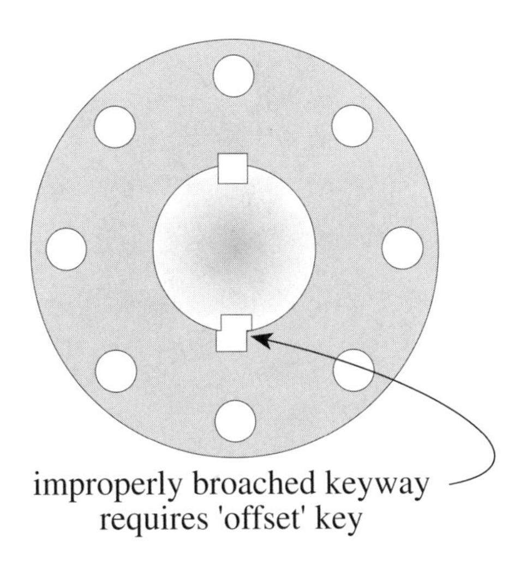

Figure 4-22. Offset key.

the shaft. To prevent the coupling hub from sliding axially along the shaft, set screws are usually locked against the key as shown in Figure 4-23.

- Insure clearance does not exceed 1 mil for shaft diameters up to 6 inches.
- Remove any burrs, clean components carefully before installation.
- If hub sticks part way on, remove it and find the problem. Do not attempt to drive it on further with a hammer.
- Install keys before the coupling is placed on the shaft.

Straight bore - interference fit with keyways

To insure that a coupling hub can be removed once it is 'shrunk' onto a shaft, proper interference fits must be adhered to. The general guideline for straight bore interference fits are found

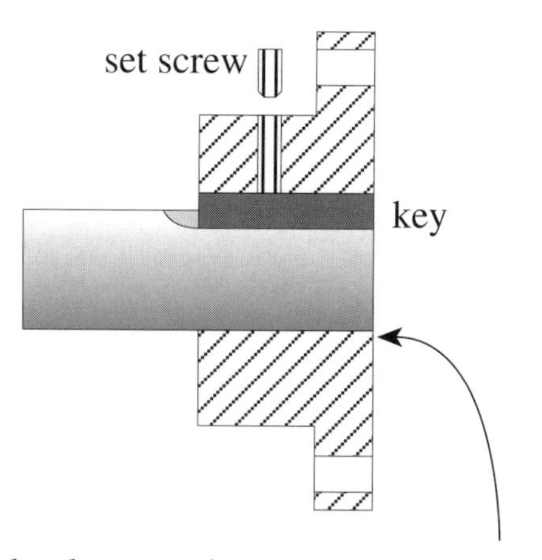

Figure 4-23. Straight bore coupling hub with clearance fit.

Nominal shaft diameter		Key Width - Set Screw Sizes	
over	to (inclusive)	Key Width	Set Screw Diameter
0.31250	0.43750	0.09375	No. 10
0.43750	0.56250	0.12500	No. 10
0.56250	0.87500	0.18750	0.25000
0.87500	1.25000	0.25000	0.31250
1.25000	1.37500	0.31250	0.37500
1.37500	1.75000	0.37500	0.37500
1.75000	2.25000	0.50000	0.50000
2.25000	2.75000	0.62500	0.50000
2.75000	3.25000	0.75000	0.62500
3.25000	3.75000	0.87500	0.75000
3.75000	4.50000	1.00000	0.75000
4.50000	5.50000	1.25000	0.87500
5.50000	6.50000	1.50000	1.00000

All dimensions are given in inches.

Table 4-3. Set screw - key size directory.

Shaft Diameter (in.)	Interference Fit (in.)
1/2 to 2	0.0005 to 0.0015 in.
2 to 6	0.005 to 0.0020 in.
6 and up	0.0001 to 0.00035 inches per inch of shaft diameter

Table 4-4. Guidelines for coupling hub shrink fits on shafts.

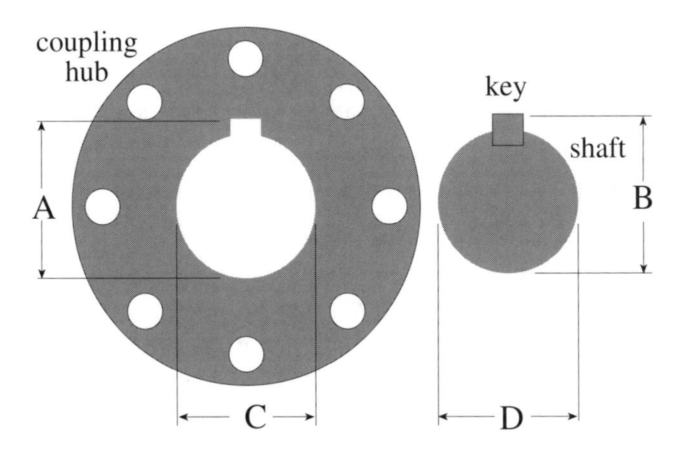

compare A to B and C to D

Figure 4-24. Measuring a coupling hub for proper key fits.

in Table 4-4.

The coupling is installed by heating in an oil bath or an oven to approximately 200 - 250 degrees F, and in some cases cooling the shaft simultaneously with a dry ice pack. Do not exceed 300 degrees F or use any direct heat such as propane or oxy-acetylene torches to expand the hub since the material properties of the shaft can be altered by direct, high temperature concentration. Once the interference fit has been determined by measuring the shaft diameter and coupling hub bores, the temperature increase needed to expand the coupling hub to exceed the shaft diameter by 2 mils (to allow for a slide on fit) is found in equation

$$\Delta T = \frac{i}{\alpha\,(d - 0.002)} \qquad \text{Equation 4.1}$$

where :
ΔT = rise in coupling hub temperature from ambient (degree F)
i = interference fit (mils or thousandths of an inch)
α = coefficient of thermal expansion
d = coupling hub bore diameter (in.)

4-1.

Removal of the coupling hub is accomplished by pulling the hub off the shaft with an acceptable puller mechanism and at times cooling the shaft with a dry ice pack. Shrink fit coupling hubs should always have fine threaded puller holes (preferably four) in the end of the coupling as shown in Figure 4-25.

Bearing type pullers that 'push' the hub off from the backside are not recommended since there is a great possibility that the puller can twist or pitch slightly, preventing a straight axial draw on the hub. For larger shaft diameters with tight interference fits, it may be necessary to apply gentle heating to the coupling for removal.

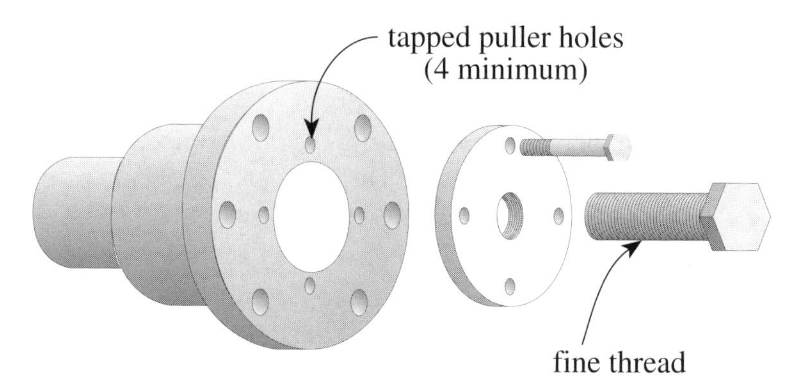

Figure 4-25. Coupling hub puller tool.

Splined shaft with end lock nut or locking plate

A splined shaft and coupling arrangement is shown in Figure 4-26. There should be a slight interference fit (0.0005") to prevent backlash or rocking of the hub on the shaft.

Figure 4-26. Splined shaft end.

Tapered bore - interference fit with keyways

Tapered shaft ends are generally used where high torques and speeds are experienced on rotating machinery necessitating a tight coupling hub to shaft fit up. The shaft end is tapered to provide an easier job of removing the coupling hub.

The degree of taper on a shaft end is usually expressed in terms of its slope (inches per foot). The amount of interference fit is expressed in inches per inch of shaft diameter. The general rule for interference fits for this type of shaft arrangement is 1 mil per inch of shaft diameter. The distance a coupling hub must travel axially along a shaft past the point where the hub is just touching the shaft at ambient temperatures is found in equation 4-2.

$$HT = \frac{12\,I}{ST} \qquad \text{Equation 4.2}$$

where :
HT = distance coupling hub must travel to provide an interference fit equal to I (mils)
I = interference fit (mils)
ST = shaft taper (in/ft)

Procedure for mounting a tapered coupling hub with keys

1. Mount bracket firmly to coupling hub and slide hub onto shaft end to lightly seat the hub against the shaft. Insure all surfaces are clean.
2. Measure hub travel gap HT with feeler gauges and lock nut down against bar. Use Equation 4-2 to determine the correct axial travel needed to obtain the required interference fit onto the shaft.
3. Remove the coupling hub and puller assembly and place in

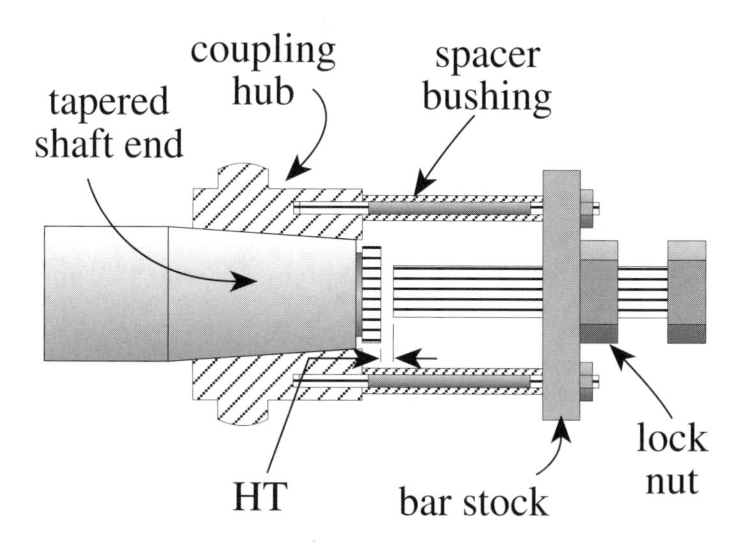

Figure 4-27. Measuring tool for insuring proper interference fit on tapered shaft end.

an oven or hot oil bath to desired differential temperature. Refer to Equation 4-1 to determine the required temperature rise needed to expand the coupling hub.

4. Set key in keyway and insure all contact surfaces are clean and burr free.
5. Carefully slide the heated coupling hub onto the shaft until the center measurement bolt touches the shaft end and hold in place until hub has cooled sufficiently.

Coupling hub to shaft surface contact

One extremely important and often overlooked consideration when working with tapered shafts and coupling hubs is the amount of surface contact between the shaft and hub. Due to slight machining inaccuracies, coupling hubs may not fully contact the shaft resulting in a poor fit when the hub is shrunk or pressed on in final assembly.

To check the surface contact, apply a thin coat of Prussian blue paste to the inner bore of the coupling hub with your finger or a soft cloth. Slide the coupling hub axially over the tapered shaft end until contact is made and rotate the coupling hub about 15 degrees to transfer the paste to the shaft. Draw the coupling hub off and observe the amount of Prussian blue paste that transferred from the hub to the shaft (not how much blue came off the inside bore of the coupling hub). If there is not at least 80% contact, the fit is not acceptable. If the bore discrepancies are slight, it is possible to lap the surfaces with a fine grit lapping compound. Apply the compound around the entire surface contact area of the tapered shaft end, lightly pushing the coupling hub up the taper and rotating the coupling hub alternately clockwise and counterclockwise through a 45 degree arc. Check the surface contact after 10 or 12 lapping rotations. Continue until surface contact is acceptable. However if a 'ridge' begins to develop on the shaft taper before good surface contact is made, start making preparations for remachining of the shaft and the coupling hub. It's better to bite the bullet now than try to heat the hub and put it on only to find out that it does not go on all the way or to pick up the pieces of a split coupling hub after the unit ran for a short period of time.

Keyless taper bores

After working with shafts having keyways to prevent slippage of the coupling hub on the shaft, it seems very unnerving to consider attaching coupling hubs to a shaft with no keys. Keyless shaft fits are quite reliable and installing hubs by hydraulic expansion methods prove to be fairly easy if installation and removal steps are carefully adhered to. Since the interference fits are usually 'tighter' than found on straight bores or tapered and keyed systems, determining a proper interference fit will reviewed.

Proper interference fit for hydraulically installed coupling hubs

The purpose of interference fits are twofold:

1) prevent fretting corrosion that occurs from small amounts of movement between the shaft and the coupling hub during rotation and
2) prevent the hub from slipping on the shaft when the maximum amount of torque is experienced during a start up or during high running loads.

For rotating shafts, the relation between torque, horsepower, and speed can be expressed as:

$$T = \frac{63000\ P}{n} \qquad \text{Equation 4.3}$$

where :
P = horsepower
T = torque (in-lbs)
n = shaft speed (rpm)

The maximum amount of shearing stress in a rotating shaft occurs in the outer fibers (i.e. the fibers at the outside diameter) and is expressed :

$$\tau\ max = \frac{T\ r}{J} = \frac{16\ T}{\pi\ d^3} \quad \text{Equation 4.4}$$

where :
τ max = maximum shear stress (lb/in)
T = torque (in-lbs)
r = radius (in)
d = diameter (in)
J = polar moment of inertia $J = \frac{\pi\ r^4}{2} = \frac{\pi\ d^4}{32}$

The accepted 'safe allowable' torsional stress for the three commonly used types of carbon steel for shafting can be found in Table 4-5.

AISI #	τ max allowable torsional stress (psi)
1040	5000
4140	10000
4340	11000

Table 4-5. Allowable torsional stresses for shafts.

Therefore the torsional holding requirement for applied torques is expressed as:

$$T = \frac{\tau_{max} \; \pi \, d^3}{16} \qquad \text{Equation 4.5}$$

The amount of torque needed to cause a press fit hub to slip on its shaft is given by:

$$T = \frac{\mu \, \pi \, p \, L \, d^2}{2} \qquad \text{Equation 4.6}$$

where :
μ = coefficient of friction (0.12)
p = contact pressure (lbs/sq. in.)
L = length of coupling hub bore (in)

The amount of contact pressure between a shaft and a coupling hub is related to the amount of interference and the outside diameters of the shaft and the coupling hub, and is expressed as:

$$p = \frac{i \, E \, (DH^2 - DS^2)}{2 \, (DH^2) \, (DS)} \qquad \text{Equation 4.7}$$

where :
p = contact pressure (lbs/sq. in.)
i = interference fit (mils)
E = modulus of elasticity (30×10^6 lb/in for carbon steel)
DH = outside diameter of coupling hub (in.)
DS = outside diameter of shaft (in.)

Since the shaft is tapered, dimension DS should be taken on the largest bore diameter on the coupling hub, where the contact pressure will be at its minimum value as shown in Figure 4-28. Therefore to find the proper interference fit between a coupling hub and a tapered shaft to prevent slippage from occurring :

1) Determine the maximum allowable torque value for the shaft diameter and the shaft material.
2) Determine the contact pressure needed to prevent slippage from occurring based on the maximum allowable torque value found in step 1.
3) Calculate the required interference fit (solve for p in Eq. 4-6 & i in 4-7).

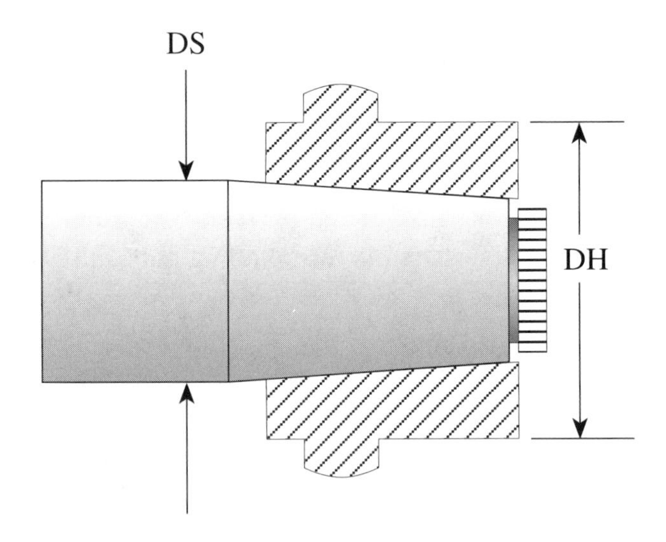

Figure 4-28. Shaft and coupling hub outside diameter measurement locations.

Installation of keyless coupling hubs using hydraulic expansion

Installing keyless taper hubs requires some special hydraulic expander and pusher arrangements to install or remove the coupling hub onto the shaft end. Figure 4-29 shows the general arrangement used to expand and push the hub onto the shaft.

Procedure for installation of coupling hub using hydraulic expander and pusher assembly

1. Check for percentage of surface contact between coupling hub and shaft (must have 80% contact or better).
2. Insure all mating surfaces are clean and that oil passageways are open and clean.

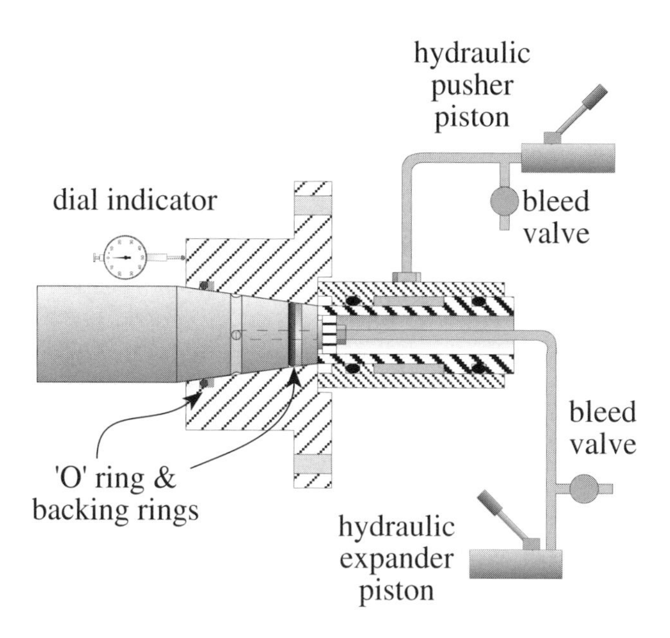

Figure 4-29. Hydraulic expander and pusher assembly.

3. Install 'O' rings and backup rings in coupling hub and shaft insuring that backup ring is on on 'outside' of 'O' ring with respect to hydraulic oil pressure. Lightly oil the 'O' rings with hydraulic fluid. Place coupling hub (and hub cover) onto shaft.
4. Install expander pump supply line to shaft end. Install the pusher piston assembly onto the the end of the shaft, insuring that the piston is drawn back as far as possible. Hook up the expander pump and begin pumping hydraulic oil through supply line to bleed any air from expansion ports and expansion groove in shaft. Once the oil has begun to seep through the coupling hub ends, lightly push the coupling hub against the shaft taper and begin to pump oil into the pusher piston assembly to seat the piston against the coupling hub.
5. Place a dial indicator against the backside of the coupling hub and zero the indicator.
6. Begin applying pressure to the pusher piston assembly forcing the hub up the taper (approximately 2000 to 4000 psig).
7. Slowly increase the pressure on the expander pump supply line until the coupling hub begins to move. The hydraulic pressure on the pusher piston assembly will begin to drop off as the hub begins to move. Maintain sufficient pressure on the pusher piston to continue to drive the hub onto the shaft. If the pusher piston pressure drops off considerably when the expansion process is underway, there is a great potential for the 'O' rings to 'blow out' the ends of the coupling hub, immediately seizing the hub to the shaft.
8. Continue forcing the hub up the shaft until the desired amount of hub travel and interference fit is attained. The expansion pressure will have to attain the required holding pressure as defined in Eq. 4-7.
9. Once the correct hub travel has been achieved, maintain

sufficient hydraulic pressure on the pusher assembly to hold the coupling hub in position and bleed off the pressure in the expansion system. Allow 15-20 minutes to elapse while bleeding to insure that any trapped oil has had a chance to escape before lowering the pusher piston pressure.

10. Remove the pusher and expander assemblies.

Removal of the coupling hub is achieved by reversing the installation process. The key to successful installation is to take your time and not try to push the hub up the shaft end all in one move.

Bibliography

Serrell, John J., "Flexible Couplings", Machinery, Oct. 1922, pgs. 91-93.

Gooding, Frank E.,, "Types and Kinds of Flexible Couplings", Industrial Engineer, Nov. 1923, pgs. 529-533.

Shaw, G.V., "Employing Limited End Play Couplings Saves Motor Bearings - Sometimes", Power, Nov. 1955, pgs. 120-121.

Zurn, Frank W., "Crowned Tooth Gear Type Couplings", Iron and Steel Engineer, Aug., 1957, pgs. 98-116.

Spector, L.F., "Flexible Couplings", Machine Design, Oct., 1958, Vol. 30, no. 22, pgs. 102-128.

Gensheimer, J.R., "How to Design Flexible Couplings," Machine Design, 33 (19), pgs. 154-159, Sept 14, 1961.

Anderson, J.H., "Turbocompressor Drive Couplings", Journal of Engineering for Industry, A.S.M.E., paper no. 60-WA-212, Jan. 1962. pgs. 115-123.

Miller, F.F., "Constant-Velocity Universal Ball Joints," Mech. Des., 37 (9), pgs. 184-193, April 15, 1965.

Moked, I., "Toothed Couplings - Analysis & Optimization", Journal of Engineering for Industry, A.S.M.E., Aug. 1968, pgs. 425-434.

DeRocker, D.E., Kaufman, S., Renzo, P.C., "Gear Couplings", Journal of Engineering for Industry, A.S.M.E., Aug. 1968, pgs. 467-474.

Weatherford, W.D., Jr., et al., "Mechanisms of Wear in Misaligned Splines," J. Lub. Tech., Trans. ASME, 90 (1), pgs. 42-48, 1968.

Broersma, G. (Ed.), "Couplings and Bearings, H. Stam Intl., N.V., Part I Couplings", pgs. 1-82, 1968.

Potgieter, F.M., "Cardan Universal Joints Applied to Steel Mill Drives," Iron and Steel Engr., 46 (3), pgs. 73-79, March 1969.

Centrifugal Pumps for General Refinery Services, API Standard 610, American Petroleum Institute, Wash., D.C., March 1971.

Mancuso, J.R., "Moments and Forces Imposed on Power Transmission Systems Due to Misalignment of a Crown Tooth Coupling," Master Thesis, Penn State Univ., June 1971.

Hunt, K.H., "Constant-Velocity Shaft Couplings: A General Theory," J. Engrg. Indus., Trans. ASME, 95 (2), pgs. 455-464, May 1973.

Loosen, P., Prause, John J., "Frictional Shaft - Hub Connectors - Analysis and Applications", Design Engineering, Jan. 1974.

Bloch, Heinz P., "Why Properly Rated Gears Still Fail", Hydrocarbon Processing, Dec. 1974, pgs. 95-97.

Calistrat, M.M., "What Causes Wear in Gear Type Couplings", Hydrocarbon Processing, Jan., 1975, pgs. 53-57.

Wright, J., "Which Shaft Coupling is Best - Lubricated or Nonlubricated," Hydrocarbon Processing, 54 (4), pgs. 191-193, April 1975.

Calistrat, M.M., "Grease Separation Under Centrifugal Forces", A.S.M.E., paper no. 75-PTG-3, July 3, 1975.

Woodcock, J.S., "Balancing Criteria for High Speed Rotors with Flexible Couplings," Intl. Conf. Vib. Noise in Pump, Fan and Compressor Installations, Univ. of Southampton, UK, C112/75, pgs. 107-114, Sept 16-18, 1975.

Wattner, K. W., "High Speed Coupling Failure Analysis", Proceedings - Fourth Turbomachinery Symposium, Gas Turbine Labs, Texas A&M University, Oct. 1975, pgs. 143-148.

Bloch, Heinz P., "Less Costly Turbomachinery Uprates Through Optimized Coupling Selection", Proceedings - Fourth Turbomachinery Symposium, Gas Turbine Labs, Texas A&M University, Oct. 1975, pgs. 149-152.

Kamii, N. and Okuda, H., "Universal Joints Applied to Hot Strip Mill Drive Lines," Iron and Steel Engr., 52 (12), pgs. 52-56, December 1975.

Bloch, H.P., "How to Uprate Turboequipment by Optimized Coupling Selection," Hydrocarbon Processing, 55 (1), pgs. 87-90, January 1976.

Bloch, Heinz P., "Use Keyless Couplings for Large Compressor Shafts", Hydrocarbon Processing, April 1976, pgs. 181-186.

Phillips, J. and Vowles, B., "Flexible Metallic Couplings," Pump Compressors for Offshore Oil Gas Conf., Univ. of Aberdeen, UK, C135/76, pgs. 103-108, June 29-July 1, 1976.

Gibbons, C.B., "Coupling Misalignment Forces", Proceedings - Fifth Turbomachinery Symposium, Gas Turbine Labs, Texas A&M University, Oct., 1976, pgs. 111-116.

Counter, Louis F., Landon, Fred K., "Axial Vibration Characteristics of Metal-Flexing Couplings", Proceedings - Fifth Turbomachinery Symposium, Gas Turbine Labs, Texas A&M University, Oct., 1976, pgs. 125-131.

Bloch, H.P., "Improve Safety and Reliability of Pumps and Drivers, Part 2 -- Gear Coupling vs Nonlubricated Couplings," Hydrocarbon Processing, 56 (2), pgs. 123-125, Feb. 1977.

Calistrat, M.M., "Metal Diaphragm Coupling Performance," Hydrocarbon Processing, 56 (3), pgs.137-144, March 1977.

Milenkovic, V., "A New Constant Velocity Coupling," J. Engrg. Indus., Trans. ASME, 99 (2), pgs. 367-374, May 1977.

Wright, C.G., "Tracking Down the Cause of Coupling Failure," Mach. Des., 49 (14), pgs. 98-102, June 23, 1977.

Mancuso, J.R., "A New Wrinkle to Diaphragm Couplings", A.S.M.E., paper no. 77-DET-128, June 24, 1977.

Proc. Intl. Conf. on Flexible Couplings for High Powers and Speeds, University of Sussex, England, June 29-July 1, 1977.

Goody, E.W., "Laminated Membrane Couplings for High Powers and Speeds", International Conference on Flexible Couplings, High Wycombe, Bucks, England, June 29 - July 1, 1977.

Calistrat, M.M., "Extend Gear Coupling Life, Part 1," Hydrocarbon Processing, 57 (1), pgs. 112-116, Jan. 1978.

"Alignment Loading of Gear Type Couplings", Bently Nevada Applications Notes no. (009)L0048, Bently Nevada Corp., Minden, Nevada, March, 1978.

Pahl, Gerhardt, "The Operating Characteristics of Gear-Type Couplings", Proceedings - Seventh Turbomachinery Symposium, Gas Turbine Labs, Texas A&M University, Dec., 1978, pgs. 167-173.

Calistrat, M.M., "Extend Gear Coupling Life, Part 2", Hydrocarbon Processing, pgs. 115-118, Jan. 1979.

Prause, John J., "New Clamping Device for Fastening Pulley End Discs to Shafts", Skillings Mining Review, June 16, 1979, Vol. 68, no. 24.

Zirkelback, C., "Couplings - A User's Point of View," Proceedings 8th Turbomachinery Symposium, Texas A&M University, pgs. 77-81, Nov. 1979.

Brown, H.W., "A Reliable Spline Coupling," J. Engrg. Indus., Trans. ASME, 101 (4), pgs. 421-426, November 1979.

Herman, A.S., "Torsionally Resilient Couplings in Diesel Engine Drives," Natl. Conf. on Power Transm., Proc. 6th Ann. Mtg., pgs. 275-279, November 13-15, 1979.

Wright, M.D., "Torsionally Soft Flexible Couplings," Natl. Conf. Power Transm. Proc., 6th Ann. Mtg., pgs. 263-265, November 13-15, 1979.

Schwerdlin, H., "Reaction Forces in Elastomeric Couplings," Mach. Des., 51 (16), pgs. 76-79, 1979.

Oberg, Erik, Jones, Franklin D., Horton, Holbrook L., **Machinery's Handbook**, 21st edition, Industrial Press Inc., 200 Madison Ave., New York, N.Y., 1980.

Marmol, R.A., et al., "Spline Coupling Induced Nonsynchronous Rotor Vibrations," J. Mech. Design, Trans. ASME, 102 (1), pgs. 168-176, Jan. 1980.

Bannister, R.H., "Methods for Modelling Flanged and Curvic Couplings for Dynamic Analysis of Complex Rotor Constructions," J. Mech. Design, Trans. ASME, 102 (1), pgs. 130-139, Jan. 1980.

Maxwell, J.H., "Vibration Analysis Pinpoints Coupling Problems," Hydrocarbon Processing, 59 (1), pgs. 95-98, Jan. 1980.

Ohlson, J.F., "Coupling of Misaligned Shafts," US PATENT, 4187698, Feb. 1980.

Gibbons, C.B., "The Use of Diaphragm Couplings in Turbomachinery," Proc. Mach. Vib. Monit. Ann. Sero., pgs. 99-116, April 8-10, 1980.

Williams, R.H., "Flexible Coupling," US PATENT, 4203305, May 1980.

Schwerdlin, H. and Eshleman, R., "Combating Vibration with Mechanical Couplings," Machine Design, 52 (20), pgs. 67-70, Sept. 1980.

Patterson, C., et al., "Vibration Aspects of Rolling Mill Horizontal Drives with Reference to Recent Coupling Development," 2nd Intl. Conf. on Vib. Rotating Mach., Churchill College, Cambridge, UK, C297/80, pgs. 315-320, Sept. 1-4, 1980.

Eshleman, R.L., "The Role of Sum and Difference Frequencies in Rotating Machinery Fault Diagnosis," 2nd Intl. Conf. on Vib. Rotating Mach., Churchill College, Cambridge, UK, C272/80, pgs. 145-149, Sept 1-4, 1980.

Beard, C.A., "The Selection of Couplings for Engine Test Beds," 2nd Intl. Conf. on Vib. Rotating Mach., Churchill College, Cambridge, UK, C267/80, pgs. 115-118, Sept 1-4, 1980.

Finn, A.E., "Instrumented Couplings: The What, the Why, and the How of the Indikon Hot-Alignment Measurement System," Proceedings 9th Turbomachinery Symposium, Texas A&M University, pgs. 135-136, December 1980.

Schwerdlin, H. and Eshleman, R., "Improve Coupling Selection through Torsional Vibration Modeling," Power Transm. Dec.,23 (2), pgs. 56-59, Feb. 1981.

Schwerdlin, H. and Roberts, C., Jr., "Combating Heat in U-Joints," Mach. Des., 53 (17), pgs. 83-86, July 23, 1981.

McCormick, D., "Finding the Right Flexible Coupling," Des. Engrg., 52 (10), pgs. 61-66, Oct. 1981.

Schwerdlin, H. and Eshleman, R.L., "Measuring Vibrations for Coupling Evaluation," Plant Engrg., 36 (12), pgs. 111-114, June 1982.

Chander, T. and Biswas, S., "Abnormal Wear of Gear Couplings - A Case Study," Trib. Intl., 16 (3), pgs. 141-146, 1983.

Winkler, A.F., "High Speed Rotating Machinery Unbalance, Coupling or Rotor," Proc. Mach. Vib. Monit. Ann. Mtg., Houston, TX, pgs. 75-80, April 19-21, 1983.

Wireman, T., "A Fitness Plan for Flexible Couplings," Power Transm. Des., 25 (4), pgs. 20-22, April 1983.

Dewell, D.L. and Mitchell, L.D., "Detection of a Misaligned Disk Coupling Using Spectrum Analysis," J. Vib., Acoust., Stress, Rel. Des., Trans. ASME, 106 (1), pgs. 9-16, Jan. 1984.

Kirk, R.G., et al.,"Theory and Guidelines to Proper Coupling Design for Rotor Dynamics Considerations," J. Vib., Acoust., Stress, Rel. Des., Trans. ASME, 106 (1), pgs. 129-138, Jan. 1984.

Kojima, H. and Nagaya, K., "Nonlinear Torsional Vibrations of a Rotating Shaft System with a Magnet Coupling," Bull. JSME, 27 (228), pgs. 1258-1263, June 1984.

Ota, H. and Kato, M., "Even Multiple Vibrations of Rotating Shaft Due to Secondary Moment of a Universal Joint," 3rd Intl. Conf. on Vib. Rotating Mach., Univ. of York, UK, C310/84, pgs. 199-204, Sept. 11-13, 1984.

Mancuso, J.R., "The Manufacturer's World of Coupling Potential Unbalance," Proc. 13th Turbomachinery Symp., Texas A&M Univ., pgs. 97-104, Nov 1984.

Seneczko, Z., "Tailoring Drive Output with Elastomeric Couplings," Mach. Des., 57 (7), pgs. 123-125, April 11, 1985.

Nataraj, C., et al., "Effect of Coulomb Spline on Rotor Dynamic Response," Proc. Instability in Rotating Machinery, NASA CP-2409, pgs. 225-233, June 10-14, 1985.

Rivin, E.I., "Design and Application Criteria for Connecting Couplings," J. Mech., Transm., Autoto. Des., Trans. ASME, 108 (2), pgs. 96-105, 1986.

Mancuso, J.R., "Disc vs Diaphragm Couplings," Mach. Des., 58 (17), pgs. 95-98, July 24, 1986.

Mancuso, J.R., **Couplings and Joints**, Marcel Dekker, 1986, ISBN #0-8247-7400-0.

Shigley, J.E. and Mischke, C.R. (Eds.), **Standard Handbook of Machine Design**, Chap. 29, pgs. 29.1-29.38, McGraw-Hill, 1986.

Peeken, H., et al., "A New Approach to Describe the Mechanical Performance of Flexible Couplings in Drive Systems," Proc. Intl. Conf. Rotordynamics, Tokyo, Japan, pgs. 159-163, Sept 14-16, 1986.

Kato, M., et al., "Lateral-Torsional Coupled Vibrations of a Rotating Shaft Driven by a Universal Joint," JSME Intl. J., Ser. III, 31 (1), pgs. 68-74, 1988.

Bühlmann, E.T. and Luzi, A., "Rotor Instability Due to a Gear Coupling Connected to a Bearingless Sun Wheel of a Planetary Gear," Proc. Rotordynamic Instability Problems in High-Performance Turbomachinery, Texas A&M Univ., NASA CP-3026, pgs. 19-39, May 16-18, 1988.

Foszcz, J.L., "Couplings - They Give and Take," Plant Engrg., 42 (18), pgs. 53-57, Nov 23, 1988.

Calistrat, Michael, **Manual for the Installation and Maintenance of Flexible Couplings**, Michael Calistrat & Associates, Missouri City, TX, January, 1989.

Baer, L., "Tolerant Couplings," Mech. Engrg., 111 (5), pgs. 58-59, May 1989.

Langworthy, V.W. (Ed.), **Centrifugal Pumps: Design. Construction and Maintenance**, Scranton Publ. Co., Inc., pgs. 48-52.

5
Defining Misalignment – Alignment and Coupling Tolerances

What exactly is shaft alignment?

As discussed in Chapter 1, shaft misalignment occurs when the centerlines of rotation of two (or more) machinery shafts are not in line with each other. As simple as that may sound, it still causes a considerable amount of confusion to people who are just beginning to study this subject when trying to precisely define the amount of misalignment that may exist between two shafts flexibly or rigidly coupled together.

How accurate does the alignment have to be? How do you measure misalignment when there are so many different coupling designs? Where should the misalignment be measured? Is it measured in terms of mils, degrees, millimeters of offset, arcseconds, radians? When should the alignment be measured ...

when the machines are off-line or when they are running?

Measuring Angles

There are 360 degrees in a circle. Each degree can be divided into 60 parts called minutes of arc and each minute of arc can be further divided into 60 parts called seconds of arc. Therefore there are 21,600 minutes of arc and 1,296,000 seconds of arc in a circle.

Another way of expressing circles is by use of radians. All circles are mathematically related by an irrational number called pi (after the Greek letter) which is approximately equal to 3.14159. There are 2 radians in a circle. Therefore one radian is equal to 57.295828 degrees.

Does level and aligned mean the same thing?

No, it does not. The term 'level' is related to earth's gravitational pull. When an object is in a horizontal state or condition or points along the length of an object are at the same altitude, the object is considered to be level. Another way of stating this is that an object is level if the surface of the object is perpendicular to the lines of gravitational force. A level rotating machinery foundation located in Boston would not be parallel to a level rotating machinery foundation located in San Francisco since the earth's surface is curved. The average diameter of the Earth is 7908.5 miles (7922 miles at the equator and 7895 miles at the poles due to the centrifugal force causing the planet to bulge at the center). When measuring distances of arc across the earths surface, one degree of arc is slightly over 69 miles, one minute of arc is slightly over 1.15 miles, and one second of arc is slightly over 101.2 feet.

It is possible, although rare, to have a machinery drive train both level and aligned. It is also possible to have a machinery drive train level, but not aligned, and it is also possible to have a machinery drive train aligned, but not level. Since shaft alignment deals specifically with the centerlines of rotation of machinery shafts it is possible to have, or not to have, the centerlines of rotation perpendicular to the lines of gravitational force.

Machinery Type	Minimum recommended levelness	Maximum recommended levelness
general process machinery supported in antifriction bearings	10 mils per foot	30 mils per foot
general process machinery supported in sleeve bearings (up to 500 hp)	5 mils per foot	15 mils per foot
process machinery supported in antifriction bearings (500+ hp)	5 mils per foot	20 mils per foot
process machinery supported in sleeve bearings (500+ hp)	2 mils per foot	8 mils per foot
machine tools	1 mil per foot	5 mils per foot
Note : 1 mil = 0.001"		

Table 5-1. Recommended level ranges for horizontally mounted rotating machinery.

You may have noticed in Chapter 2 that there were a considerable number of patents filed from 1900 to 1950 that seemed to combine (or maybe confuse) the concept of level and aligned. A number of these alignment devices were used in the paper industry where extremely long 'line shafts' were installed to drive different parts of a paper machine. These line shafts were constructed with numerous sections of shafting that were connected end to end with rigid couplings and supported by a number of bearing pedestals along the length of the line shaft and could be up to 300 feet in length. Even if a mechanic carefully aligned each section of shafting at each rigid coupling connection along the length of the line shaft and perfectly leveled each shaft section, the centerline of rotation at each end of a 300 foot long line shaft would be out by 0.018" due to the curvature of the earths surface.

Types of misalignment

Shaft misalignment can occur two basic ways: parallel and angular as shown in figure 5-1. Actual field conditions usually have a combination of both parallel and angular misalignment so measuring the relationship of the shafts gets to be a little complicated in a three-dimensional world, especially when you try to show this relationship on a two-dimensional piece of paper. I find it helpful at times to take a pencil in each hand and position them based on the dial indicator readings to reflect how the shafts of each unit are sitting.

Commonly used misalignment terms

parallel misalignment

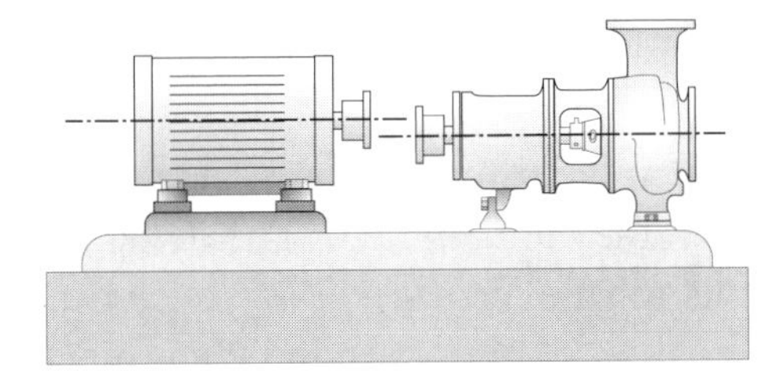

angular misalignment

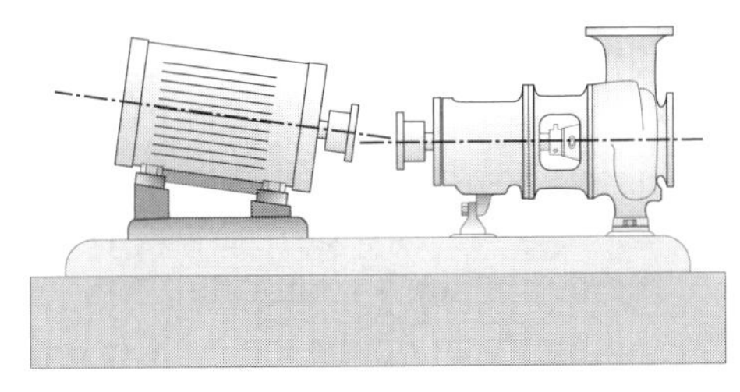

'real world' misalignment usually exhibits a combination of both parallel and angular shaft centerline positions

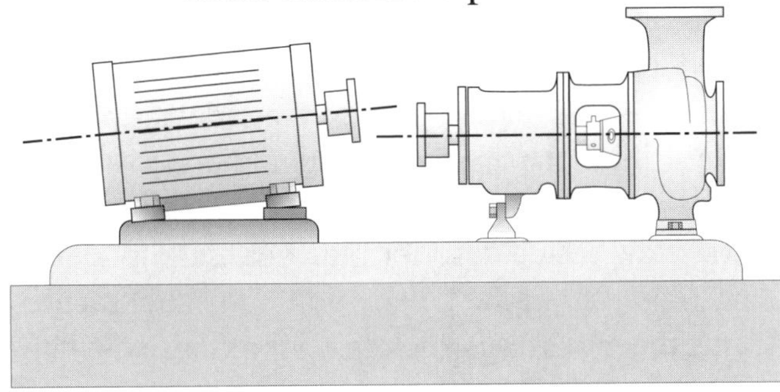

Figure 5-1. How shafts can be misaligned.

Definition of shaft misalignment

In more precise terms, shaft misalignment is the deviation of relative shaft position from a colinear axis of rotation measured at the points of power transmission when equipment is running at normal operating conditions. To better understand this definition, let's dissect each part of this statement to clearly illustrate what's involved.

Colinear means in the same line or in the same axis. If two shafts are colinear, then they are aligned. The deviation of relative shaft position accounts for the measured difference between the actual centerline of rotation of one shaft and the projected centerline of rotation of the other shaft. Figure 5-1 shows a typical misalignment situation on a motor and a pump.

For a flexible coupling to accept both parallel and angular misalignment there must be at least two points where the coupling can 'flex' or give to accommodate the misalignment condition. By projecting the axis of rotation of the motor shaft toward the pump shaft (and conversely the pump shaft rotational axis toward the motor shaft) there is a measurable deviation between the projected axis of rotation of each shaft and the actual shaft centerlines of each shaft where the power is being transmitted through the coupling from one 'flexing' point to another. Since we measure misalignment in two different planes (vertical and horizontal) there will be four deviations that occur at each flexible coupling. In a horizontally mounted drive train, two of these deviations occur in the top view describing the amount of lateral (side to side) misalignment. Two more deviations occur when viewing the drive train in the side view which describes the vertical (up & down) misalignment. The goal of the person doing the alignment is to position the machinery casings such that all of these deviations are below certain tolerance values.

Defining misalignment

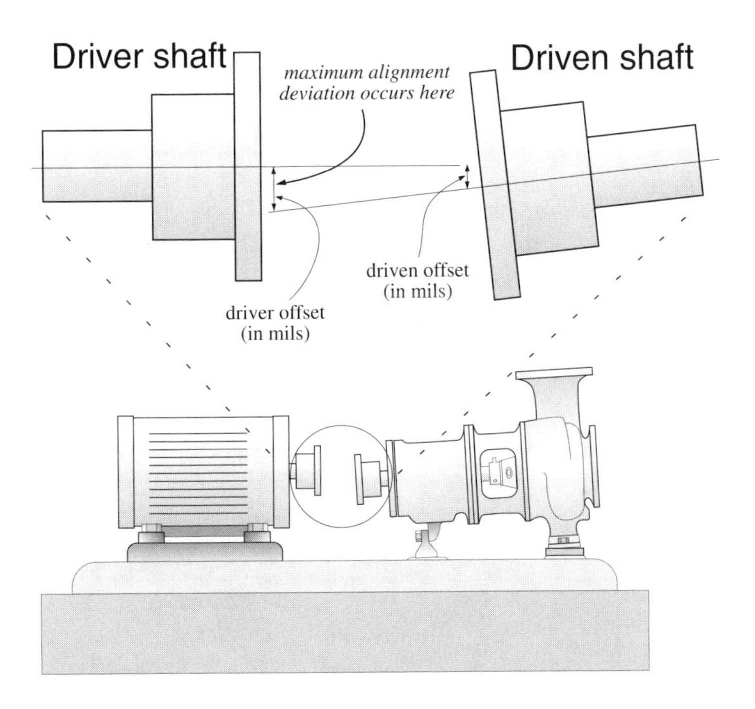

Misalignment is the deviation of relative shaft position from a colinear axis of rotation measured at the points of power transmission when equipment is running at normal operating conditions.

Figure 5-2. Definition of shaft misalignment.

How to determine your alignment tolerance

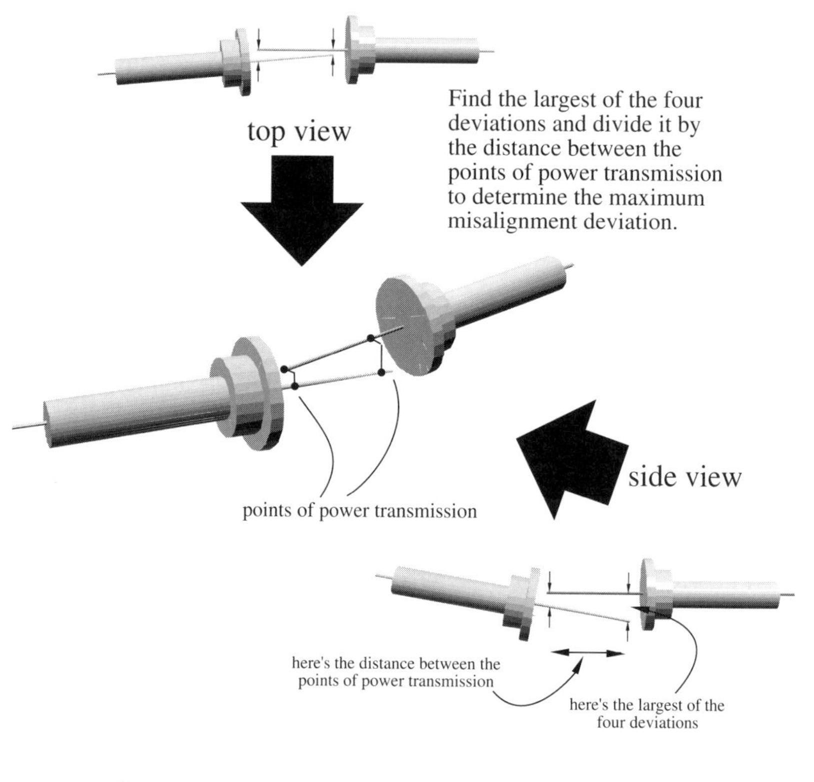

for example ...

If the largest of the four deviations is ... 6 mils (0.006")

... and the distance between the power transmission points are ... 4"

$$\frac{6 \text{ mils } (0.006")}{4"} = 1.5 \text{ mils/inch}$$

is your maximum deviation

Figure 5-3. How to determine alignment tolerances.

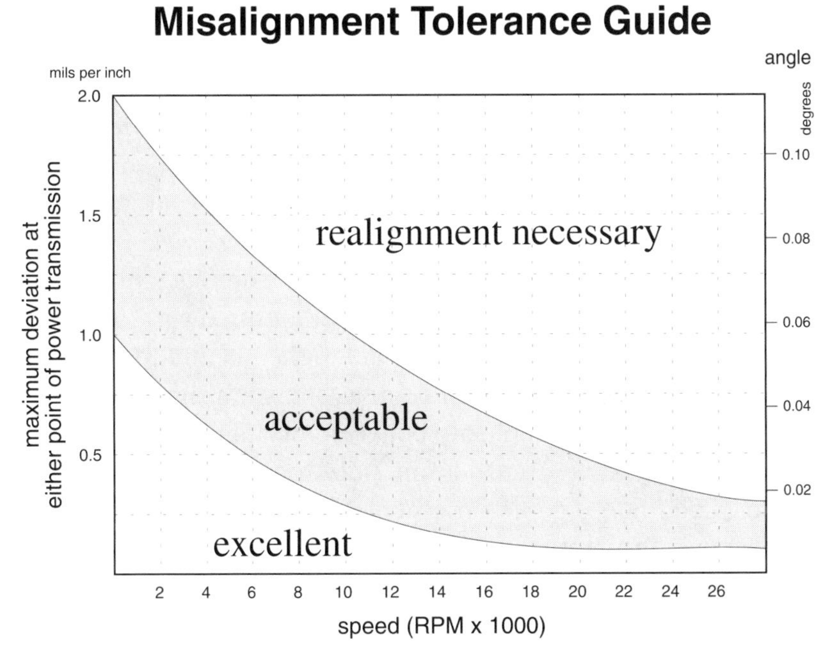

Figure 5-4. Alignment tolerance guide for flexibly coupled rotating machinery

Therefore, there are three factors that affect alignment of rotating machinery - the speed of the drive train, the maximum deviation at either flexing point or point of power transmission / power reception, and the distance between the flexing points or points of power transmission.

The last part of the definition of shaft misalignment is probably the toughest to achieve, and usually the one aspect of alignment that is most often ignored. When rotating equipment is started, the shafts will begin to move to another position. The most common cause of this movement is due to temperature changes that occur in the machinery casings and, therefore, this movement is commonly referred to as hot and cold alignment. These temperature changes are caused by friction in the bearings or by thermal changes that occur in the process liquids and gases. Movement of machinery may also be caused by process reaction moments in attached piping or counter-reactions due to the rotation of the rotor, something similar to the forces you feel when you try to move you arm around with a spinning gyroscope in your hand. Chapter 9 covers this topic in great detail.

Throughout this book misalignment conditions will be specified in mils offset occurring at each point of power transmission, with particular attention being paid to the maximum offset that occurs at either transmission point as shown in figure 5-3. Figure 5-4 shows the relationship between misalignment angles, distance between power transmission points, and offset in mils and is used as a go / no-go guide for acceptable shaft alignment for flexibly coupled rotating machinery.

Shaft vs. coupling alignment

Frequently people use the terms 'shaft alignment' and 'coupling alignment' interchangeably. Is there really a difference?

Notice in figure 5-5 that the centerline of rotation of the shaft on the left is in line with the centerline of the bore of the coupling hub on the shaft on the right, but it is not in line with the centerline of rotation of the shaft on the right.

By it purest definition, shaft alignment is when the centerlines of rotation are colinear. This is a very important point in aligning rotating machinery that a vast number of people overlook. It is possible to align the centerlines of rotation of machinery shafts that are bent or that have improperly bored coupling hubs and never know that these runout types of problems exist.

Notice in figure 5-6 that when the bent shaft is rotated, its centerline of rotation is straight but the shaft itself is not. In this situation what should we try to align the shaft on the right to? The

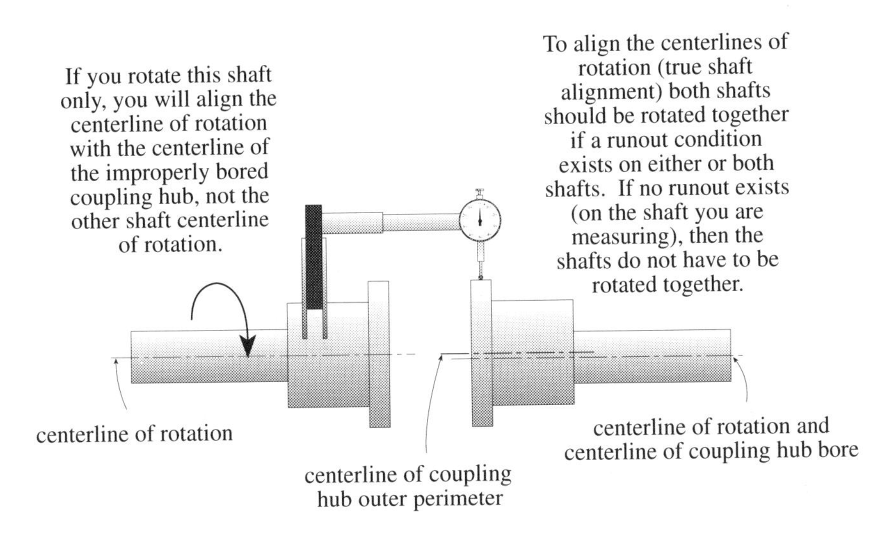

Figure 5-5. Aligning a centerline of rotation to the center of an improperly bored coupling hub.

centerline of the bore of the coupling hub at the end of the shaft or the centerline of rotation itself? The correct answer should be why would you ever try to align a piece of machinery that has a bent shaft or an improperly bored coupling hub?

How straight are rotating machinery shafts?

The assumption that many people make is that the centerlines of rotation on machinery are perfectly straight lines. In vertically oriented shafts this may indeed be true but the vast majority of rotating machinery have shafts that are horizontally mounted and the weight of the shaft and the components attached to the shaft cause the shaft to bow from its own weight. This

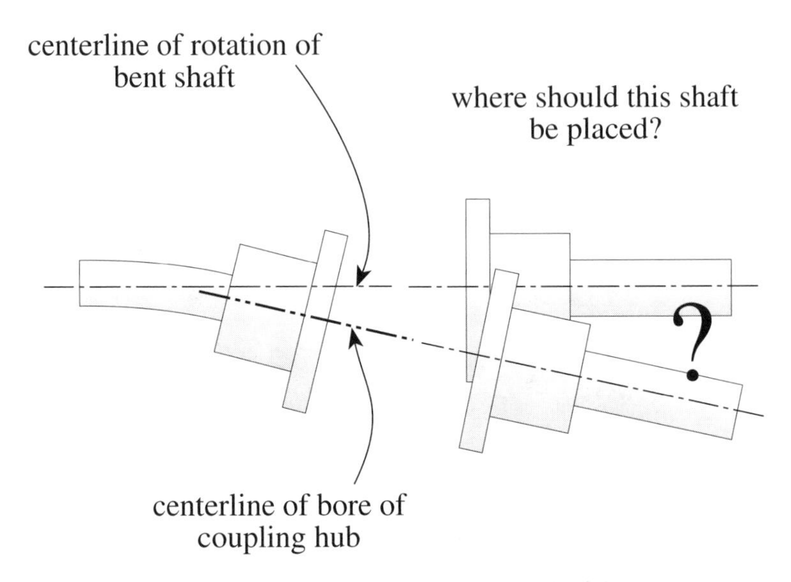

Figure 5-6. Aligning a bent shaft?

naturally occurring curvature of a machinery rotor is often referred to as a catenary curve.

Definitions :

catenary - the curve assumed by a perfectly flexible inextensible cord of uniform density and cross section suspended from two fixed points
catenoid - the surface described by the rotation of a catenary about its axis

The amount of deflection depends on several factors such as the stiffness of the shaft, the amount of weight between support points, the bearing design, and the distance between support points. For the vast majority of rotating machinery in existence, this catenary bow is negligible and for all practical purposes is ignored.

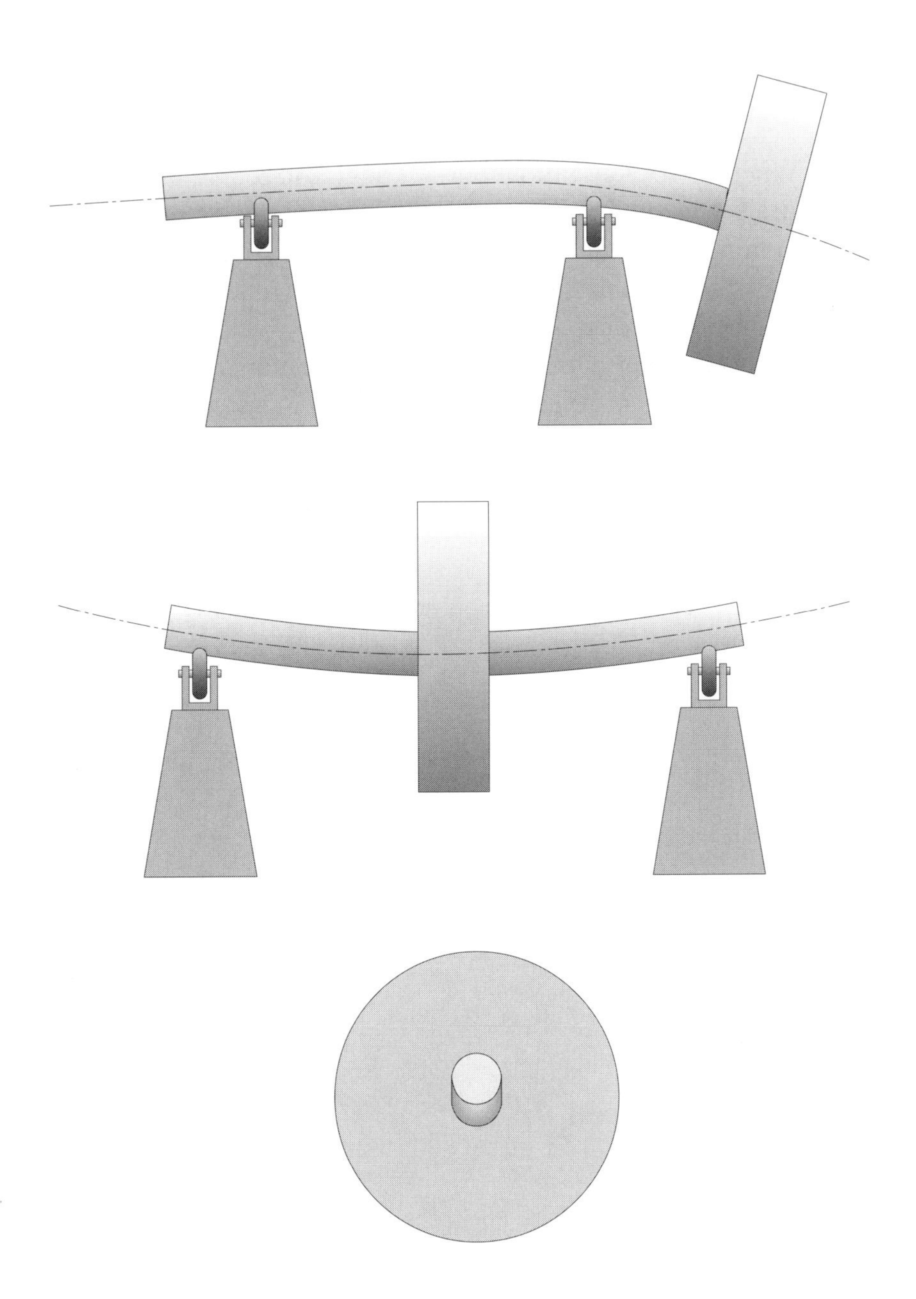

Figure 5-7. The natural bending of shafts from their own weight.

On extremely long drive trains however, (eg. turbine generators in power generating plants and MG sets in the metal industry) this catenary curve must be taken into consideration.

When a very long, flexible shaft begins to rotate, the bow tries to straighten out, but will never become a perfectly straight line. It is important to understand that the axis of rotation of a shaft could very possibly run on a curved axis of rotation. In situations where two or more pieces of machinery are coupled together, where one or more of the shafts are rotating around a catenary shaped axis of rotation, it is important to align the shafts so they maintain the curved centerline of rotation.

6
Preliminary Alignment Checks

Perhaps the most overlooked step in the process of aligning rotating machinery is this one. All too often, people who skip this step find themselves having problems measuring the off-line shaft positions accurately, adding and then removing shim stock several times from under the machinery feet, and they frequently find themselves 'chasing their tail' trying to reposition the machines laterally several times with marginal or no success. After wasting several hours in their attempt to align the machinery, they realize that something is wrong and they go back to check for many of the problems listed below.

In Chapter 1, we examined the 8 basic steps of aligning rotating machinery. As illustrated in figure 1-5, a considerable amount of time should be spent on these 'preliminary' checks. In summary, you will be trying to find and correct any problems in the following areas ...

- unstable or deteriorated foundations and baseplates (refer to Chapter 3)
- damaged or worn components on the rotating machinery (eg. bearings, shafts, seals, couplings)
- excessive runout conditions (eg. bent shafts, improperly bored coupling hubs)
- machine casing to baseplate interface problems (eg. soft foot)
- excessive piping, ductwork, or conduit forces (refer to Chapter 3)

Excessive Runout Conditions

The term 'runout' describes out of round or non-perpendicular conditions that exist on rotating machinery shafts and should be one of the first things you check on the machinery you are attempting to align. All rotating machinery shafts, coupling hubs, impellers, or other types of components rigidly attached to shafts will have runout. Some runout can be as small as ten millionths of an inch or as high as 100 mils.

'Radial' runout quantifies the eccentricity of the outer surface of a shaft or a component rigidly attached to a shaft with respect to the shafts centerline of rotation. 'Face' runout quantifies the amount of non-perpendicularity that may exist at the end of a shaft or on surfaces of components rigidly attached to a shaft. Runout conditions are typically measured with dial indicators as illustrated in figure 6-4. Runout checks should also be made at several points along the length of a rotor. Notice that the amount of face runout can vary depending on the distance from the centerline of rotation.

How much runout is acceptable? Table 6-1 can be used as a guideline for acceptable amounts of runout on rotating machinery shafts.

Dial indicator basics

bottom plunger type   back plunger type

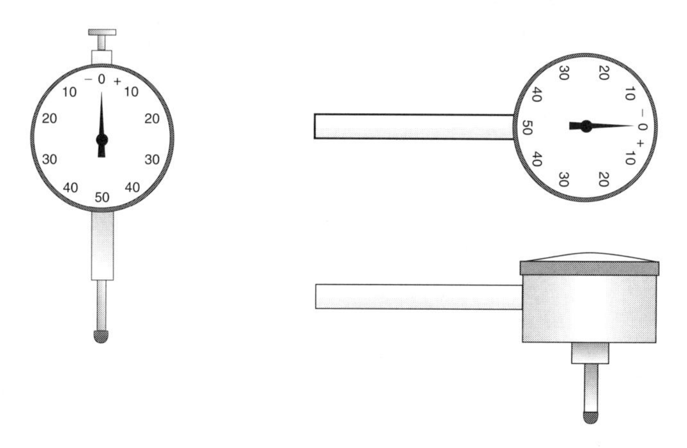

stem moves *outward* ...
needle rotates *counterclockwise*

stem moves *inward* ...
needle moves *clockwise*

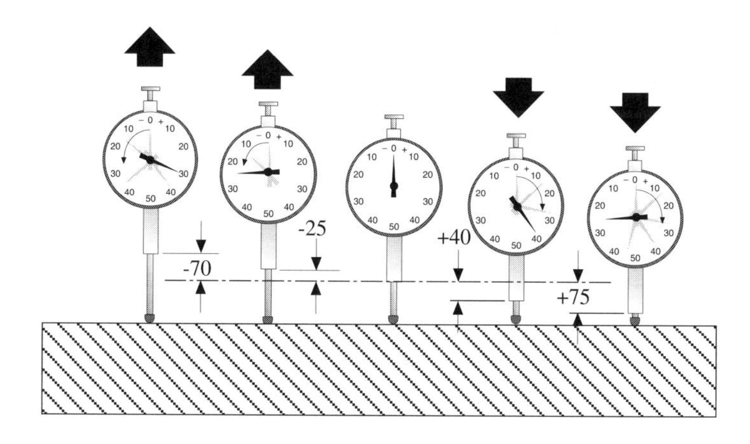

Figure 6-1. How a dial indicator works.

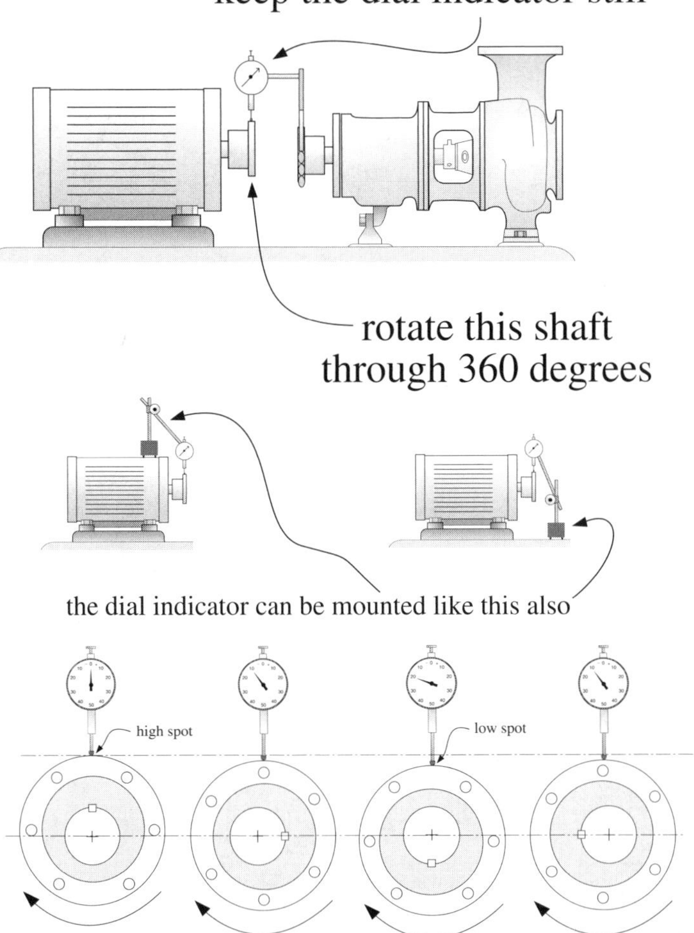

Figure 6-2. How to check for a runout condition.

'Runout' problems usually fall into one of these three catagories ...

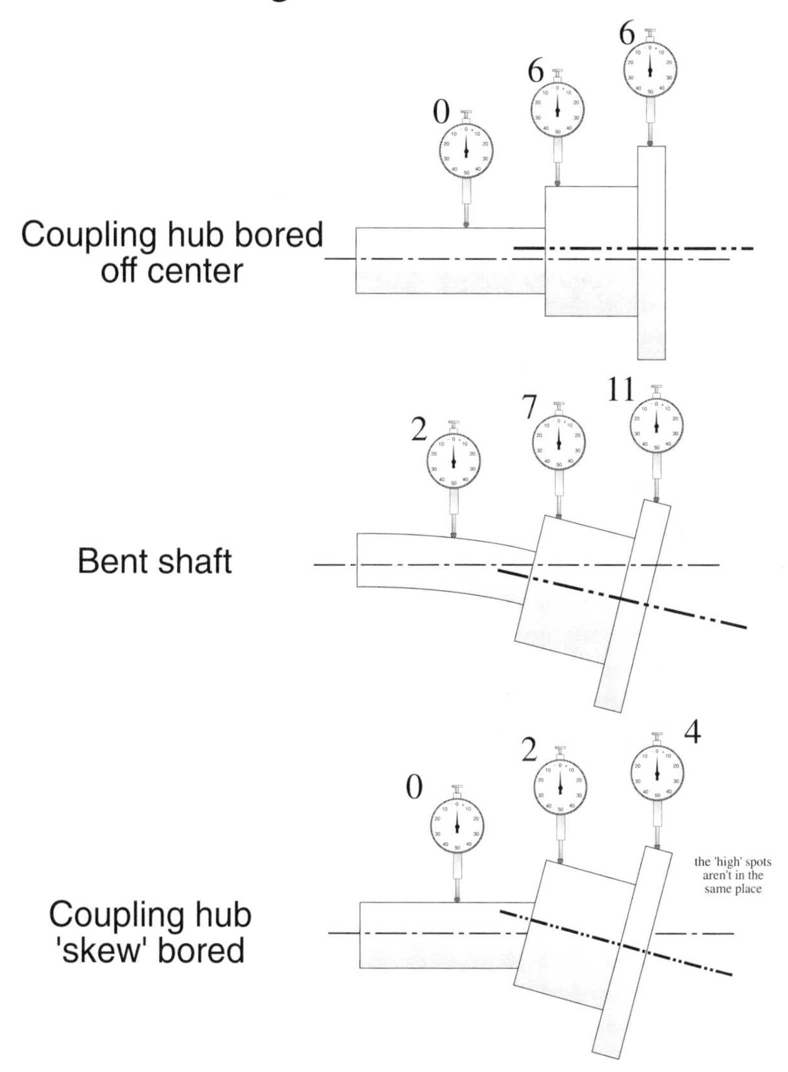

Figure 6-3. How to check for different types of runout problems.

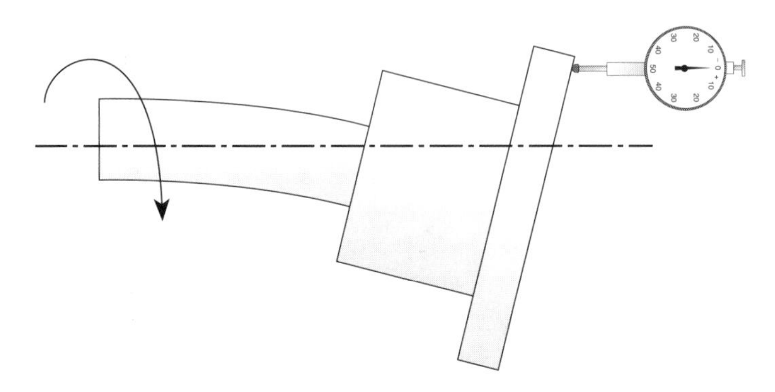

Figure 6-4. How to measure 'face' runout.

Recommended Runout Guidelines	
Machine Speed (RPM)	Maximum allowable Total Indicated Runout (T.I.R.)
0-1800	5 mils (0.005")
1800-3600	2 mils (0.002")
3600 and up	less than 2 mils

Table 6-1. Recommended maximum radial runout.

Measuring runout can be confusing at times. ‘High spots’ and ‘hills’ are not the same thing. ‘Low spots’ and ‘valleys’ are not the same thing. The high spot and low spot should occur 180 degrees apart. Hills and valleys can occur at any point or perhaps at several points around the outer surface of a coupling hub for example.

Machine casing to baseplate interface problems (aka 'Soft foot')

One of the most prevalent problems with aligning rotating machinery can be traced to machine case to baseplate interface problems. When rotating machinery is set in place on its base / frame / soleplate, one or more than one of the 'feet' are not making good contact at the 'foot points' on the frame. This can be attributed to warped or bowed frames, warped or bowed machine cases, improper machining of the equipment feet, improper machining of the baseplate, or a combination of a warped / uneven frame and warped / uneven machine case. This problem is commonly referred to as 'soft foot'. Soft foot generally describes any condition where poor surface contact is being made between the underside of the machine casing 'feet' and where they contact the baseplate or frame.

Soft foot problems seem to be worse on fabricated baseplates as opposed to cast baseplates. A fabricated baseplate is frequently made from sections of channel iron, angle iron, structural tubing, or I-beams. These pieces are then welded together to construct a machine frame. The chances of making true 45 or 90 degree cuts on the frame pieces and welding them all together and insuring that everything is flat, square, and in the same plane is very slim. However, cast baseplates are not exempt from this problem either. Even in cast baseplates where the base is sand cast and the machinery feet are machined, it is possible that during the installation process the frame was warped when it was placed on the concrete pedestal introducing a soft foot problem.

Why would we bother to worry about this phenomena? There are two important reasons why this should be corrected:

1- Depending on which sequence the foot bolts are tightened down, the centerline of rotation can be shifted into various positions causing a considerable amount of frustration when trying to align the machinery.

2- Tightening down any foot bolts that are not making good contact will cause the machine case to warp upsetting critical clearances on critical components such as bearings, shaft seals, mechanical seals, pump wear rings, compressor staging seals, motor armature/stator air gaps etc.

People often mistake this problem as being analogous to the short leg of a four legged chair. This metaphor is too simplistic and does not reflect the true nature of what really occurs on the machinery. The feet of a chair typically are making point contact on the floor. If our rotating machinery feet were only making point contact, then every soft foot problem could be corrected by using three legs such as found in a tripod.

It is important to recognize that our machinery feet are not making point contact. Instead, there are typically four (or more) supposedly flat foot surfaces on the underside of our machine case trying to mate up to four (or more) supposedly flat surfaces on the baseplate. Now, the chances of all four surface on the undersides of our machines cases being flat and in the same plane and all four surfaces on the baseplate being flat and in the same plane are rare indeed. Quite often when we try to mate the underside of a machine foot to the point of contact on the baseplate, a non-parallel, very complex, tapered, wedged shaped gap type of condition exists that cannot be corrected with a flat piece of shim stock.

In addition, it is most probable that a soft foot condition exist at all of the foot points. Now I'm not saying that our machinery is levitating in free space, just that the feet are not making good contact on the baseplate. A wide variety of different conditions can exist. Machinery can 'rock' across two diagonal corners or can rock from side to side or end to end. It is not uncommon to see three of the feet 'toe up' and the fourth foot 'toe down'. It is possible to have 'edge contact' at the inboard side of a foot and the outboard edge at that foot exhibit a gap.

Step 1. Relieve any stresses in machine cases and baseplates.

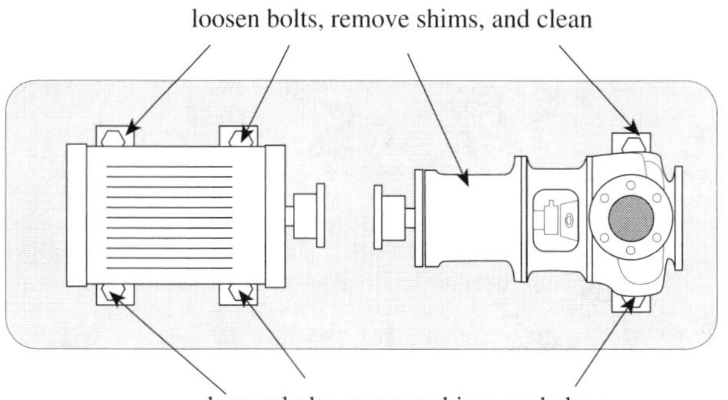

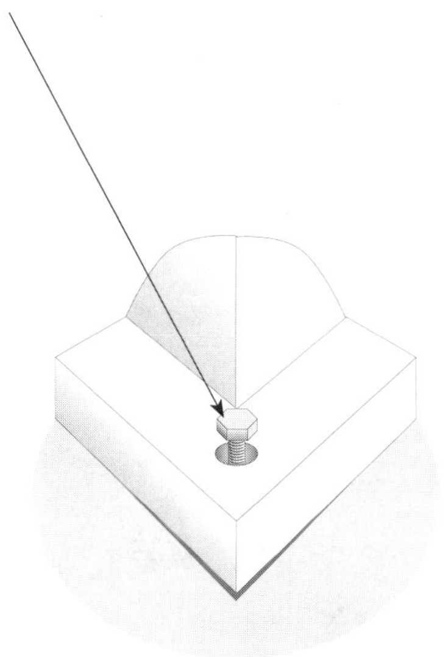

• If the machinery has been running for some time and there are shims under the feet, and you suspect that the soft foot has not been corrected, remove all of the existing shim packs and set the machine cases flat down on the baseplate.

• Clean the underside of each machinery foot and the points of contact on the baseplate. Remove any dirt, rust, or old shim stock from underneath each of the feet. If necessary, use some sandpaper or emerypaper (approximately 80 grit) to clean the surfaces on the underside of the machine 'foot' and the points of contact (sometimes called the 'pads') on the baseplate.

• Install the foot bolts but do not tighten them down. Try to 'center up' the machine cases in their bolt holes and 'rough' align both units.

Figure 6-5. Soft Foot check - step 1.

Step 2. Check for 'rocking case' conditions and measure the gap conditions around all of the foot bolts.

• With the foot bolts completely removed, or very loose in the bolt holes, check to see if the machine can be rocked from corner to corner, or end to end, or side to side. If it can, determine if the machine case seems to mate to the baseplate better in one position than the other. Hold the machine in that position by 'finger tightening' one (or more) of the bolts and measure four points around that bolt point. Then measure four points around each of the remaining bolt holes with a set of feeler gauges and record the readings at each remaining bolting point on the machine case.

• With the foot bolts in place but not tightened down, measure four points around each bolt hole with a set of feeler gauges and record the readings.

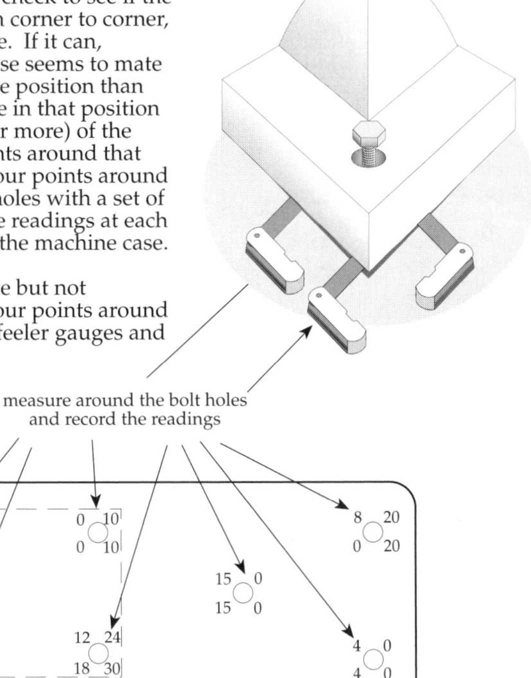

Figure 6-6. Soft Foot check - step 2.

Step 3. Correcting the soft foot conditions.

• Eliminate the 'soft foot' condition under each foot by installing complete U-shaped shims (if you have an even gap at all four points around the bolt hole) or by constructing a 'stair stepped shim wedge' with L-shaped or J-shaped shims or shims 'strips' and installing the special wedges under each foot that needs correction. See shim sections below.

• If you have to build a custom 'shim wedge' with L-shaped or J-shaped shims or shim 'strips', as much as possible, try to maintain the outline of a complete U-shaped shim when you stack the shim pieces together. Later on, you may have to install additional shims under that foot to change the height or 'pitch' of the machine case when you align the machinery. If the soft foot shim packs are neatly fabricated and stacked together in a U-shape, you can easily remove the soft foot shim pack, place the additional shims on top or underneath the pack and reinstall the entire assembly without disorienting the soft foot shim arrangement.

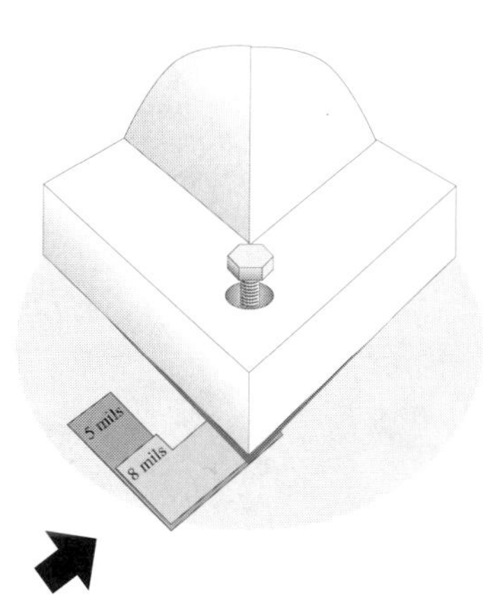

Note : After you install the shim correction under a foot, it might be helpful to 'feel' if the soft foot has been eliminated. To do this, initially finger tighten the bolt, put a wrench on the bolt head and try to tighten it all the way. If the bolt tightens very quickly (e.g. you only have to turn the wrench an 1/8 turn or less) the soft foot is probably corrected. If however, you have to make a 1/4 or 1/2 turn on the wrench and the foot feels 'spongy', the soft foot probably still exists and you have to try another shim pack.

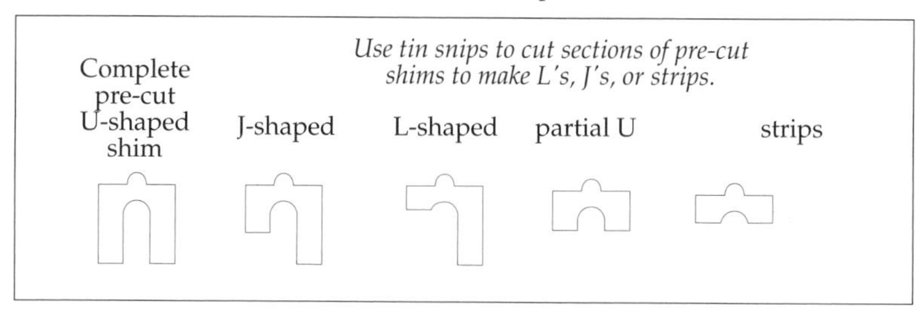

Figure 6-7. Soft Foot check - step 3.

Step 4. Verifying that the soft foot has been corrected.

Verify that the 'soft foot' or machine case / frame warpage problem has been eliminated by one of the following methods:

- Multiple bolt - multiple indicator method
- Multiple bolt - single indicator method
- Shaft movement method
- Single bolt - single indicator method

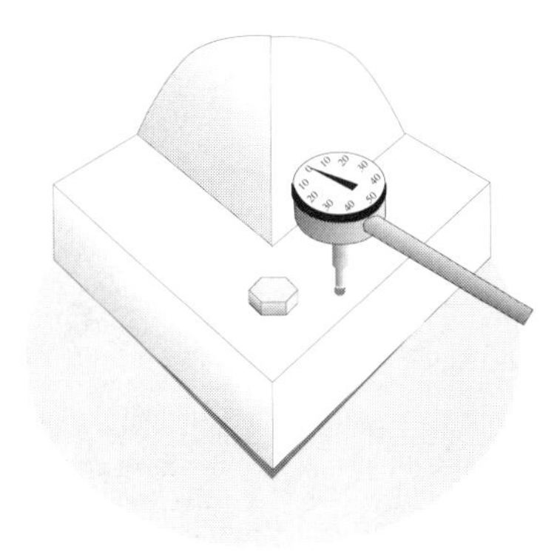

Figure 6-8. Soft Foot check - step 4.

There are four basic steps to detecting and correcting a soft foot problem as illustrated in figures 6-5 to 6-8.

The real trick to correcting soft foot is to insure that contact crosses through the axis of each bolt hole. Ideally it would be great to achieve full annular contact all the way around each bolt hole, but this would require fabricating compound wedge shaped shims that would tax the patience of many people bereft with this problem. However, there is a small minority of people who are willing to spend the time to do this to insure that a quality job is done.

Verifying that the 'Soft Foot' has been eliminated

There are several methods for checking whether a soft foot condition has been eliminated. Due to the sometimes complex nature of machine case and frame warpage, the best method requires that several points on the machine case be monitored for movement. If this is impractical, one point on the machine case can be monitored while several bolts are loosened. Another method is to monitor the movement of the machinery shafts while loosening one or more foot bolts.

Multiple Bolt - Multiple Indicator Method (preferred method)

1 - Tighten all of the foot bolts holding the machine in place.
2 - Place one dial indicator at each bolt location holding the machine case in place. Anchor the dial indicators to the frame or base and place the dial indicator stems as close as possible to the bolt holes, insure that the stems are touching the top of the 'feet', and zero the indicators.
3 - Loosen the bolt where the worst soft foot condition existed, watching the indicator at that foot for any movement. If more than 2-3 mils of movement is detected, there is probably some soft foot still remaining at that foot, but do not do anything yet. Leave the bolt loose.
4 - Loosen another bolt watching the indicator at that foot for any movement and also watch the indicator at the first foot for any additional movement. If more than 2-3 mils of movement is detected when this bolt was loosened and if more than 2-3 mils of additional movement is detected at the indicator on the first bolt that is loosened, there is probably some warpage occurring across those two bolts, but do not do anything yet. Leave both bolts loose.
5 - Continue loosening each of the remaining bolts holding the machine case watching the indicators at every loosened

bolt observing any additional movement. Carefully watch the indicator since each corner may raise or lower as each bolt is loosened giving you clues as to whether diagonal frame warpage or warpage along a side is occurring.

6 - Once all of the bolts have been loosened, review what you observed when each bolt was loosened. If only one indicator showed more than 2-3 mils of movement, then there is probably a soft foot condition at that foot only. Remove any soft foot shims under that foot and re-measure four points around that bolt hole with feeler gauges and install a flat shim or shim wedge to correct the observed condition. If more than 2-3 mils of movement was noticed at several bolt locations, then there is probably a soft foot condition at each one of those feet. Remove any soft foot shims under those feet and re-measure four points around those bolt holes with feeler gauges and install flat shims or shim wedges to correct the observed condition.

7 - Repeat the procedure if additional corrections are required.

Multiple Bolt - Single Indicator Method (second choice)

1 - Tighten all of the foot bolts holding the machine in place.

2 - Place one dial indicator at the bolt location where the worst soft foot condition was noticed. Anchor the dial indicator to the frame or base and place the dial indicator stem as close as possible to the bolt hole, insure that the stem is touching the top of the 'foot', and zero the indicator.

3 - Loosen the bolt where the indicator is located watching the indicator at that foot for any movement. If more than 2-3 mils of movement is detected, there is probably some soft foot still remaining at that foot, but do not do anything yet. Leave the bolt loose.

4 - Loosen another bolt watching the indicator at the first foot for any movement. If more than 2-3 mils of additional movement is detected when this bolt was loosened, there is probably some warpage occurring across those two bolts but do not do anything yet. Leave both bolts loose.
5 - Continue loosening each of the remaining bolts holding the machine case watching the indicator at the first bolt, observing any additional movement. Carefully watch the indicator since each corner may raise or lower as each bolt is loosened giving you clues as to whether diagonal frame warpage or warpage along a side is occurring.
6 - Once all of the bolts have been loosened, review what you observed when each bolt was loosened. If more than 2-3 mils of movement occurred when just one of the bolts was loosened, then there is probably a soft foot condition at that foot only. Remove any soft foot shims under that foot and re-measure four points around that bolt hole with feeler gauges and install a flat shim or shim wedge to correct the observed condition. If more than 2-3 mils of movement was noticed when several of the bolts were loosened, then there is probably a soft foot condition at each one of those feet. Remove any soft foot shims under those feet and re-measure four point around those bolt holes with feeler gauges and install flat shims or shim wedges to correct the observed condition.
7 - Repeat the procedure if additional corrections are required.

Shaft Movement Method (third choice)

1 - Tighten all of the foot bolts holding the machine in place.
2 - Attach a bracket to one shaft, place a dial indicator on the top side of the other shaft, and zero the indicator (similar to the piping stress check shown in Chapter 3).

3 - Sequentially loosen one foot bolt at a time observing for any movement at the indicator when each bolt is loosened.
4 - If there was more than 2-3 mils of movement when only one of the bolts were loosened, then there is probably a soft foot condition at that foot only. Remove any soft foot shims under that foot and re-measure four point around that bolt hole with feeler gauges and install a flat shim or shim wedge to correct the observed condition. If more than 2-3 mils of movement was noticed when several of the bolts were loosened, then there is probably a soft foot condition at each one of those feet. Remove any soft foot shims under those feet and re-measure four point around those bolt holes with feeler gauges and install flat shims or shim wedges to correct the observed condition.
5 - Repeat the procedure if additional corrections are required.

Single Bolt - Single Indicator Method (last choice)

1 - Tighten all of the foot bolts holding the machine in place.
2 - Place a dial indicator at one of the feet on the machine case. Anchor the dial indicator to the frame or base and place the dial indicator stem as close as possible to the bolt hole, insure that the stem is touching the top of the 'foot', and zero the indicator.
3 - Loosen the bolt where the indicator is located watching the indicator at that foot for any movement. If more than 2-3 mils of movement is detected, there is probably some soft foot still remaining at that foot. Remove any soft foot shims under that foot and re-measure four point around that bolt hole with feeler gauges and install a flat shim or shim wedge to correct the observed condition. Re-tighten the bolt.
4 - Sequentially move the indicator to each one of the feet,

loosening that bolt and watching the indicator for any movement. If more than 2-3 mils of movement is detected when each bolt was loosened, there is probably some soft foot still remaining at that foot. Remove any soft foot shims under that foot and re-measure four point around that bolt hole with feeler gauges and install a flat shim or shim wedge to correct the observed condition. Re-tighten each bolt.

5 - Repeat the procedure if additional corrections were required.

Once the soft foot has been corrected, the shims will stay there for the rest of the alignment process. We may be adding more shims later on to change the height or 'pitch' of the machine case, but the shims used to correct the soft foot condition will remain in place.

Other methods for correcting soft foot problems

For many of you who are reading about this problem for the first time there is a great tendency to disbelieve that this malady actually exists! Be forewarned, this is a time consuming, frustrating process that frequently can consume more time than actually aligning the rotating machinery itself!

It does seem rather silly to cut U-shaped shims into strips, L-shapes, J-shapes, or shortened U-shapes to correct this problem but pre-cut, U-shaped shims are commonly used in industry to adjust the position of rotating machinery in the process of aligning equipment. Again, you cannot correct a wedge shaped gap condition with a flat piece of shim stock. Since many people only have this pre-cut shim stock available to them, then the only way to construct a wedge is to 'stair-step' pieces of shims together to construct the wedge that is needed.

Despite the fact that virtually every piece of rotating ma-

Soft Foot Example

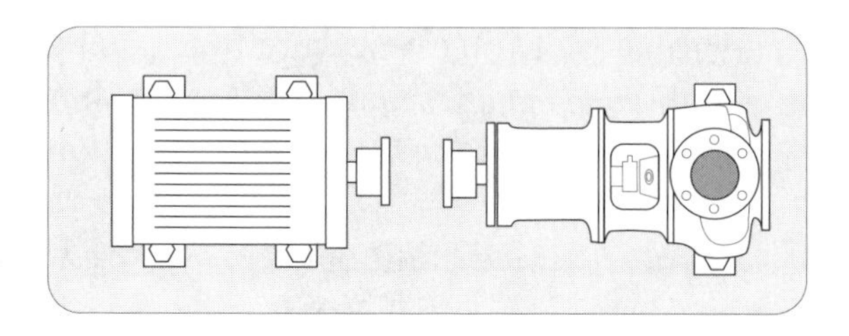

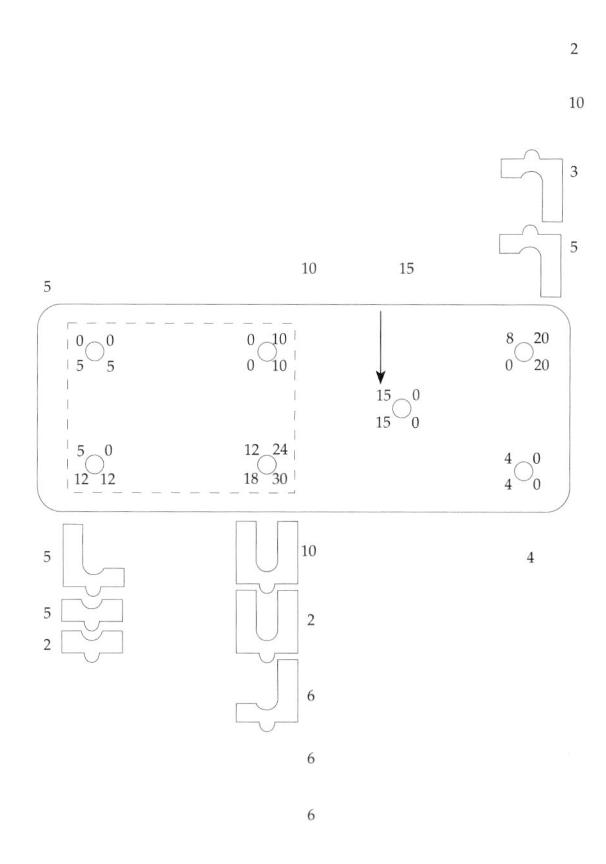

Figure 6-9. Soft Foot example for motor pump arrangement.

chinery has a soft foot problem, very few solutions have been forwarded on how to correct this problem easily. Whatever mechanism is used, there are six problems that need to be addressed:

- the vast majority of soft foot problems are non-parallel gap situations
- one or more than one machine foot may not be making contact whether parallel or non-parallel conditions exist
- it is possible that a slight soft foot condition can be introduced when adding more shims under one end of a machine case than the other end when attempting to correct a misalignment condition
- thermal warpage of a machine base or frame can occur during operation that would alter the soft foot condition observed off-line
- the device must be 'thin' enough to fit under all of the currently installed rotating machinery without major frame, machine case, or piping alterations
- maintain its shape and form for long periods of time
- be relatively inexpensive

Under these circumstances, cutting 'U' shaped shims into sections only addresses some of these situations. To address all of the concerns, the soft foot device must be able to correct for non-parallel conditions, have an adjustable height capability, and be able to change its shape in the event the conditions between the underside of the machinery feet and the base change during operation. One possible solution is a machinery positioner / soft foot correction device as shown in figure 6-10.

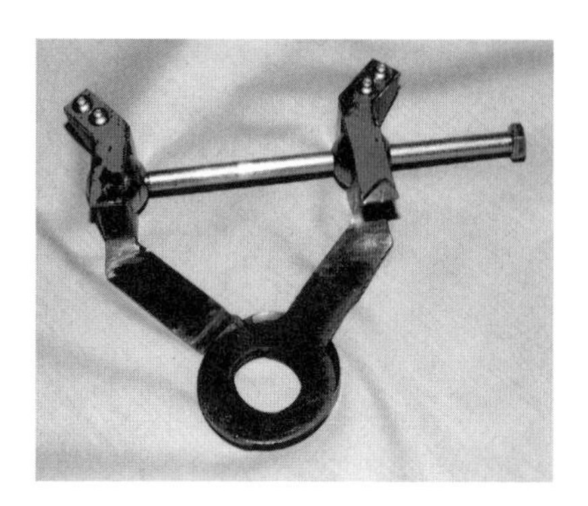

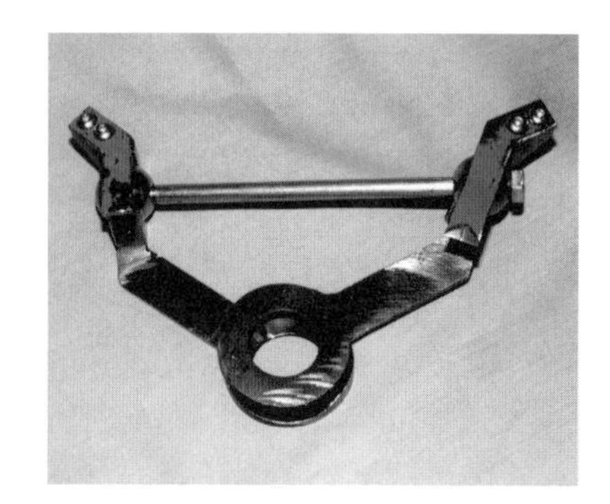

Figure 6-10. Prototype machinery positioner that automatically corrects for non-parallel contact and has adjustable height capabilities. Photo courtesy Turvac Inc., Cincinnati, OH (patent pending).

7
Shaft Alignment Techniques and Measuring Tools

The techniques and tools shown in this chapter illustrate the wide variety of methods used to measure the position of one shaft's centerline of rotation with respect to another shaft's centerline of rotation when the machinery is off-line. An alignment 'expert' is someone who knows how to perform these shaft positional measurements in a variety of different ways. There are advantages and disadvantages to each one of these techniques. There is no one method or measuring device that will solve every alignment problem that one can possibly encounter on the various types of rotating machinery drive systems in existence. Understanding each one of these techniques will enable you to select the best measurement method for the alignment situation confronting you. In many cases two (or more) different techniques could be used to make shaft centerline positional measurements.

Frequently people who capture a set of readings using one of these techniques or measurement tools will run across a situation where the data just doesn't seem to make any sense. Knowing how to perform shaft positional measurement a different way can verify whether the data from the initial technique in doubt is valid. Since the machinery shafts can only be in one position at any point in time, the data from two or more measurement methods should indicate the same shaft positional information. For example, if you have captured a set of readings with a laser alignment system and you don't believe what the system is telling you, take a set of reverse indicator readings. If the two sets of readings agree, then the measurement data is probably correct. If it is not, then it would be wise to determine why there is a discrepancy between the two methods before you continue on and incorrectly position the machinery based on bad measurement data.

Since shaft alignment is primarily concerned with the application of distance measurement, this chapter will begin by covering the wide variety of tools available to measure dimensions. Next, five different dial indicator techniques commonly employed for rotating machinery shafts connected together with flexible couplings will be shown along with two traditional dial indicator / feeler gauge / inside micrometer techniques used for rotating machinery shafts connected together with rigid couplings. It is important to understand each of these basic measurement techniques since every alignment measurement system in existence utilizes one or more of these methods regardless of the measurement sensor used to capture the shaft position information.

Finally, a comprehensive review of the wide variety of shaft position measurement tools that utilize a myriad of different measurement sensors will be shown to give you an indication of what's available on the market today.

Keep in mind that this chapter covers one small but important facet of shaft alignment: measuring the relative positions of two rotating machinery shafts. In other words, these methods will show you how to find the positions of two shaft centerlines when the machinery is not running (Step 5 in Chapter 1). Once you have determined the relative positions of each shaft in a two element drive train, the next step is to determine if the machinery is within acceptable alignment tolerances. If the tolerance is not yet acceptable, the machinery will have to be repositioned. Chapter 8 will discuss a very useful and powerful technique where the data collected from these techniques can be used to construct a graph or model of the relative shaft positions to assist you in determining how you might want to move the machinery to achieve acceptable tolerances.

Dimensional Measurement

As illustrated in Chapter 2, the task of accurately measuring distance was one of the first problems encountered by man. The job of 'rope stretcher' in ancient Egypt was a highly regarded profession and dimensional measurement technicians today can be seen using laser interferometers capable of measuring distances down to the sub-micron level.

It is important for us to understand how all of these measurement tools work since new tools rarely replace old ones, they just augment them. Despite the introduction of laser shaft alignment measurement systems in the mid 1980's for example, virtually all manufacturers of these systems still include a standard tape measure for the task of measuring the distances between the hold down bolts on machinery casings and where the measurement points are being captured on the shafts.

The two common measurement systems in worldwide use today are the English and metric systems. Without going into a lengthy dissertation of English to metric conversions, the easiest one most people can remember is this

$$25.4 \text{ mm} = 1.00 \text{ inch}$$

By simply moving the decimal point three places to the left it becomes obvious that

0.0254 mm = 0.001" = 1 mil (one-thousandths of an inch)

Standard tape measures and rulers

Perhaps one of the most common tools used in alignment are standard rulers or tape measures as shown in figure 7-1. The tape measure is typically used to measure the distances between machinery hold down bolts (commonly referred to as the machinery 'feet') and the points of measurement on the shafts or coupling hubs. Graduations on tape measures are usually as small as 1/16" to 1/32" (1mm on metric tapes) which is about the smallest dimensional measurement capable of being discerned by the unaided eye. The straight edge is often used to 'rough align' the units as shown in figure 7-2.

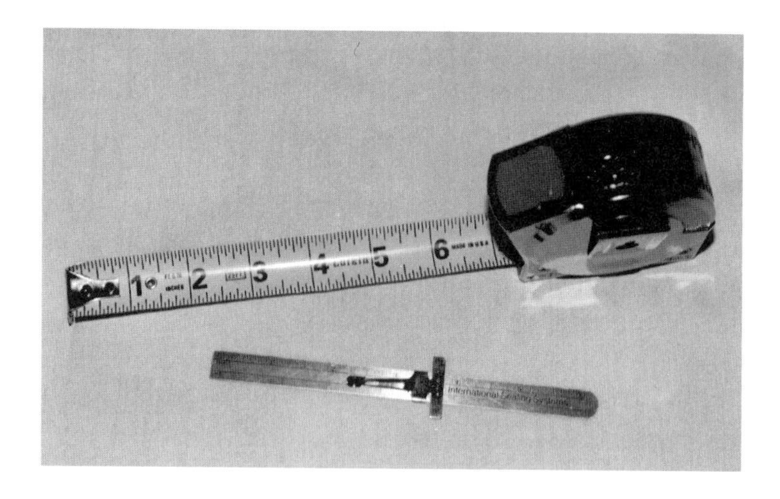

Figure 7-1. Standard rulers.

'Rough' alignment methods

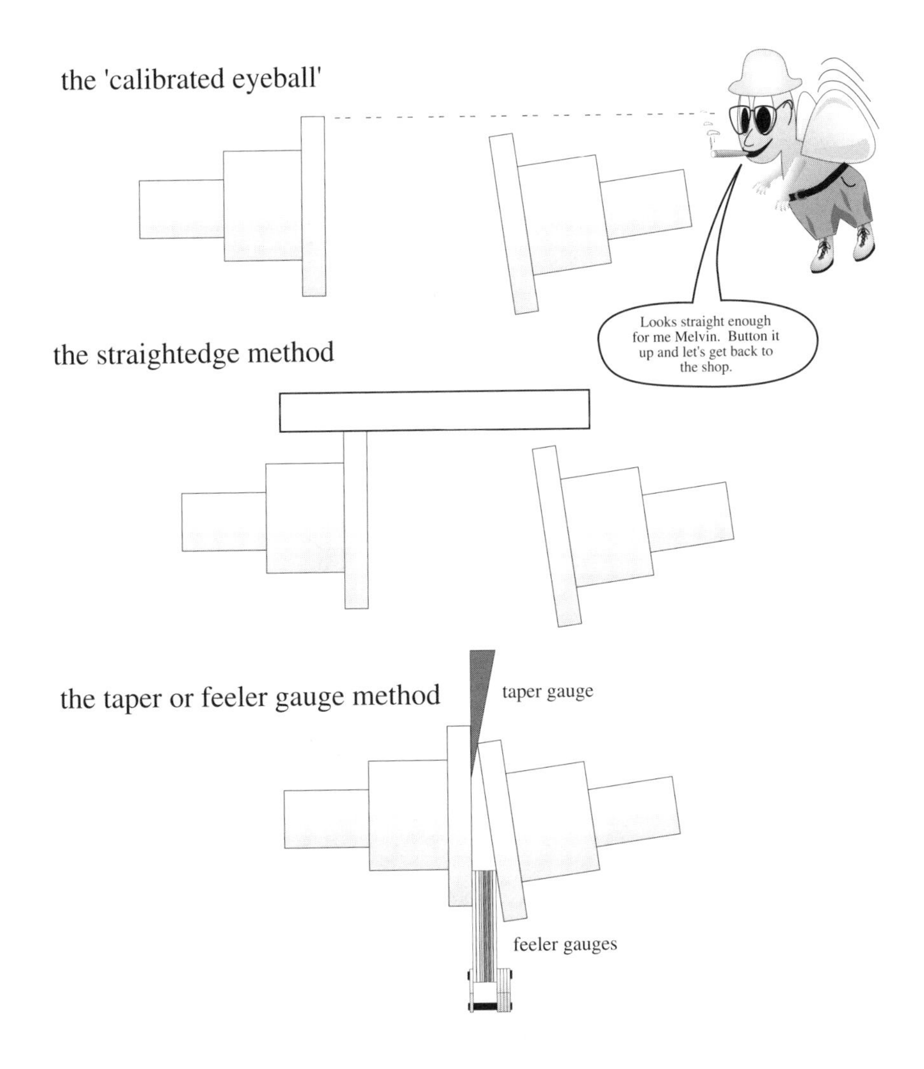

Figure 7-2. Rough alignment methods using straightedges or feeler gauges.

Feeler and taper gauges

'Feeler gauges' are simply strips of metal shim stock arranged in a 'fold out fan' type of package design. They are used to measure 'soft foot' gap clearances, closely spaced shaft end to shaft end distances, rolling element to raceway bearing clearances, and a host of similar tasks where fairly precise (+/- 1 mil) measurements are required.

Taper gauges are precisely fabricated wedges of metal with lines scribed along the length of the wedge that correspond to the thickness of the wedge at each particular scribe line. They are typically used to measure closely spaced shaft end to shaft end distances where accuracies of +/- 10 mils is required.

Slide Caliper

As indicated in Chapter 2, the slide caliper has been used to measure distances with an accuracy of 1 mil (0.001") for the last

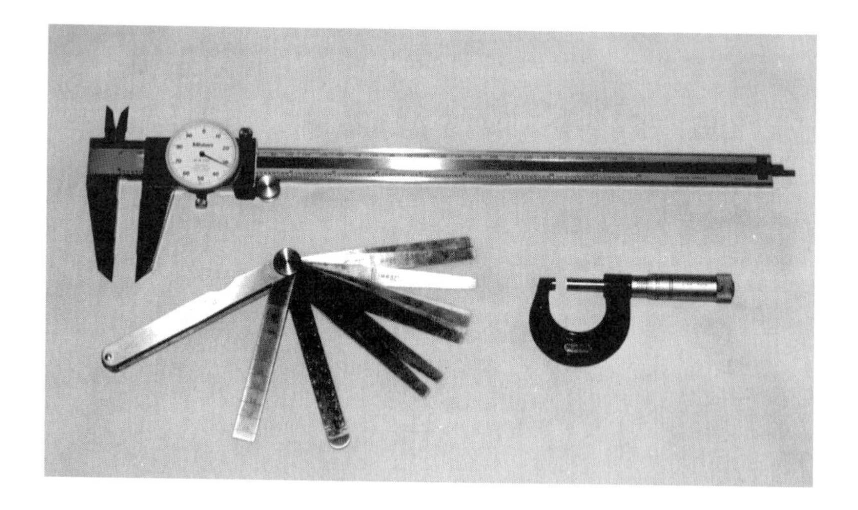

Figure 7-3. Feeler gauges, slide caliper, and micrometer.

four hundred years. The primary scale looks like a standard ruler with divisions marked along the scale at increments of 0.025". The secondary, or sliding scale, has a series of 25 equally spaced marks where the distance from the first to the last mark on the sliding scale is 1.250" apart. The jaws are positioned to measure a dimension by positioning the sliding scale along the length of the

How to read a Vernier Caliper

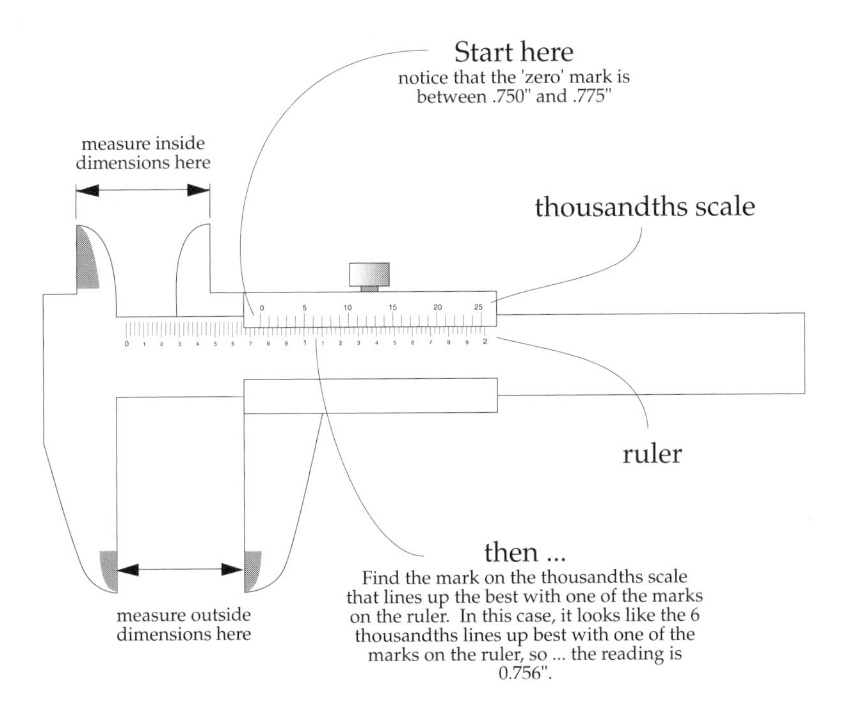

Figure 7-4. How to read a slide caliper.

primary scale. A reading is then taken by observing where the position of the '0' mark on the sliding scale aligns with the divisions on the primary scale to obtain the inch and quarter of an inch reading. The thousandths of an inch reading is obtained by observing which mark aligns most closely on the secondary with one on the primary scale as illustrated in figure 7-4.

Micrometers

Although the micrometer was originally invented by William Gascoigne in 1639, its use did not become widespread until 150 years later when Henry Maudslay invented a lathe capable of accurately and repeatably cutting threads. That of course brought about the problem of how threads should be cut (number of threads per unit length, thread angles, thread depth, etc.), which forced the emergence of thread standards in the Whitworth system (principally abandoned) and the current English and Metric standards.

The micrometer is still in prevalent use today. The micrometer is typically used to measure shaft diameters, hole bores, shim or plate thicknesses, ad infinitum and is almost a 'must' for the person performing alignment jobs.

Dial Indicators

The dial indicator came from the work of 19th century watchmakers in New England. John Logan of Waltham, Massachusetts filed a US patent on May 15th, 1883 for what he termed 'an improvement in gages'. Its outward appearance was no different than the dial indicators of today but the pointer (needle) was actuated by an internal mechanism consisting of a watch chain wound around a drum (arbor). The arbor diameter determined the amplification factor of the indicator. Later, Logan developed a rack and pinion assembly that is currently in use today on most mechanical dial indicators.

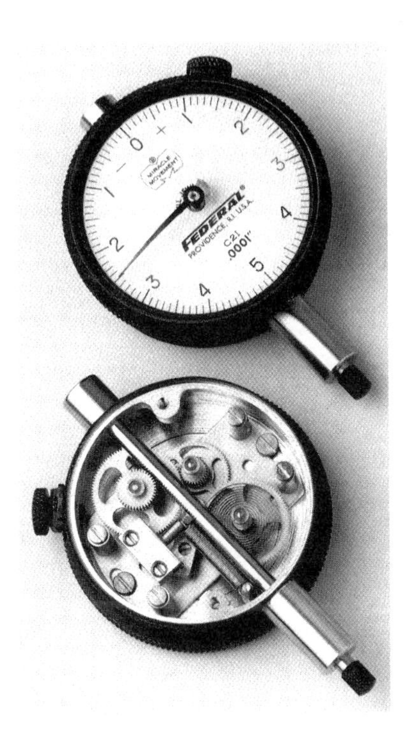

Figure 7-5. Dial indicator. Photo courtesy Federal Products, Providence, RI.

The full range of applications of this device was not recognized for another 15 years when one of Logan's associates, Frank Randall, another watchmaker from E. Howard Watch Co. in Boston, bought the patent rights from Logan in 1896. He then formed a partnership with Francis Stickney and began manufacturing dial indicators for industrial use. A few years later B. C. Ames also began manufacturing dial indicators for general industry.

German professor Ernst Abbe established the measuring instrument department at the Zeiss Works in 1890, and by 1904 had developed a number of instruments, which included a dial indicator, for sale to industry.

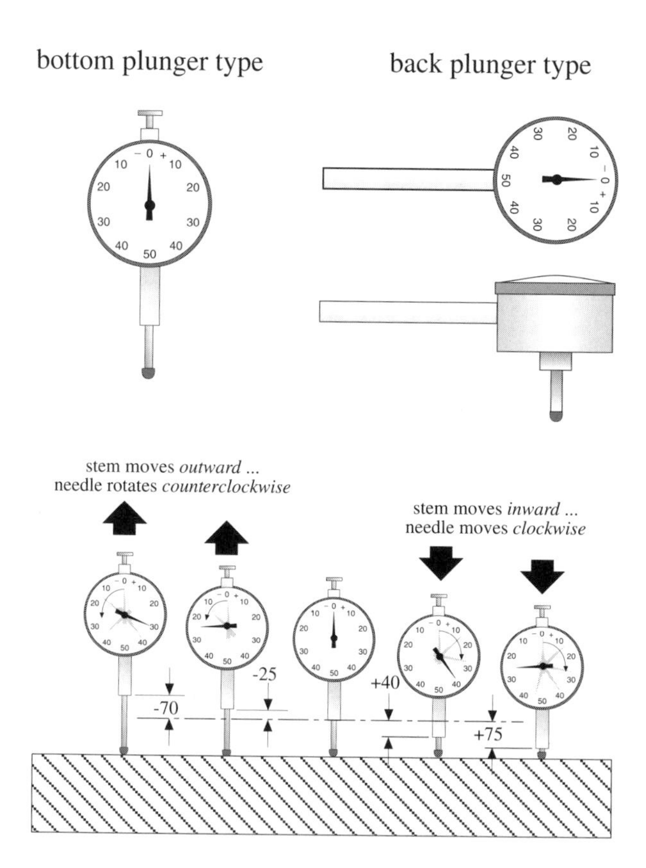

Figure 7-6. Dial indicator basics. Photo courtesy Federal Products, Providence, RI.

Typically, AGD Group 1 or 2 (AGD is an acronym for American Gauge Design, Group 1 specifies that the dial face is approximately 1" in diameter, Group 2, 2" in diameter) dial indicators are used for shaft positional measurements although any size dial face could be used. The AGD Group 1 indicators can be purchased with 200 - 250 mils (0.200" - 0.250") of total stem travel and are best suited for shaft alignment purposes. AGD Group 2 indicators with up to 2" of total stem travel are also frequently used but contribute to additional 'bracket sag' problems due to their weight (see bracket sag in this Chapter).

Optical Alignment Tooling

Optical alignment tooling consists of devices that combine low power telescopes with accurate bubble levels and optical micrometers for use in determining precise elevations (horizontal slices through space) or plumb lines (vertical slices through space). They are not to be confused with theodolite systems that can also measure the angular pitch of the line of sight. Optical tooling methods date back to the earliest times as covered in Chapter 2.

Figure 7-7. Optical tilting level and jig transit.

Optical alignment systems are perhaps one of the most versatile tools available for a wide variety of applications such as leveling foundations, checking for squareness on machine tools or frames, aligning bearing pedestal locations of lengthy rotating machinery drive trains, measuring off-line to running machinery movement (covered in Chapter 9), checking for roll parallelism in paper and steel manufacturing plants, and on and on. If you have a considerable amount of rotating machinery in your plant, it is highly recommended that someone examine potential applications for this extremely useful and accurate tooling.

Proximity Probes

Proximity probes (aka inductive pickups) are basically non-contacting, electronic dial indicators and are therefore devices used to measure distance or displacement. Their use in measuring vibration (i.e. shaft motion) dates back to the mid-1960's and are typically used as permanently mounted vibration sensors to measure shaft motion, but they are also used to measure shaft position in both the radial and axial directions.

Although they have been proposed for use as shaft alignment measuring devices (see figure 2-17 in Chapter 2), no company currently offers such a system. Proximity probes have been used extensively to measure off-line to running machinery movement in some very innovative ways as explained in Chapter 9. Their primary limitation is the range of useful distance measurement (approximately 50 -150 mils) that can be attained with standard probes.

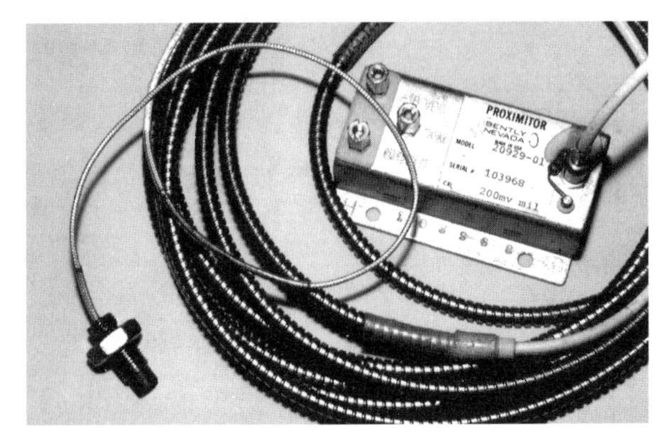

Figure 7-8. Proximity probe and oscillator - demodulator.

typical target sensitivity is 100 or 200 mv/mil

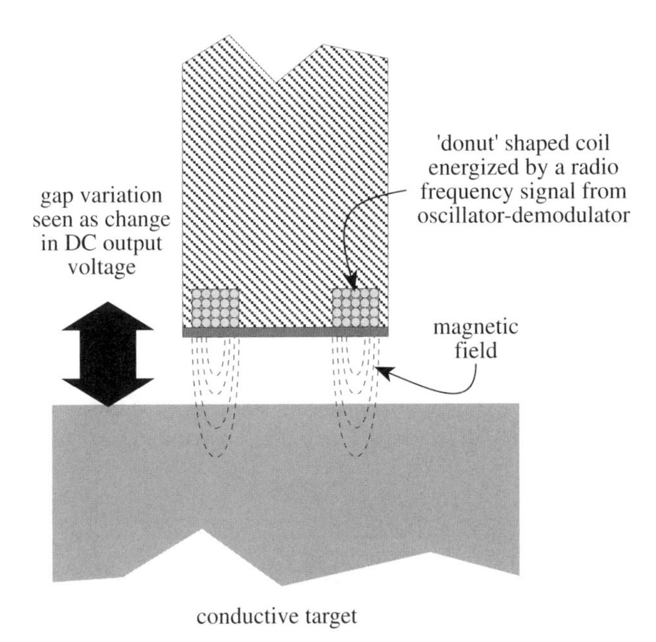

Figure 7-9. Basic operation of a proximity probe.

Linear Variable Differential Transformers (LVDT)

These devices are also called variable inductance transducers that output an AC signal proportional to the position of a core that moves through the center of the transducer as illustrated in

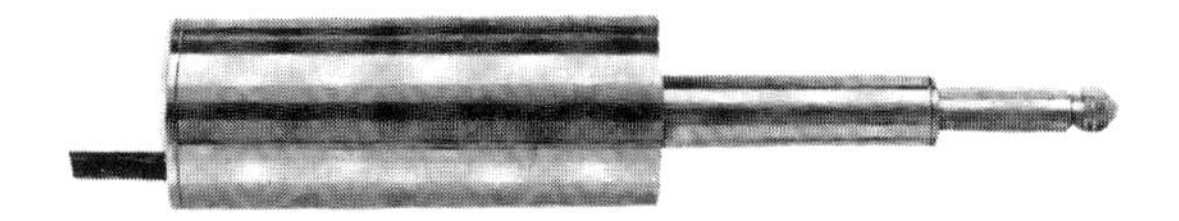

Figure 7-10. Typical LVDT sensor.

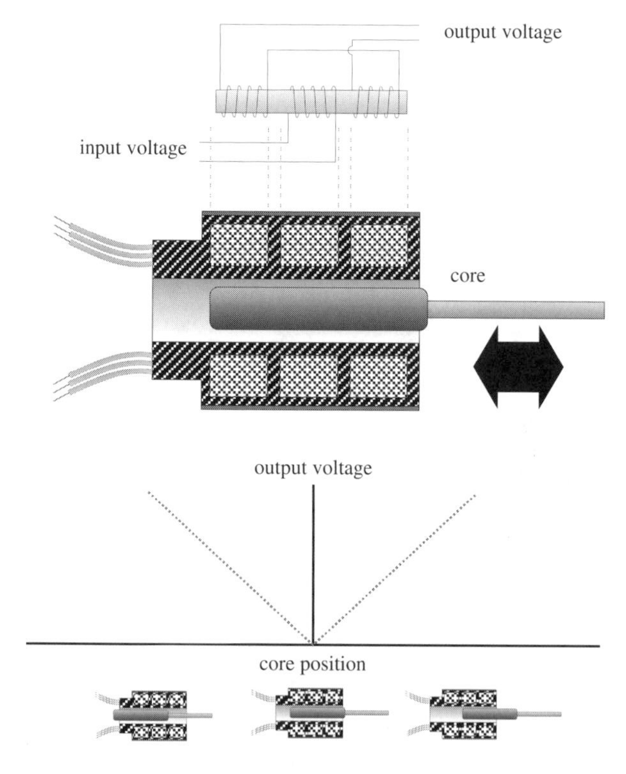

Figure 7-11. LVDT sensor basic operation.

figure 7-11. These devices can attain accuracies of +/- 1% of full scale range with stroke ranges available from 20 mils to over 20 inches. No current manufacturer of alignment measurement systems use this type of transducer for shaft alignment purposes.

Optical Encoders

Optical encoders are essentially pulse counters. They are most frequently used to measure shaft speed or shaft position and are therefore sometimes called shaft or rotational encoders. A

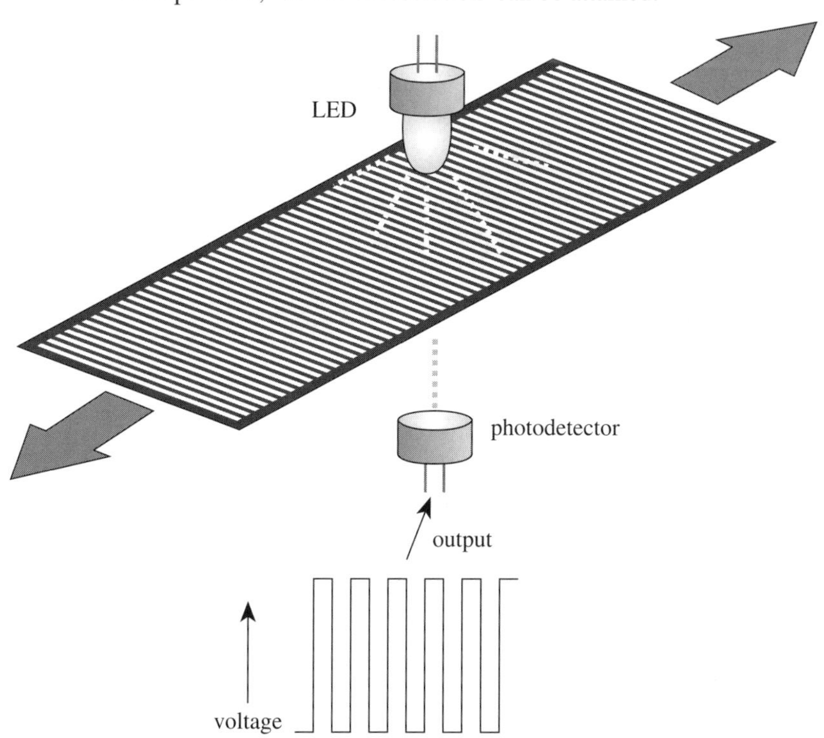

Figure 7-12. Optical encoder operation.

series of slots are etched on a disk or flat strip. A light source (typically an LED) aims at the disk or flat strip and as the disk or strip is moved or rotated, a photodetector on the other side of the disk / strip counts the number of slots that are seen. One manufacturer currently uses this type of sensor for shaft alignment measurement.

Charge Couple Devices (CCD)

The charge-coupled device was originally proposed by Boyle and Smith in 1970 as an electrical equivalent to magnetic bubble digital storage devices. The basic principle of their device was to store information in the form of electrical 'charge packets' in potential wells created in the semiconductor by the influence of overlying electrodes separated from the semiconductor by a thin insulating layer. By controlling voltages applied to the electrodes, the potential wells and hence the 'charge packets' could be shifted through the semiconductor.

The potential wells are capable of storing variable amounts of charge and can be introduced electrically or optically. Light impinging on the surface of the charge-coupled semiconductor generates charge carriers which can be collected in the potential wells and afterward clocked out of the structure enabling the CCD to act as an image sensor.

A considerable amount of effort was put forth in the 1960's in developing optical imagers that utilized matrices of photodiodes that effectively became undone by the development of the CCD. The rate of progress in CCD design through 1974 was so astonishing that R. L. Rodgers demonstrated a 320x512 bit CCD sensor that could be used for 525 line television imaging just four years after the CCD was invented. Charge coupled devices have found their way into everyday life in video cameras and in high technology fields such as astronomy where large area CCD's capture images in telescopes both in orbit and on earth.

With the recent pace of introducing electronic measurement sensors in the arena of alignment it seems odd that no one has incorporated the CCD as a measurement sensor. The only known applications of CCD's for use in alignment was investigated in a doctoral thesis by Brad Carman and a research project at the University of Calgary (see figures 7-14 and Chapter 9).

A CCD is a multilayered silicon chip. In one layer, an array of electrodes divides the surface into pixels. Each electrode is connected to leads, which carry a voltage. The image forms on the silicon substrate. Light particles pass through the CCD freeing electrons in the silicon substrate. The voltage applied to the leads draws freed electrons together in special areas in the silicon substrate, called photo sites. The number of gathering electrons at the photo site is dependant on the intensity of the light striking in that area. The CCD transfers captured electrons, one by one, to an analog to digital converter, which assigns each site a digital value corresponding to the number of electrons a site holds. The number of electrons at each site determines how light or dark each pixel in the image is.

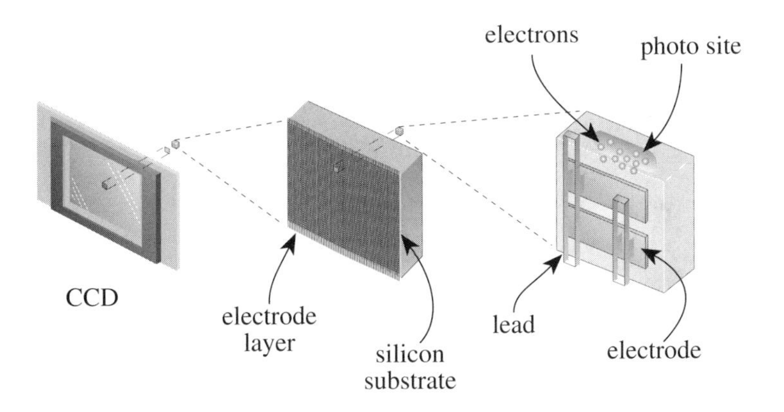

Figure 7-13. How a CCD works.

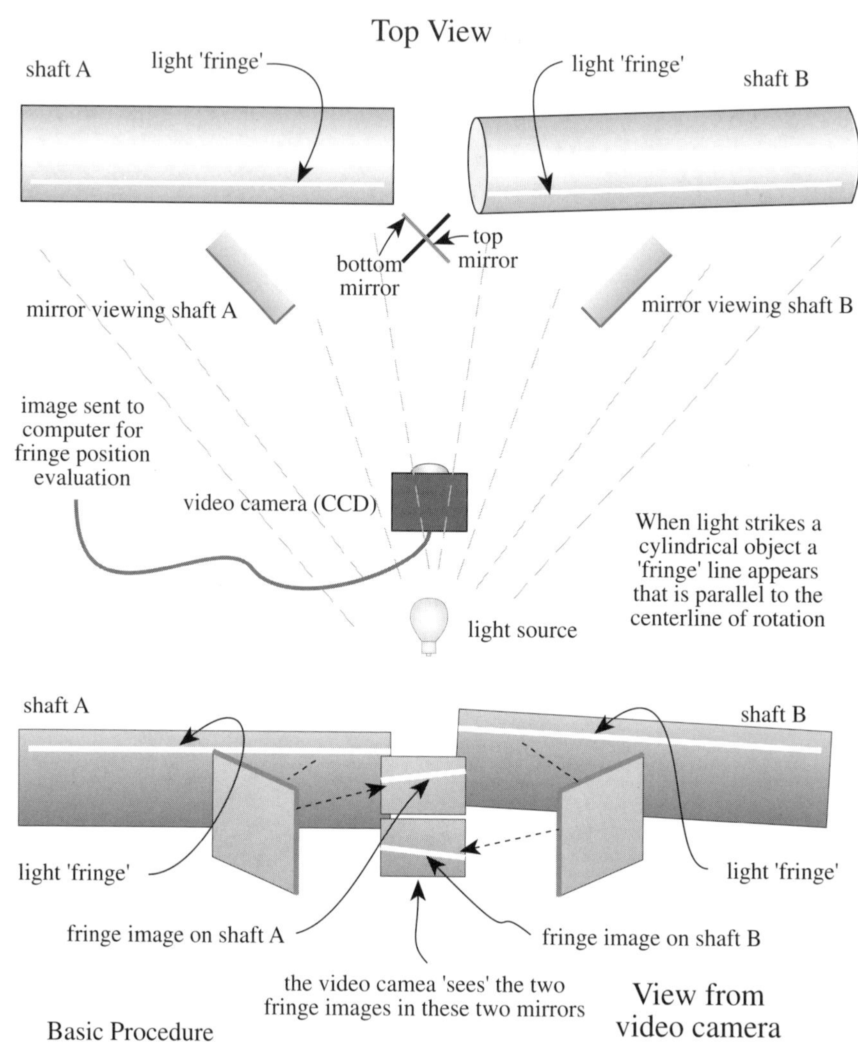

Since the relative mirror positions are unknown during set-up, a reference line is obtained by placing a straight cylinder (posessing a continuous straight fringe line) in front of the two mirrors establishing a reference line (and reference fringe positions). An image of the actual shaft fringes are then compared to the reference fringe lines to determine the relative displacements and slopes between the two shafts.

Figure 7-14. Using a CCD to detect light fringes on shafts.

Lasers and Detectors

From relying on the accuracy of our eyesight to using straightedges and feeler gauges and then eventually to shaft brackets and dial indicators, the art of measuring machinery shaft positions has been continually refined to improve accuracy and reduce the amount of time required to achieve acceptable machinery alignment.

It was inevitable, however, that all was not to stop here, particularly in light of the technological explosion in electronics. With the advent of the microprocessor chip, the semiconductor junction laser, and silicon photodiodes, new inroads have been forged in the process of measuring rotational centerlines that utilize these new electronic devices instead of the mechanical measuring instruments. Since the first useable laser shaft alignment measurement system was introduced by Prüftechnik in Germany in 1984, a host of manufacturers have introduced other laser shaft alignment systems. At the time of this printing, there are six companies that currently offer laser shaft alignment measurement systems. Since some of the manufacturers have taken slightly different approaches to using lasers and detectors, it will be beneficial to initially discuss some of the basic theory of operation of photonic semiconductors and how they are applied to mechanical measurements.

Useful Terms -

photonics - field of electronics that involves semiconductor devices that emit and detect light.

semiconductors - typically silicon 'crystal' doped (i.e. made impure) with other elements such as phosphorus (n-type due to 5 electrons in outer shell) or boron (p-type due to 3 electrons in outer shell). Depending on certain conditions, semiconductors can act as insulators or conductors.

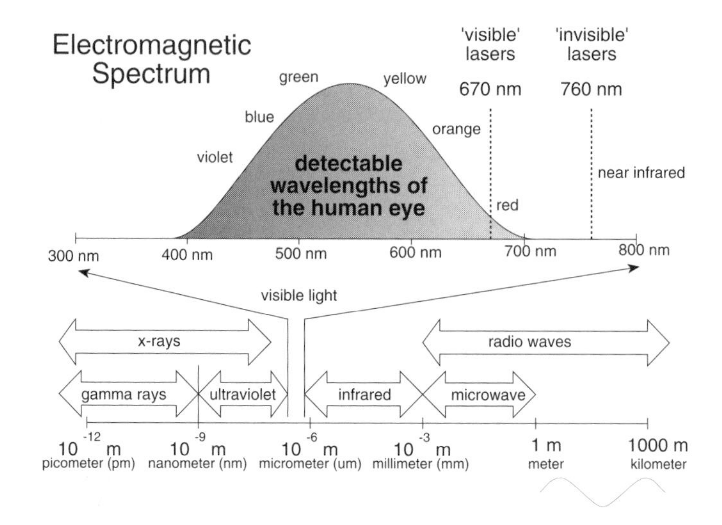

The two 'faces' of electromagnetic energy

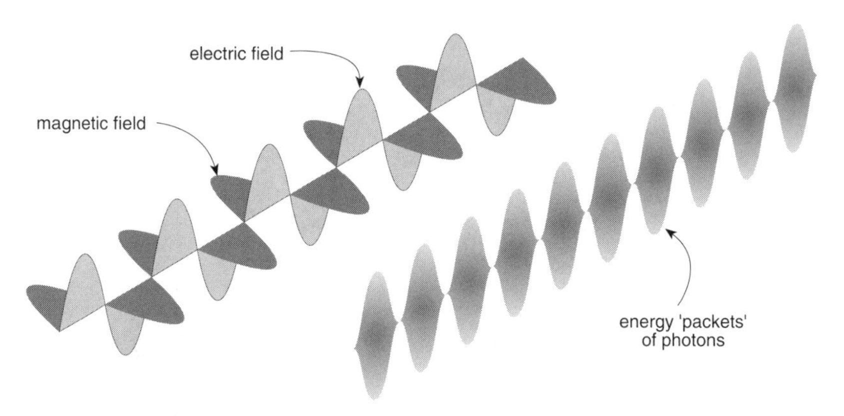

The photon is the key behind controlling an atoms orbital energy. Absorption occurs when electrons go from a lower to a higher orbital level (shell). Emission occurs when electrons go from a higher to a lower orbital level.

Figure 7-15. The electromagnetic spectrum.

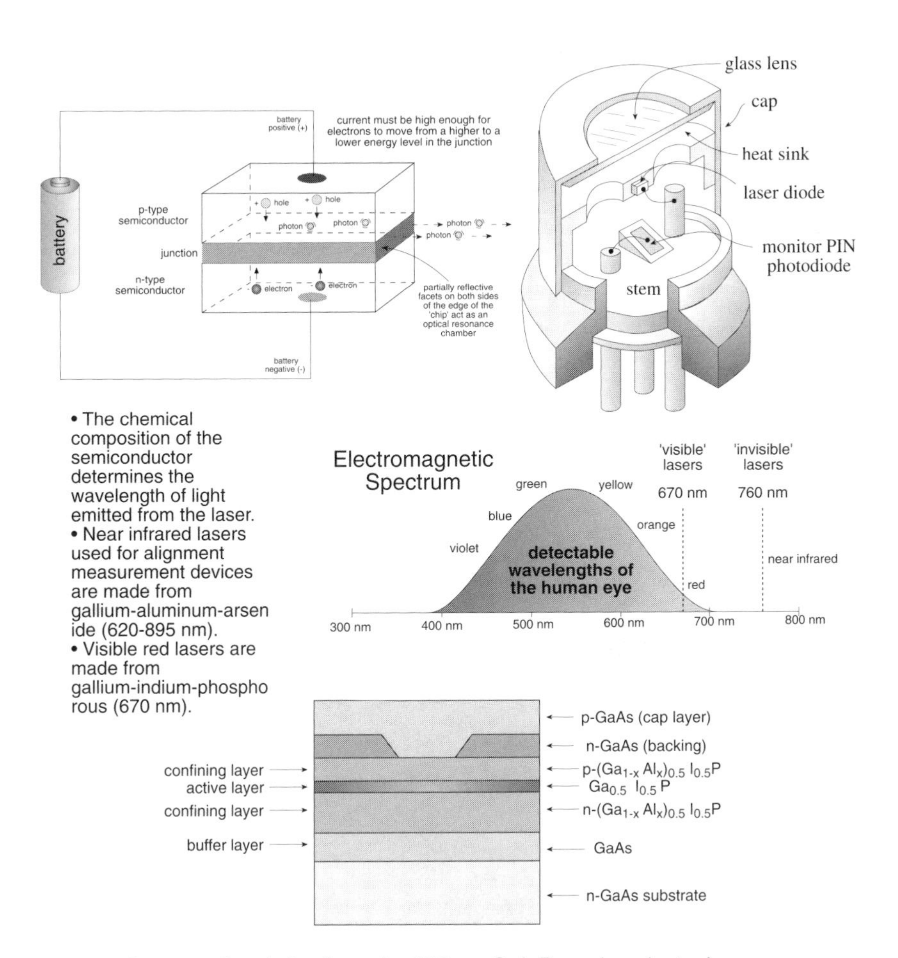

Cross sectional structure of a 670 nm GaInP semiconductor laser.

Figure 7-16. How semiconductor laser diodes work.

How Laser Detectors Work

Laser detectors are semiconductor photodiodes capable of detecting electromagnetic radiation (light) from 350 to 1100 nm. When light strikes the surface of the photodiode, an electrical current is produced. Most manufacturers of laser - detector shaft alignment systems use 10mm x 10mm detectors (approximately 3/8 sq. inch) a few may use 20mm x 20mm detectors. Some manufacturers of these systems use bi-cell (uni-directional) or quadrant cell (bi-directional) photodiodes to detect the position of the laser beam. When light strikes the center of the detector, output currents from each cell are equal. When the beam moves across the surface of the photodiode, a current imbalance occurs indicating the off-center position of the beam.

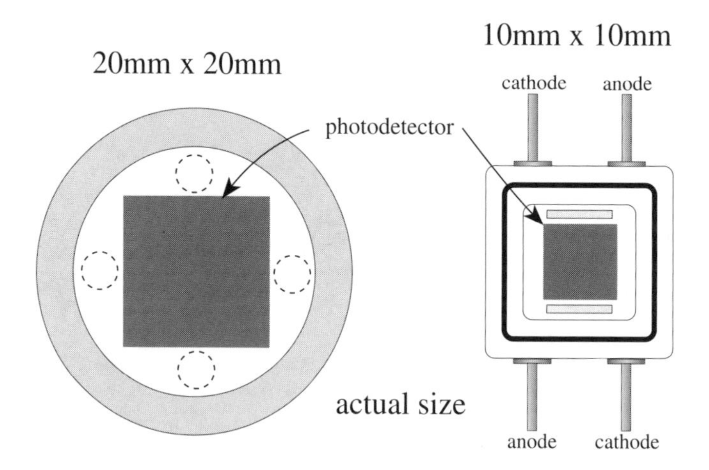

Figure 7-17. How photodiodes work..

LASER - acronym for Light Amplified by Stimulated Emission of Radiation.

LED - Light Emitting Diode. All diodes emit some electromagnetic radiation when forward biased. When the forward current attains a certain level, called the threshold point, lasing action occurs in the semiconductor. Gallium-

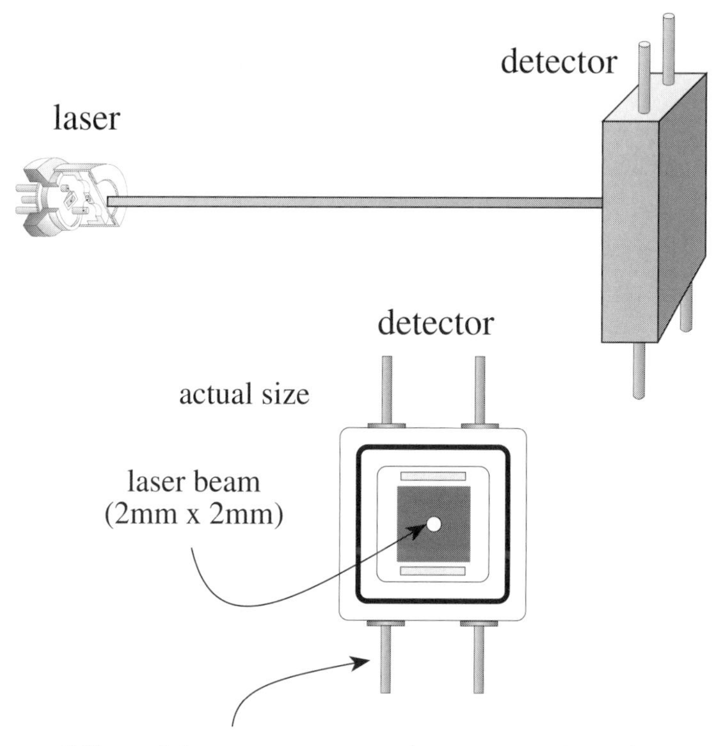

Figure 7-18. Laser photodiode operation.

arsenide-phosphide diodes emit much more radiation that silicon type diodes and are typically used in semiconductor junction diode lasers.

photodiode - all diodes respond when subjected to light, (electromagnetic radiation). Silicon diodes respond very well to light and are typically used to detect the presence or position of light as it impedes on the surface of the diode.

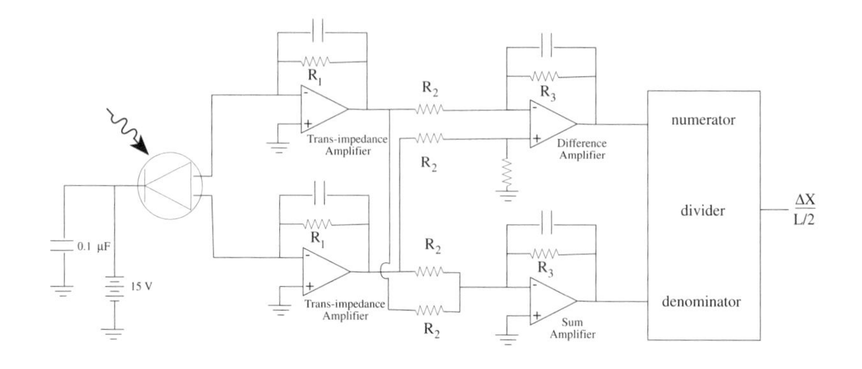

Figure 7-19. Typical single axis photodiode circuit.

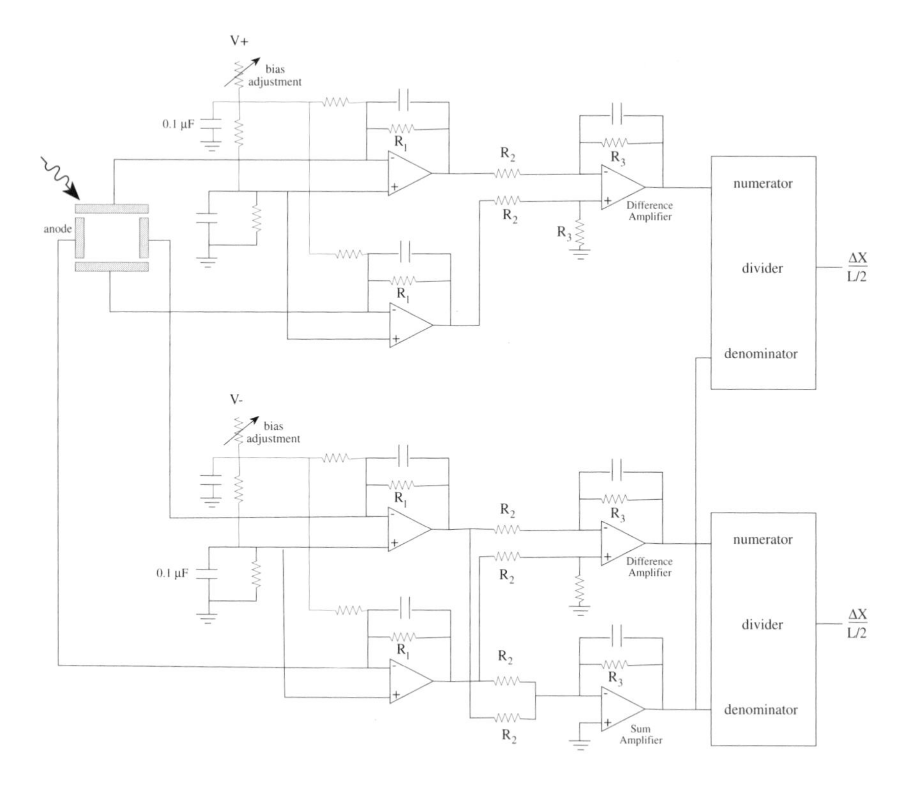

Figure 7-20. Typical dual axis photodiode circuit.

Measuring shaft centerline positions using dial indicators

As we have seen so far, man has invented many ingenious ways to capture dimensional measurements. Some of these inventions have been employed to measure shaft centerline positions, and some others are just waiting for someone to adapt the device for this purpose.

For the past forty years the most common tool that has been used to accurately measure shaft misalignment is the dial indicator. There are some undeniable benefits of using a dial indicator for alignment purposes.

- One of the preliminary steps of alignment is to measure runout on shafts and coupling hubs to insure that eccentricity amounts are not excessive. As we have seen in Chapter 6, the dial indicator is the measuring tool typically used for this task and is, therefore, usually one of the tools that the alignment expert will bring to an alignment job. Since a dial indicator is being used to measure runout, why not use it to also measure the shaft centerline positions?
- The operating range of dial indicators far exceeds the range of many other types of sensors used for alignment. Dial indicators with total stem travels of .200" (5 mm) are traditionally used for alignment but indicators with stem travels of 3 inches or greater could also be used if the misalignment condition is moderate to severe when you first begin to 'rough in' the machinery.
- The cost of a dial indicator (around $70 to $110 US) is far less expensive than many of the other sensors used for alignment. You could purchase over 140 dial indicators for the average cost of some other alignment tools currently on the market.

- Since the dial indicator is a mechanically based measurement tool, there is a direct visual indication of the measurement as you watch the needle rotate.
- They are very easy to test for defective operation.
- They are much easier to find and replace in virtually every geographical location on the globe in the event that you damage or lose the indicator.
- Batteries are not included since they are not needed.
- The rated measurement accuracy is equivalent to the level of correction capability (i.e. shim stock cannot be purchased in thicknesses less than 1 mil).

Figure 7-21. Dial indicator and bracket arrangement taking rim readings on a large flywheel.

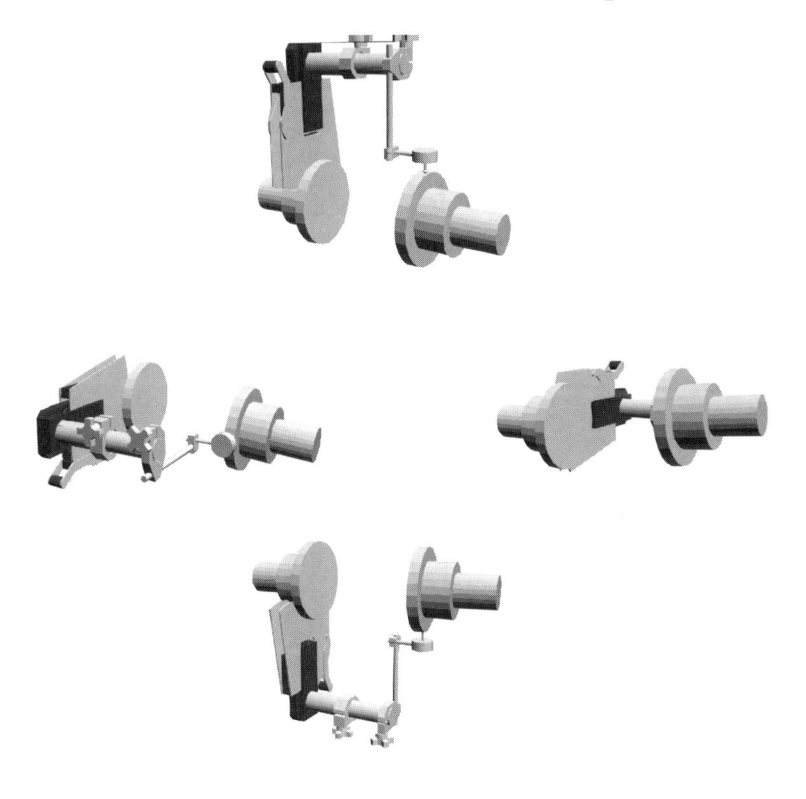

Figure 7-22. Rotate both shafts through 90 degree increments when capturing shaft centerline position information.

Rotating both shafts to override any runout conditions

Chapter 6 covered information on determining if runout conditions exist from coupling hubs bored off center or skew bored and bent shafts. How do you align the centerlines of rotation if the surfaces you are capturing readings on are eccentric?

To override any runout problems that exist on coupling hubs or shafts, you must rotate the machinery shafts together (i.e. take the readings at the same points on the shaft surface). Figure 7-23 illustrates what will happen if you have a runout condition and you only rotate one of the shafts when taking readings.

Keep in mind that taking readings at the same angular position on the shafts will override any runout that exists. This procedure does not suggest that runout conditions should be ignored. If excessive amounts of runout exist, they should always be corrected before you ever begin to align the equipment.

Face and Rim Method

Perhaps the 'oldest' dial indicator technique used to align rotating machinery shafts is the face and rim or face-peripheral method shown in figure 7-24. It is not entirely clear who first used this technique to align rotating machinery shafts but this method is frequently found in machinery installation manuals and flexible coupling installation instructions and is still used by a large number of mechanics to align machinery.

Figures 7-25 and 7-26 show some variations to this method.

Advantages -

- This is a good technique to use in situations where one of the machinery shafts cannot be rotated or it would be difficult to rotate one of the machinery shafts (see also Double Radial Technique).

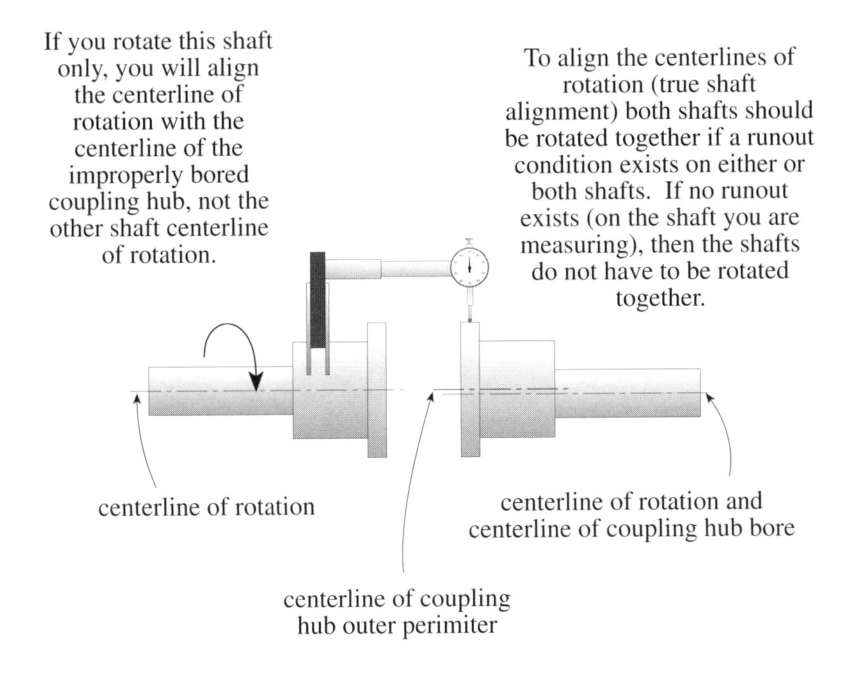

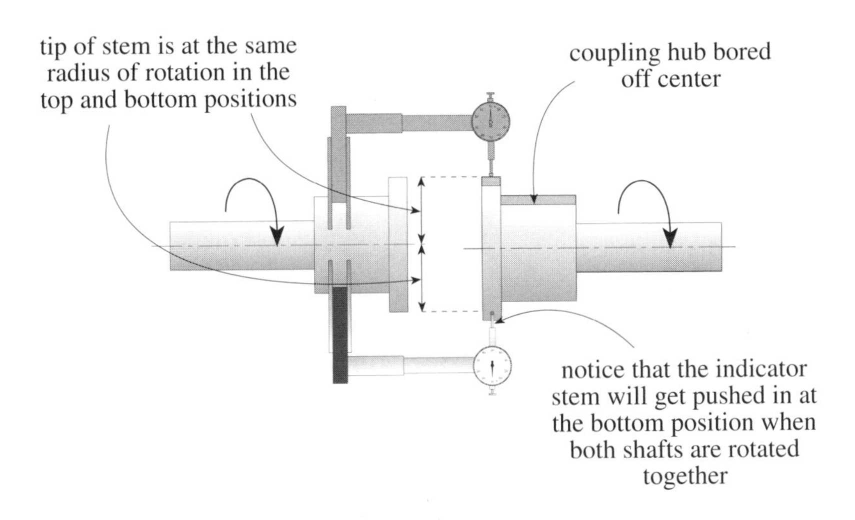

Figure 7-23. Why you should rotate both shafts to override a runout condition.

Face - Peripheral technique

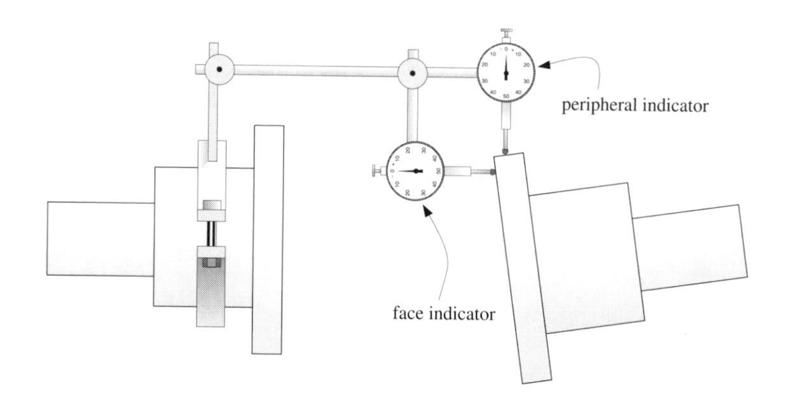

Procedure ...

1- Attach the alignment bracket firmly to one shaft and position the indicator(s) on the face and perimeter of the other shaft.
2 - Zero the indicators at the 12 o'clock position.
3 - Slowly rotate the shaft and bracket arrangement through 90 degree intervals stopping at the 3, 6, and 9 o'clock positions. Record each reading (plus or minus).
4 - Return to the 12 o'clock position to see if the indicator(s) re-zero.
5 - Repeat steps 2 through 4 to verify the first set of readings.

indicator readings log

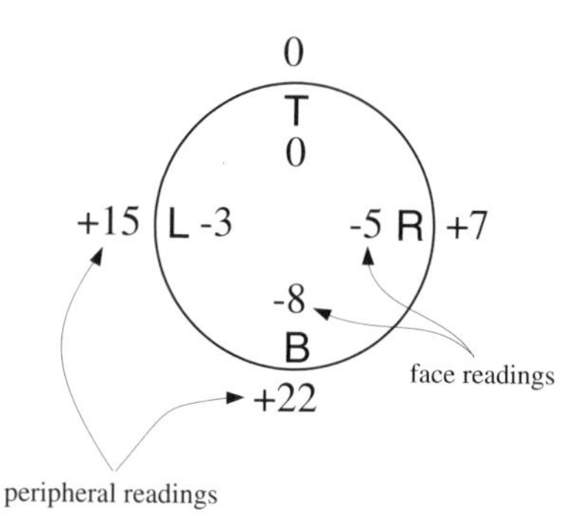

Figure 7-24. Face - peripheral (i.e. face - rim) technique.

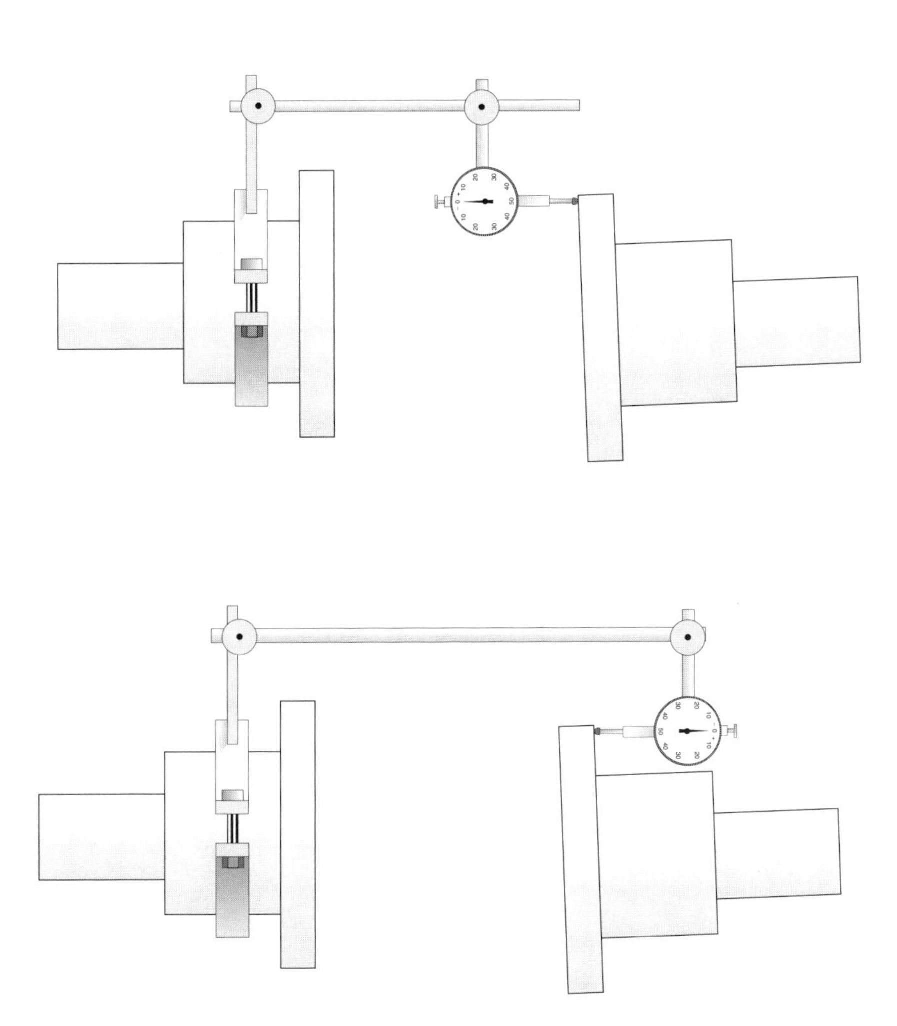

Figure 7-25. Face readings can be taken on the front or back sides.

If both shafts can be rotated, the diameter the face readings are being captured at can be extended by using a post of another bracket attached to the shaft or coupling hub or ...

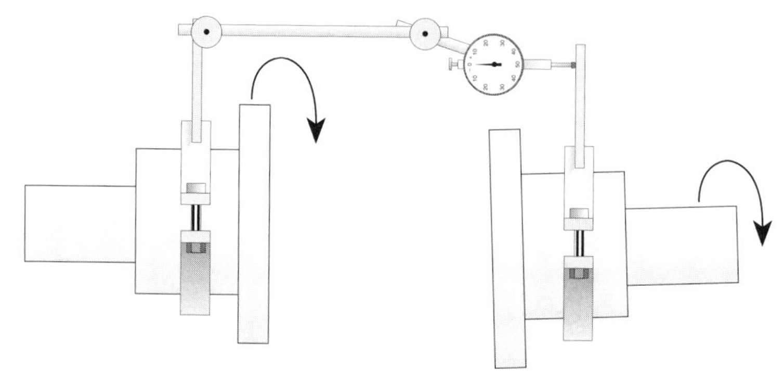

... to a section of flat steel stock clamped or bolted to the 'face'.

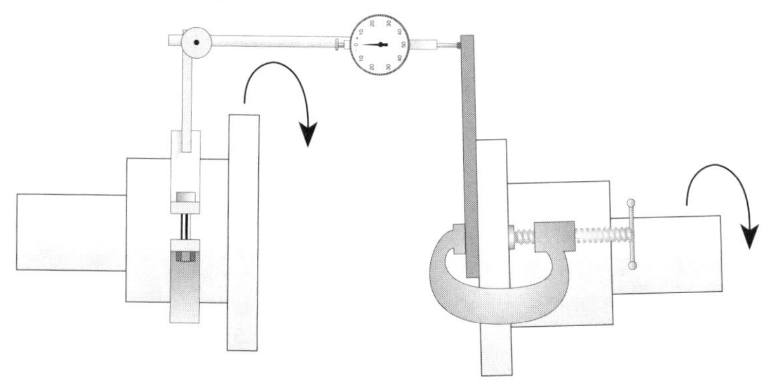

Figure 7-26. Face readings can be captured on any device rigidly attached to a shaft (assuming the shafts are rotated together).

Measure the face diameter

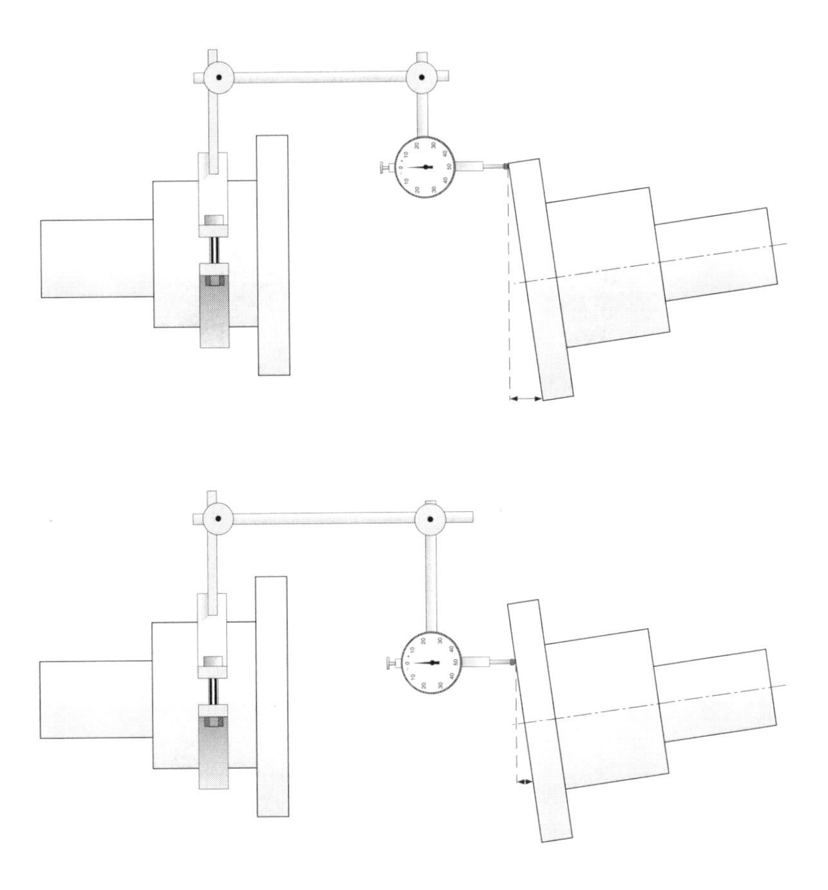

The face readings are different even though the shaft positions are the same.

Figure 7-27. Taking face readings on different diameters will result in different readings even though the shafts are in the same angular position.

- Many people who use this method understand that the rim (or perimeter) dial indicator shows centerline offset or parallel misalignment and the face indicator indicates angular misalignment.
- A good method to use when the face readings can be captured on a fairly large diameter (e.g. 8" or greater). This method begins to approach the accuracy of the Reverse Indicator Technique when the diameter the face readings are being captured on equals or exceeds the span from the bracket location to the point where the rim indicator readings are being captured.

Disadvantages -
- Not as accurate as the reverse indicator method if both shafts can be rotated and particularly if the face measurements are being taken on diameters less than 8".
- If the machinery shaft(s) are supported in sliding (plain/sleeve) bearings, it is very easy to axially 'float' the shafts toward or away from each other when rotating the shaft(s) resulting in bad or inaccurate face readings (see Validity Rule).
- Bracket sag must be measured and compensated for.

Reverse Indicator Method

This method is also often called the Indicator Reverse method or the Double Dial method. It seems to have originated around the mid to late 1950's in the United States and again it is not clear who first developed this technique. Clark Brothers (now Dresser-Rand) adopted this technique for use in aligning their rotating machinery and Don Cutler (currently working for Thomas-

Reverse Indicator technique

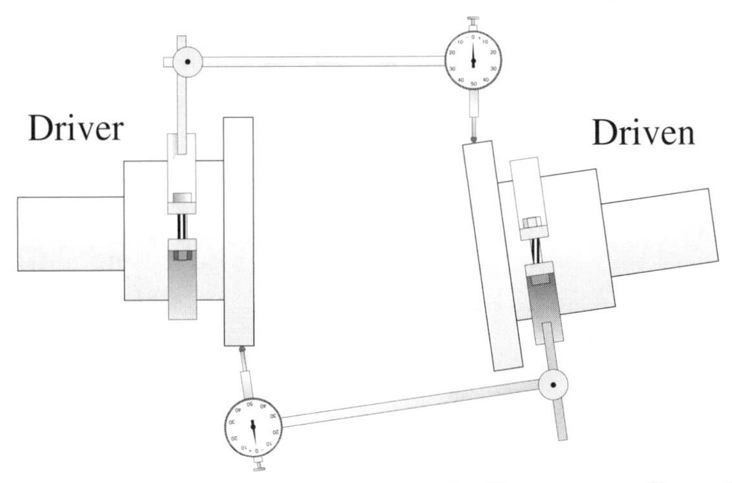

Procedure ...

1- Attach the alignment bracket(s) firmly to one (both) shaft(s) and position the indicator(s) on the perimeter of the other shaft.
2 - Zero the indicator(s) at the 12 o'clock position.
3 - Slowly rotate the shaft and bracket arrangement through 90 degree intervals stopping at the 3, 6, and 9 o'clock positions. Record each reading (plus or minus).
4 - Return to the 12 o'clock position to see if the indicator(s) re-zero.
5 - Repeat steps 2 through 4 to verify the first set of readings.
6 - If one bracket was used, mount the bracket onto the other shaft and repeat steps 1 through 5.

indicator readings log

Driver

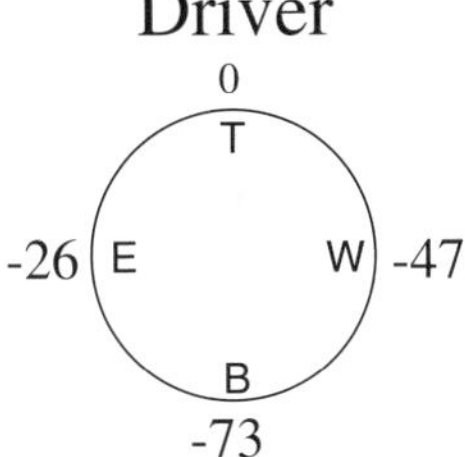

Driven

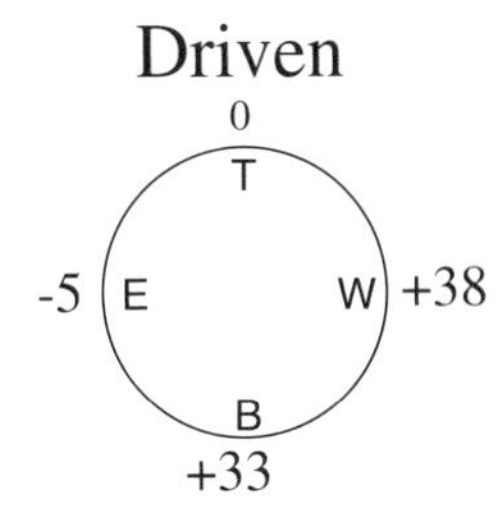

Figure 7-28. Reverse indicator technique.

The 'traditional' method of capturing reverse indicator readings is to clamp a bracket on one shaft, span over to the other shaft with a bar that holds a dial indicator used to measure the circumference (or rim / perimeter) of the other shaft.

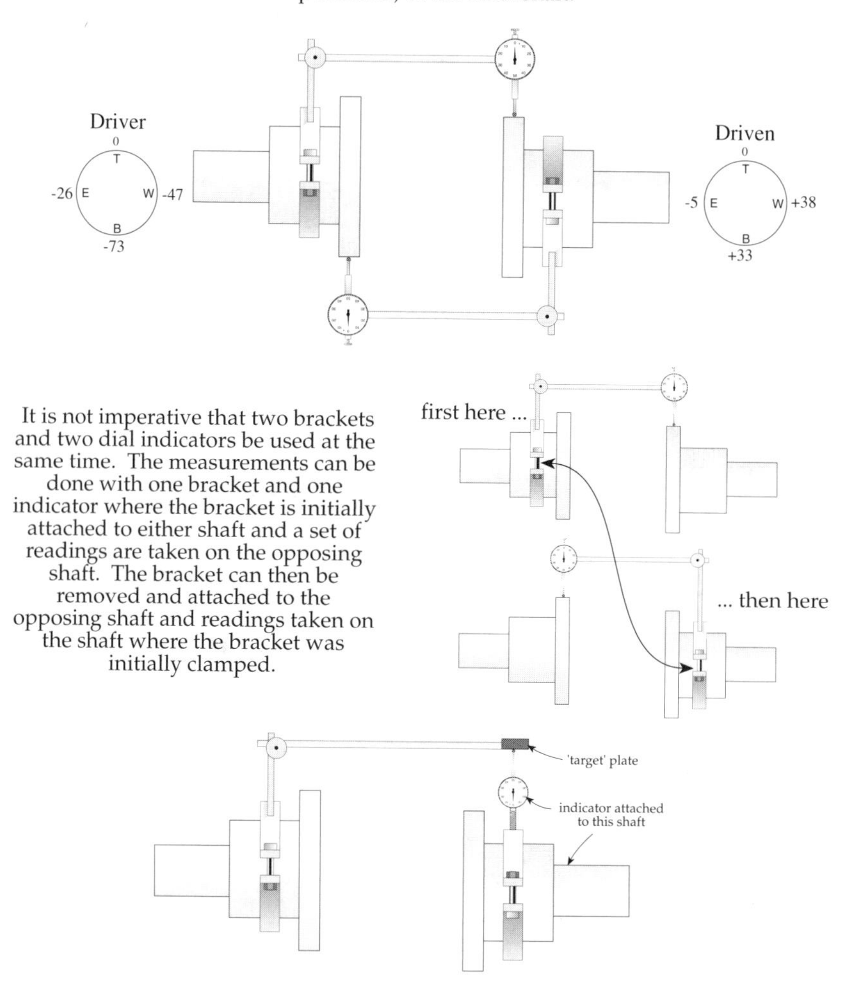

An alternative method is to clamp a bracket onto a shaft that supports a rod extending over to the other shaft. An indicator is then attached to the shaft that reads the 'underside' of the bar.

Figure 7-29. Reverse indicator set up variations.

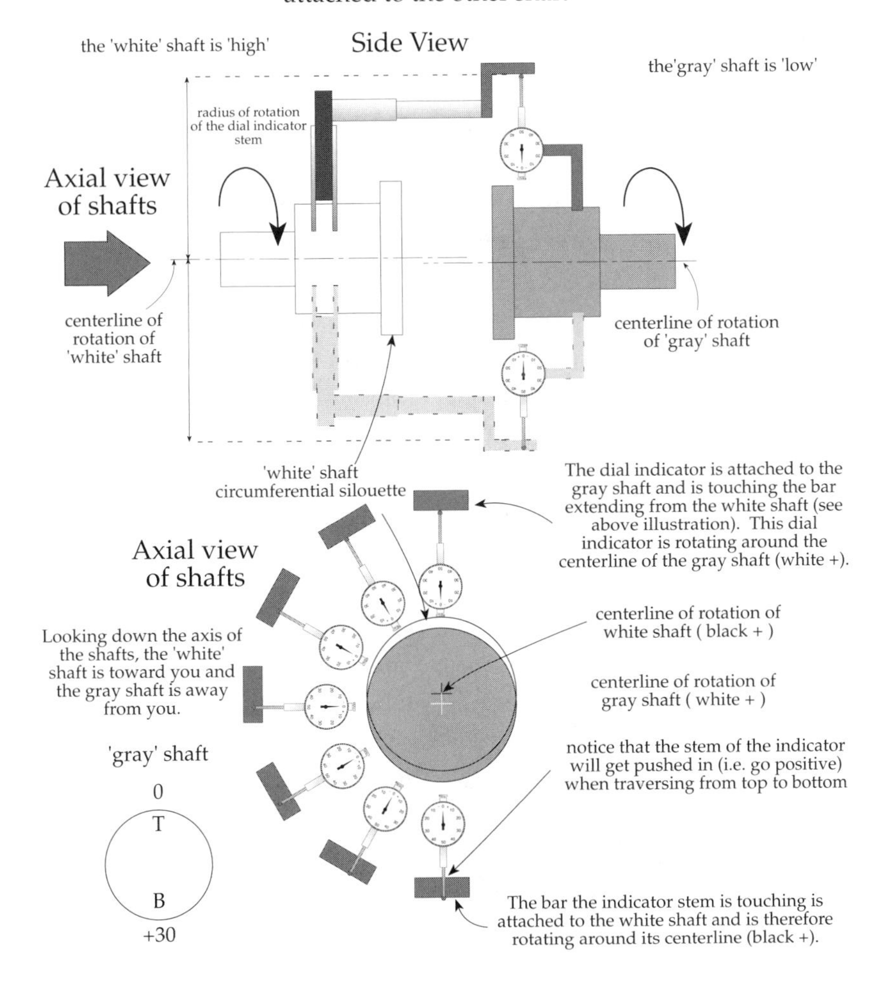

Figure 7-30. Rim readings can also be taken on an inside diameter.

Rexnord) developed the line to point graphing method while working for Clark (refer to Chapter 8 for more information on modeling / graphing techniques).

The reverse indicator method can be used on 60-70% of the rotating machinery in existence and is the currently preferred dial indicator method for measuring rotating machinery shafts. It is best suited for use when the distances between measuring points on each shaft range from 3 to 30 inches. Although figure 7-28 shows using two brackets and two dial indicators at the same time, there is no reason why one bracket / dial indicator set up could not be used where a set of readings are captured on one shaft first and then reversing the bracket / indicator to capture a set of readings on the other shaft. In fact, it may be wise to use just one bracket at a time to insure that readings are being taken correctly and minimize the confusion that could result from trying to observe two indicators simultaneously. Additionally, how much time does it take to set up two brackets and two indicators versus setting up one bracket and one indicator twice?

Advantages -

- Typically more accurate than the face-rim method since the distance from the mounting point of the bracket to the point where the indicators capture the readings on the shafts is usually greater than the distance a face reading can be taken.
- If the machinery is supported in sliding type bearings and the shafts are 'floating' back or forth axially when rotating the shaft to capture readings, there is virtually no effect on the accuracy of the readings being taken.
- Can be performed with the flexible coupling in place (see precautions later on in this chapter).

Disadvantages -

- Both shafts must be rotated.

Figure 7-31. Reverse indicator technique employed across gear coupling with spacer.

Figure 7-32. Reverse indicator technique employed across two different diameter hubs with spool piece removed.

- Difficult to visualize the positions of the shafts from the dial indicator readings being collected.
- Bracket sag must be measured and compensated for.

Double Radial Method

Although not frequently used, this method has some distinct advantages compared to some of the other methods discussed in this chapter. This method should only be used if there is at least a 3 inch or greater separation between reading points. The accuracy of this technique increases as the distance between reading points increases. The disadvantage of this method is that there is usually not enough shaft exposed to be able to spread the indicators far enough apart to merit using the method, except for very special circumstances.

Advantages -

- This is a good technique to use in situations where one of the machinery shafts cannot be rotated or it would be difficult to rotate one of the machinery shafts.
- Can be set up to measure inner circular surfaces such as the bore of a barrel.
- A good method to use when the dial indicator readings can be captured across a large span (e.g. 8" or greater). This method begins to approach the accuracy of the Reverse Indicator Technique when the distance between the two sets of dial indicator readings being captured on one shaft equals or exceeds the span between reading points from shaft to shaft.

Disadvantages -

- Not enough shaft surface is exposed to spread the readings far enough apart for acceptable accuracy.
- If the machinery is supported in sliding type bearings and the shafts are 'floating' back or forth axially when rotating the shaft to capture readings,

Double radial indicator technique

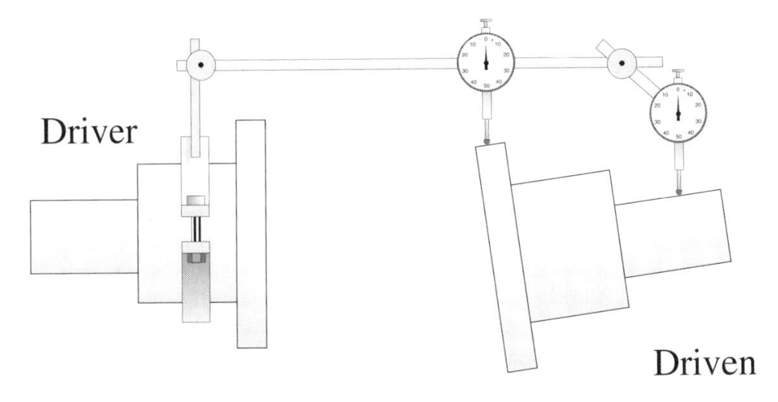

Procedure ...

1- Attach the alignment bracket firmly to one shaft and position the two indicators at different axial locations on the perimeter of the other shaft.
2 - Zero the indicator(s) at the 12 o'clock position.
3 - Slowly rotate the shaft and bracket arrangement through 90 degree intervals stopping at the 3, 6, and 9 o'clock positions. Record each reading (plus or minus).
4 - Return to the 12 o'clock position to see if the indicator(s) re-zero.
5 - Repeat steps 2 through 4 to verify the first set of readings.
6 - If one bracket was used, mount the bracket onto the other shaft and repeat steps 1 through 5.

indicator readings log

Near indicator

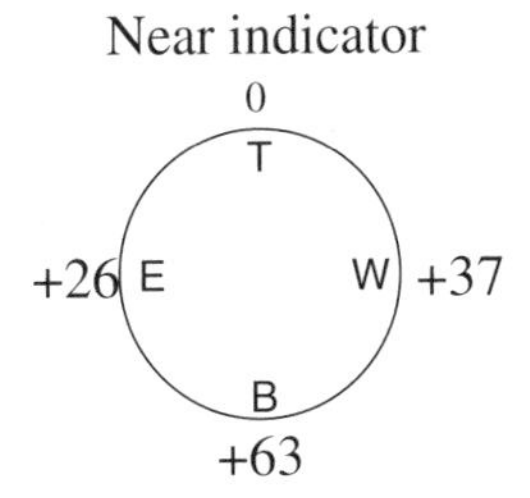

Far indicator

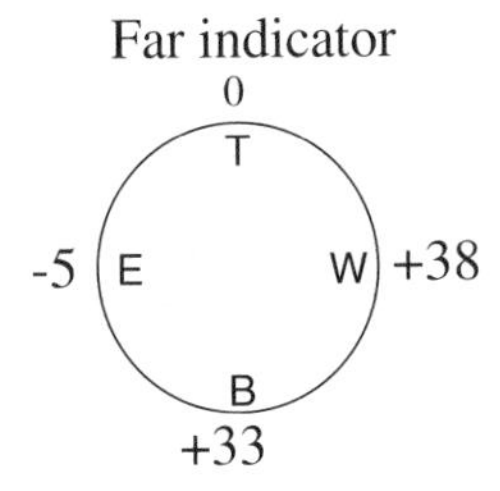

Figure 7-33. Double radial technique.

there is virtually no effect on the accuracy of the readings being taken.

- Bracket sag must be measured and compensated for.

Shaft to Coupling Spool Method

There are situations where two pieces of rotating machinery are positioned a considerable distance apart where trying to employ any of the former dial indicator techniques would prove to be cumbersome due to the extreme distances between the ends of each shaft. When the distance between shaft ends begins to exceed 30 - 40 inches, this technique is recommended to measure the positions of the shafts. This method can be applied to : cooling tower fan drive systems, press roll drives in the paper industry with universal joints, dryer can drive sections, vehicular drive shaft systems, gas/ power turbine to generators or compressors for example.

For acceptable accuracies when using this technique, the distance from the flexing points at each end of the coupling to the point where the dial indicator is capturing readings should be at least 4 inches or greater. A good rule of thumb when setting up the tooling for this technique is to maintain at least a 1:15 ratio of flex point to reading location distance compared to the distance between flex points. For example, if the distance between flexing points in the coupling is 120 inches, the distance from the flex points to where the indicators are capturing readings should be at least 8 inches.

Advantages -

- Perhaps the most accurate measurement technique where extreme distances occur between shaft ends.
- Relatively easy to set up and capture readings.

Disadvantages -

- Since coupling spool (aka 'jackshaft') must be kept in place, both shafts must be rotated together.

Shaft to coupling spool technique

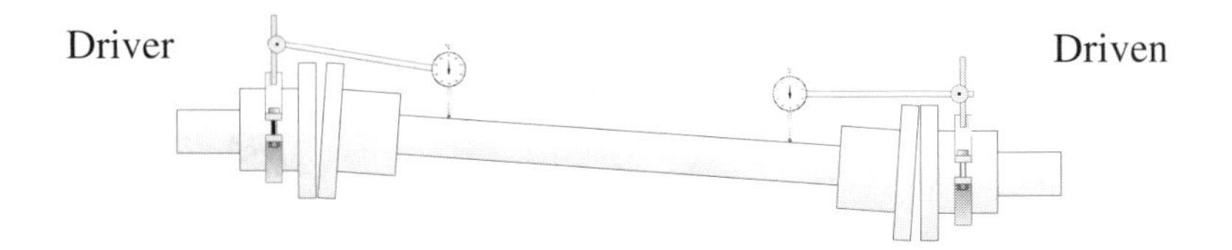

Procedure ...

1- Attach the alignment bracket(s) firmly to one (both) shaft(s) and position the indicator(s) at some point along the coupling spool with the indicator(s) touching the outside diameter of the spool.
2 - Zero the indicator(s) at the 12 o'clock position.
3 - Slowly rotate the shaft and bracket arrangement through 90 degree intervals stopping at the 3, 6, and 9 o'clock positions. Record each reading (plus or minus).
4 - Return to the 12 o'clock position to see if the indicator(s) re-zero.
5 - Repeat steps 2 through 4 to verify the first set of readings.
6 - If one bracket was used, mount the bracket onto the other shaft and repeat steps 1 through 5.

indicator readings log

Driver to spool

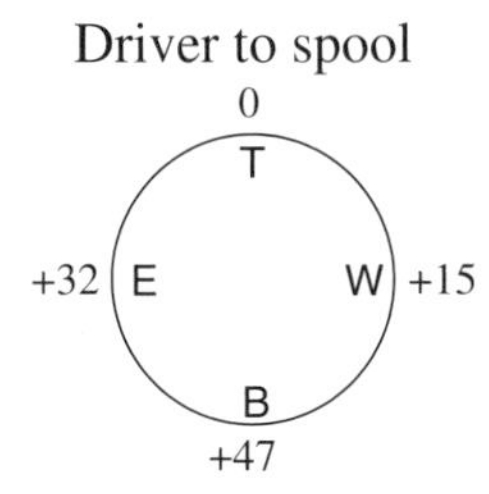

Driven to spool

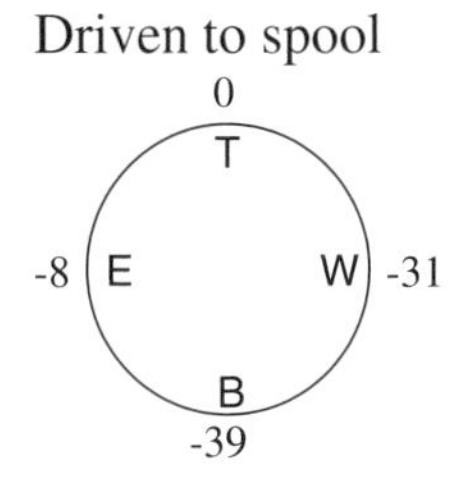

Figure 7-34. Shaft to coupling spool technique.

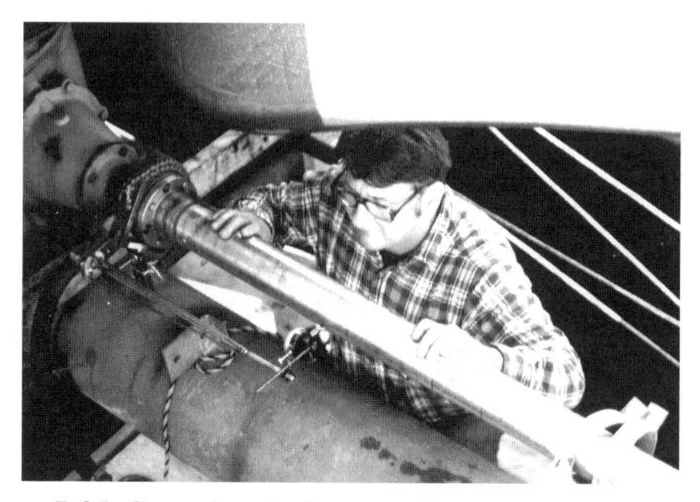

Figure 7-35. Capturing shaft to coupling spool readings on a cooling tower fan drive system.

Face - Face Coupling Spool Method

Another method used to measure shaft centerline position data typically for long spans between shaft ends is the face-face technique shown in figure 7-36. Note that two sets of face readings are taken across each flexing point in the coupling. One of the first shaft alignment patents filed in the United States embodied this technique although relatively few people use this method in the field.

Although usually not as accurate as the shaft to coupling spool method, there are occasions where this method must be used instead. The accuracy of this technique is increased as the diameter of the face readings are increased. There are some interesting applications of this technique when adapted to measure off-line to running machinery movement as explained in Chapter 9.

Advantages -

• Better set up if the brackets cannot be attached to the machinery

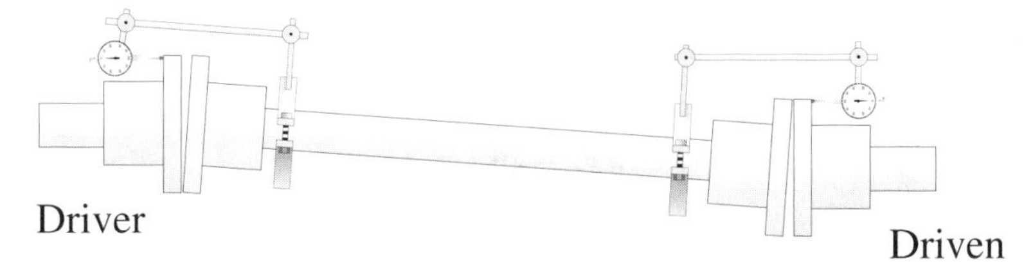

indicator readings log

Procedure ...

1- Attach the alignment bracket(s) firmly to the coupling spool and position the indicator(s) at some point on the 'face' of the coupling hub rigidly attached to the shaft(s).
2 - Zero the indicator(s) at the 12 o'clock position.
3 - Slowly rotate the shaft and bracket arrangement through 90 degree intervals stopping at the 3, 6, and 9 o'clock positions. Record each reading (plus or minus).
4 - Return to the 12 o'clock position to see if the indicator(s) re-zero.
5 - Repeat steps 2 through 4 to verify the first set of readings.
6 - If one bracket was used, mount the bracket onto the other shaft and repeat steps 1 through 5.

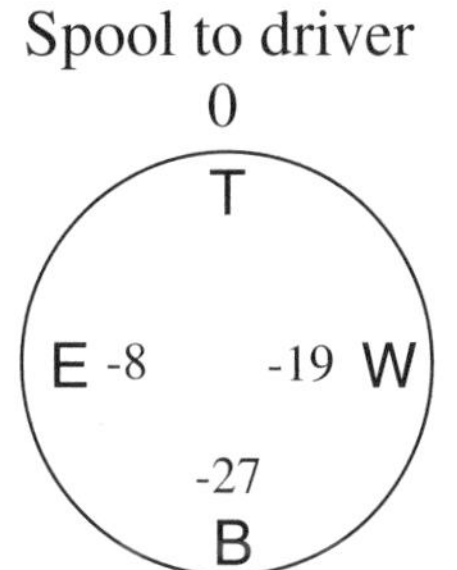

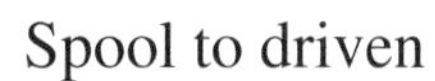

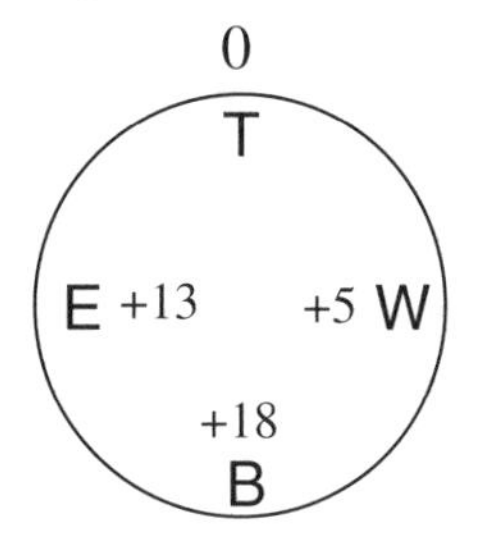

Figure 7-36. Face - face coupling spool technique.

shafts (as in the shaft to coupling spool method) but can be attached to the spool piece.

Disadvantages -

• Not as accurate as the shaft to coupling spool method assuming the readings were taken on a relatively small diameter.

Sixteen Point Method

The 16 point method is frequently used on rotating machinery connected together by rigid rather than flexible couplings. The general procedure is illustrated in figure 7-37.

It is typically used where both shafts are supported in two bearings. The flange bolts are loosened, the shafts separated slightly, and a series of face readings are taken at four points around the flange faces at the 12, 3, 6, and 9 o'clock positions. The assumption made when performing this technique is that there is pure angular alignment present (i.e. no centerline offset) and that the flange faces are perpendicular to the centerlines of rotation. In many cases, these assumptions are not true, resulting in improper shaft alignment.

This method has also been used where one shaft is supported in two bearings and the other shaft is supported in one bearing on the outboard end. The coupling flanges have a recessed (rabbeted) fit. The flange bolts are loosened, the shafts separated just slightly, insuring that the flange faces are still indexed in the recess, and a series of face readings are taken at four points around the flange faces at the 12, 3, 6, and 9 o'clock positions.

Twenty Point Method

The 20 point method is also frequently used on rotating machinery connected together by rigid rather than flexible couplings. The general procedure is illustrated in figure 7-38. Again, it is typically used where both shafts are supported in two bearings. The flange bolts are loosened, the shafts separated slightly, and a series of face readings are taken at four points around the flange faces at the 12, 3, 6, and 9 o'clock positions along with a rim (circumferential) reading typically taken with a dial indicator. It is effectively the face and rim technique explained earlier except the face readings are usually taken with feeler gauges.

Rigid Coupling Alignment Techniques

The Sixteen Point Alignment Measurement Method

This technique is typically used for rigid couplings with spigot (recessed) fits commonly found on machinery where one rotor is supported in two bearings and the other rotor is supported by one bearing.

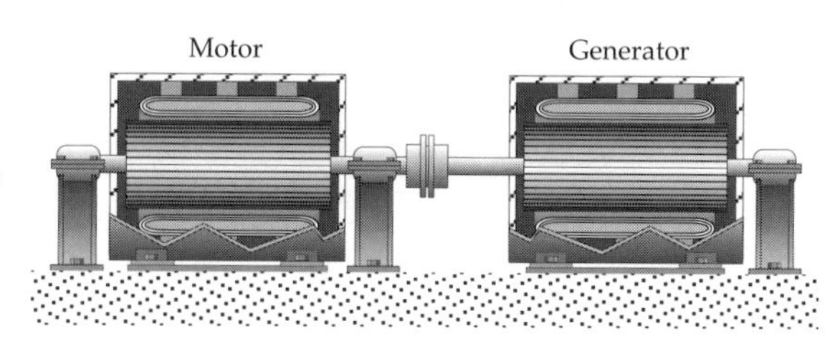

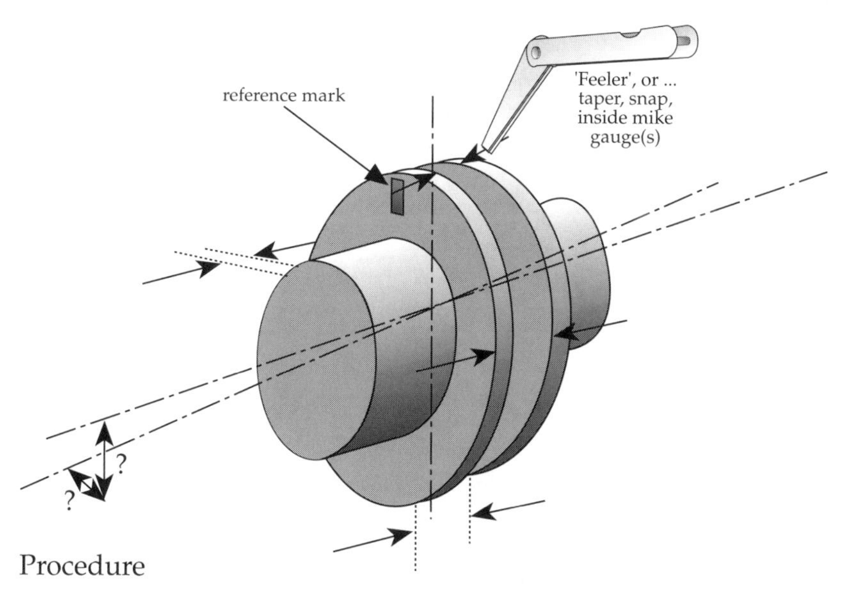

Procedure

1. Insure the coupling bolts are loose and there is a slight separation (around 20 mils) between the coupling hub faces to prevent any stress or binding force interaction from one shaft to another.
2. Place a reference mark on one (or both) of the shafts, usually at 12 o'clock.
3. Accurately mark off 90 degree increments on the coupling hubs from the 12 o'clock reference.
4. Use feeler, or taper gauges capable of measuring to 0.001" (1 mil) to measure the 'gaps' between the coupling hub 'faces' at these 90 degree intervals (ie. both sides, the top, and bottom).
5. Measure the diameter of the coupling hubs where the 'gaps' were captured.
6. Record each gap reading and rotate both shafts 90 degrees.
7. Capture another set of readings and rotate the shafts 90 degrees again.
8. Repeat step 7 until the reference mark has returned to its original position at 12 o'clock.

Figure 7-37. 16 point technique.

Rigid Coupling Alignment Techniques

The Twenty Point Alignment Measurement Method

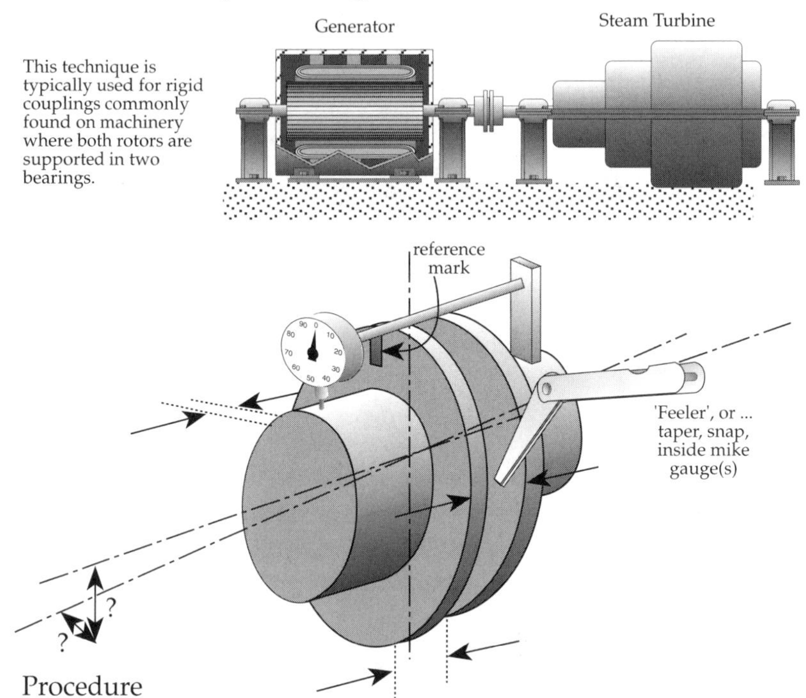

Procedure

1. Insure the coupling bolts are loose and there is a slight separation (around 20 mils) between the coupling hub faces to prevent any stress or binding force interaction from one shaft to another.
2. Place a reference mark on one (or both) of the shafts, usually at 12 o'clock.
3. Accurately mark off 90 degree increments on the coupling hubs from the 12 o'clock reference.
4. Attach a bracket or fixture to one shaft and span over to the other shaft to place a dial indicator on the perimeter or rim of the coupling. Zero the indicator at the 12 o'clock position.
5. Use feeler, or taper gauges capable of measuring to 0.001" (1 mil) to measure the 'gaps' between the coupling hub 'faces' at these 90 degree intervals (i.e. both sides, the top, and bottom).
6. Measure the diameter of the coupling hubs where the 'gaps' were captured.
7. Record each gap reading and rotate both shafts 90 degrees.
8. Capture another set of feeler gauge readings and note the reading on the dial indicator that is now on the side of the coupling hub. Rotate the shafts 90 degrees again.
9. Capture another set of feeler gauge readings and note the reading on the dial indicator that is now on the bottom of the coupling hub. Rotate the shafts 90 degrees again.
10. Capture another set of feeler gauge readings and note the reading on the dial indicator that is now on the other side of the coupling hub. Rotate the shafts 90 degrees again returning the reference mark back to 12 o'clock.

Figure 7-38. 20 point technique.

Validity Rule

Due to the geometry of taking readings around the circumference of a circular shaft, a pattern emerges which is commonly referred to as the 'validity rule'. The validity rule states that when the two measurements taken 90 degrees on each side of the 'zeroing' point are added together, they will equal the measurement taken 180 degrees from the 'zeroing' point. Examples of the validity rule are shown in figure 7-39, and as you may have noticed, in all of the example readings shown in the various dial indicator techniques covered in the previous pages. The validity rule applies to both radial (i.e. perimeter / circumferential) and face readings.

The validity rule is important for two reasons:
1 - To insure that you are capturing an accurate set of readings when measuring the off-line positions of machinery shafts. A considerable amount of time will be wasted attempting to reposition machinery based on inaccurate measurements.

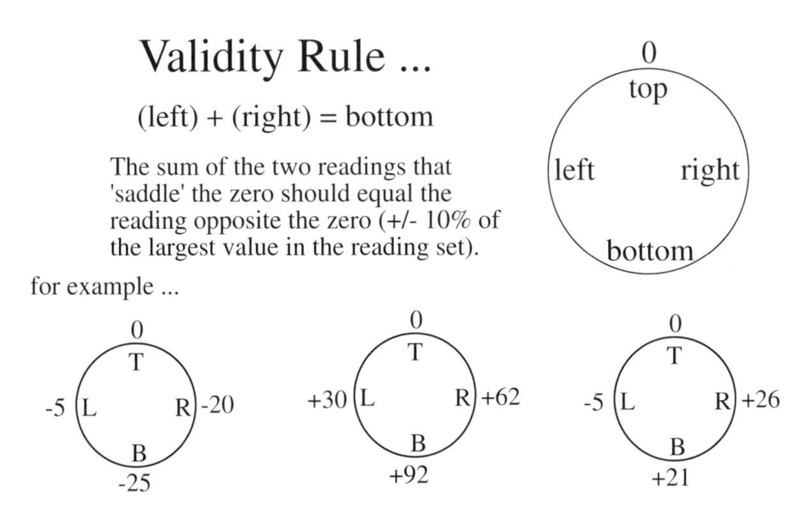

Figure 7-39. The validity rule.

2 - You do not have to rotate all the way around a shaft to determine the position of its centerline. If you can capture readings through a 180 degree arc (i.e. three out of four readings) you can determine what the other reading would be without actually having to take the measurement at that position. This comes in very handy when a physical restriction (e.g. lube oil pipe, the baseplate, a coupling guard stand) prevents sweeping through a full circle.

In fact, it is possible to determine the position of a shaft's centerline with readings captured on less than a 180 degree sweep. These types of readings are called 'partial arc' readings. However, there are inherent inaccuracies when attempting to determine centerline positions from partial arc readings.

Why measurements are taken at 90 degree intervals

Invariably the question arises on why readings are taken at the 12, 3, 6, and 9 o'clock positions on rotating machinery shafts.

In horizontally mounted rotating equipment, adjustments are made to the machinery cases to align the shafts in two planes, the up and down plane (i.e. vertical movement) and the side to side plane (i.e. lateral movement). Vertical adjustments made to rotating machinery casings are based on the 12 and 6 o'clock measurements. Lateral adjustments are made to rotating machinery casings based on the 3 and 9 o'clock measurements. In other words, when you are adjusting the height or pitch of the machinery cases, the side readings (3 and 9 o'clock) don't mean anything, only the top and bottom readings indicate the vertical position. Likewise, when you are adjusting the side to side positions of the machinery casings, the top and bottom readings aren't regarded, only the side readings are considered. Shaft positional measurements are taken in the planes that define the directions of movement the machinery casings will undergo to correct the misalignment condition.

In vertically oriented rotating machinery, however, it becomes obvious that there is no 'top' and 'bottom'. In this case,

Problem ...
The dial indicators are not stopping exactly at the 3, 6, and / or 9 o'clock positions.

Solution ...
accuartely measure 90 degree angles or use a twin spirit level

Problem ...
the dial indicator stem is not perpendicular to the surface

Solution ...
insure that indicator is perpendicular to the reading surface

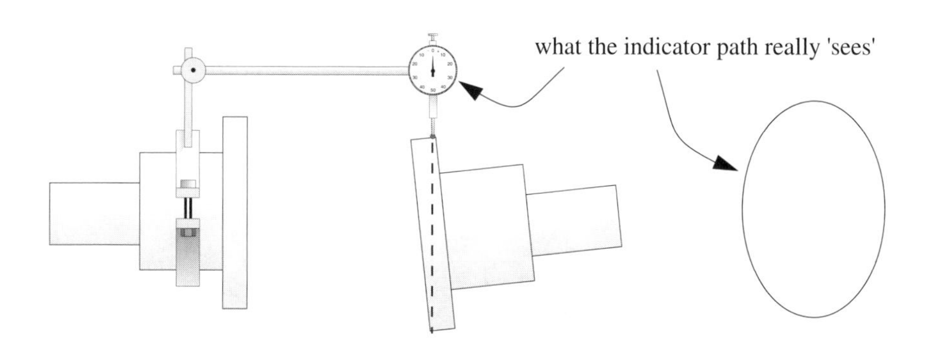

there's not much you can do about this

Figure 7-40. Reasons for deviations to the validity rule .

Tips for getting good readings

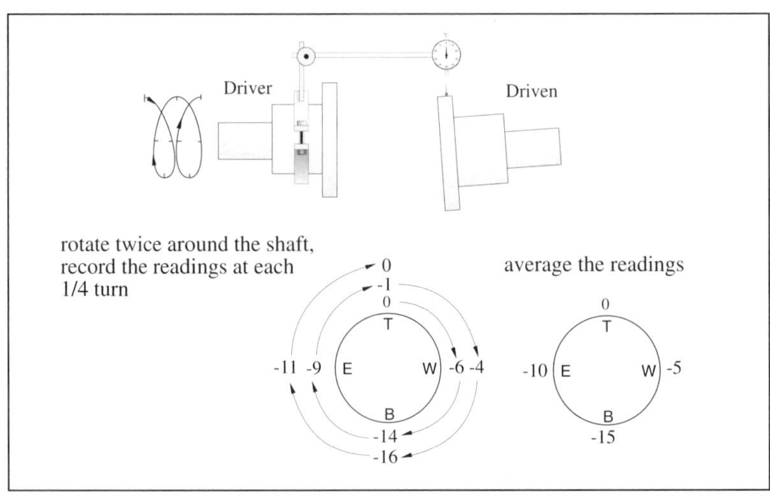

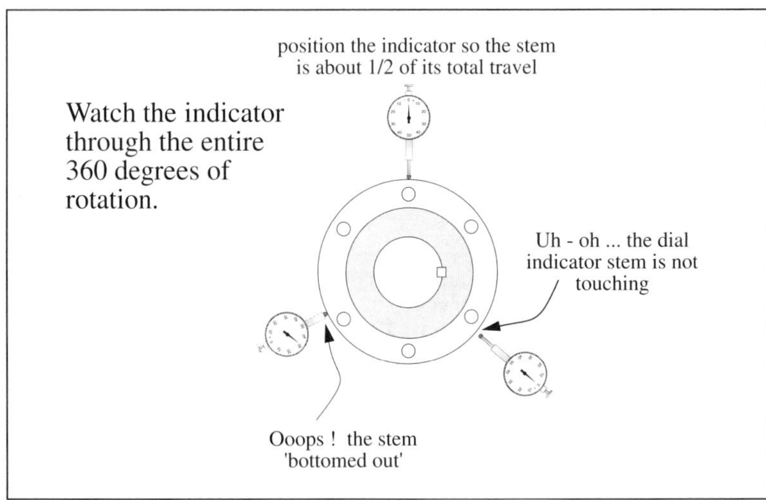

Figure 7-41. Some tips to insure that you're getting good measurement data.

you must determine what the planes of movement / translation will be on the machinery cases and capture the measurements in those planes. Refer to Chapters 12 and 13 for additional information on this subject.

Bracket / Bar Sag

Whenever mechanical brackets and dial indicators are used to measure shaft positions, 'bracket' or 'bar' sag must be measured and compensated for. The span bar is an overhung cantilever beam that bends under its own weight and the weight of devices attached to the end of the bar from the gravitational pull of the earth. This is similar to the phenomena discussed in Chapter 5 on rotor bow (i.e. the ‘catenary’ curve).

Bracket sag is somewhat confusing (and surprising) to people who observe it for the first time. Not only does this phenomena affect radial / circumferential measurements, but it also affects face measurements. Attempting to align machinery with measurements that have not been compensated for bracket sag will result in incorrect shim changes in an attempt to rectify the vertical misalignment condition. This is one of the classic mistakes made by people who align rotating machinery and cannot be stressed enough.

There are four factors that affect the amount of sag that you will get with any mechanical bracket arrangement:

1 - The amount of overhung weight (the weight of the dial indicator at the end of the span bar and the incremental weight of the span bar itself).
2 - The span of the bar.
3 - The stiffness of the span bar.
4 - Clamping force of the bracket to the shaft.

Typically, when you set out to align a piece of rotating machinery, there are several things that you don’t know until you set your measuring system onto the shafts. You don’t know what the diameter of the shafts are that you will be clamping on to, nor do you know what the height of the span bar needs to be from the point of contact on each shaft, nor do you know the distance you will be spanning from shaft to shaft.

Perhaps the best way to approach this due to all of these variables, is to perform the following procedure when taking alignment readings:

1 - Set up the bracket / span bar / indicator on the machinery being aligned.

2 - Take a set of shaft to shaft readings and record the data. These are referred to as the 'field' readings.

3 - Remove the bracket / span bar / indicator set up being very careful not to disturb the span bar length and the arrangement of the bracket / span bar configuration. Use the same dial indicator you used to capture the readings.

4 - Find a rigid piece of piping, conduit, or bar stock of sufficient length to clamp the bracket / span bar / indicator set up onto. Try to select a rigid piece of pipe close to the diameter of the shaft the bracket was clamped to when you measured the shaft to shaft positions.

5 - Set the indicator in the top position insure that the stem is loaded part way in its travel and zero the indicator.

6 - Hold the assembly in the horizontal position and rotate the entire pipe / bracket / span bar / dial indicator assembly through 90 degree arcs and note the readings at each position (particularly at the bottom) and record what you see. These are referred to as the 'sag' readings. Usually, the readings on each side are half of the bottom reading and all of the readings have a negative value (uh .. typically, but not always!). Refer to figure 7-42.

7 - Calculate what the readings would have been if you had been using a bracket that had no sag in it. These are referred to as the 'compensated' readings. Refer to the example shown in figure 7-43.

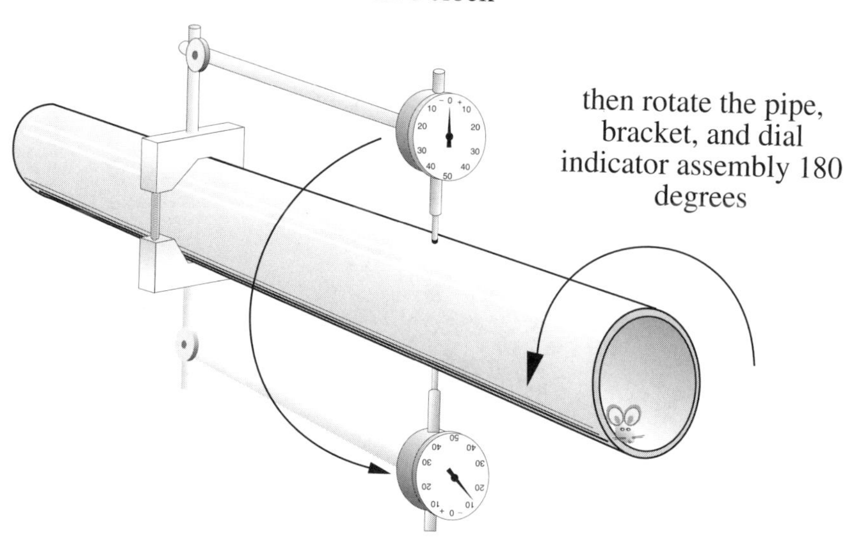

Figure 7-42. How to measure bracket sag (aka bar sag) in mechanical shaft alignment brackets that clamp around the outer diameter of a shaft.

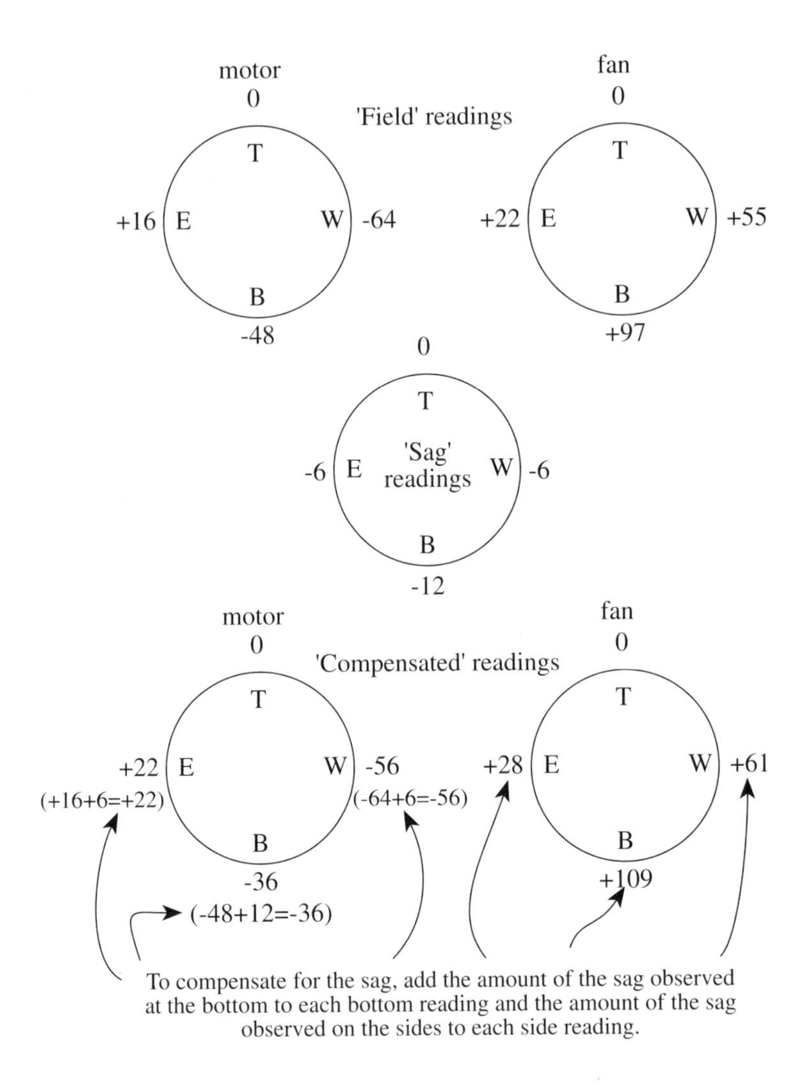

Figure 7-43. Sample problems illustrating how to compensate for bracket sag (aka bar sag) in mechanical shaft alignment brackets.

It is extremely important to note that when the machinery shafts are in perfect alignment, the dial indicator readings will be exactly what the sag readings are. In other words, if you are 'spinning zeros', the shafts are misaligned. When using mechanical brackets, you virtually never want to get a series of zeros when you're done (see figure 7-44 and the section in Chapter 9 on 'shoot-for' readings).

As mentioned previously, if you do not compensate the 'field' readings for bracket sag, and attempt to calculate shim changes to adjust the positions of the machinery for realignment, you will come up with wrong amounts for the shims.

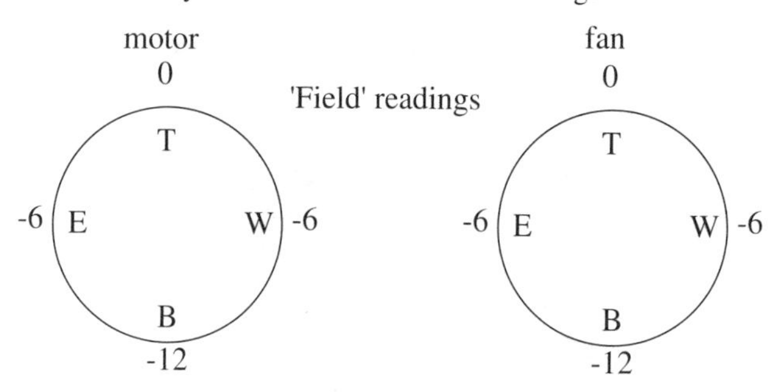

Figure 7-44. These are the ideal set of reverse indicator readings that must be obtained for the motor and fan shafts to be perfectly in line with each other from the sample problems illustrated in figure 7-43.

Figure 7-45. X-mas tree type bracket.

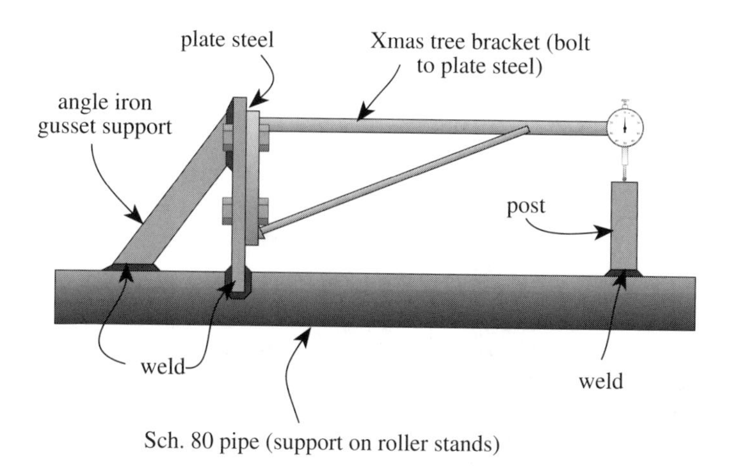

Figure 7-46. How to measure bracket sag in X-mas tree type brackets.

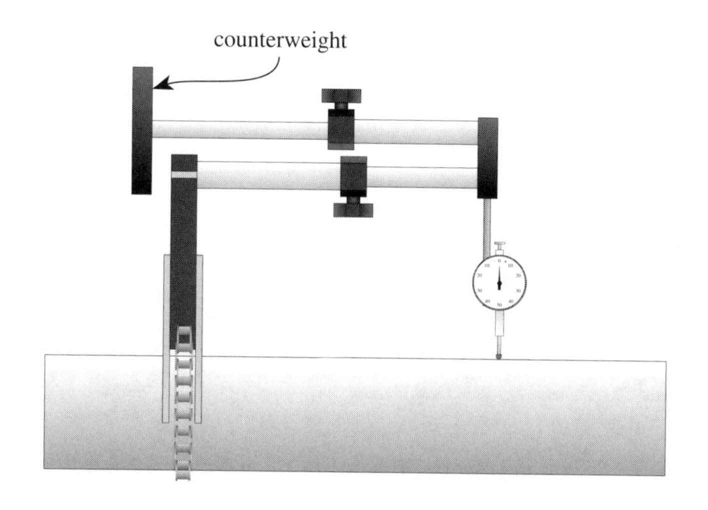

Figure 7-47. Zero sag compensation device.

Figure 7-48. Sag compensation device. Photo courtesty Murray & Garig Tool Works, Baytown, TX.

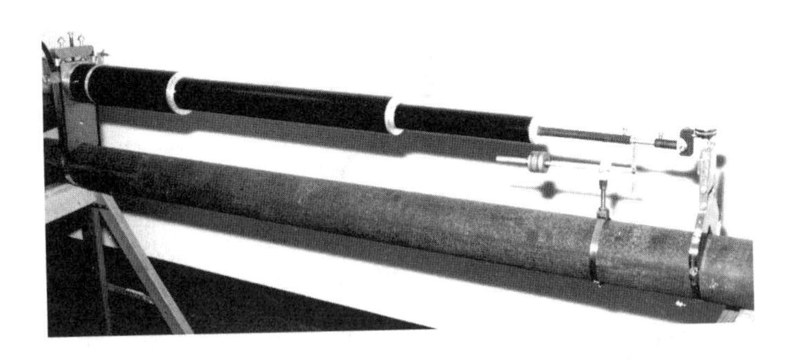

Figure 7-49. Sag compensation device on long span bracket manufactured from carbon-carbon tubing. Photo courtesty Murray & Garig Tool Works, Baytown, TX.

Circumferential (Rim) Readings are Twice the Amount of Offset

Whenever measurements are taken 180 degrees around the perimeter or circumference of a shaft or coupling hub, the measured value is twice the amount of centerline to centerline offset as illustrated in figures 7-50 and 7-51. This 'doubling' of the measurement must be accounted for when calculating the required

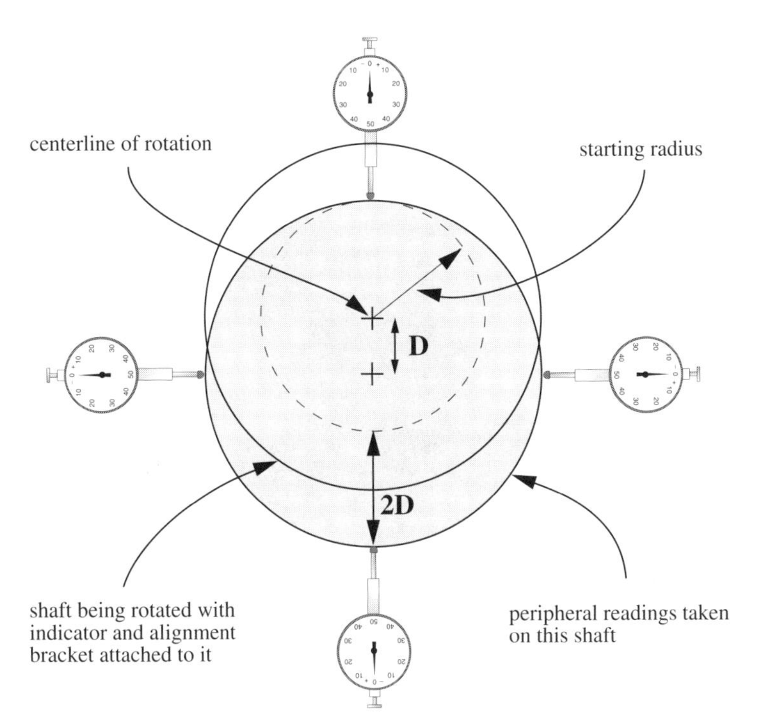

Figure 7-50. Why rim or circumferential readings are measuring twice the amount of centerline offset.

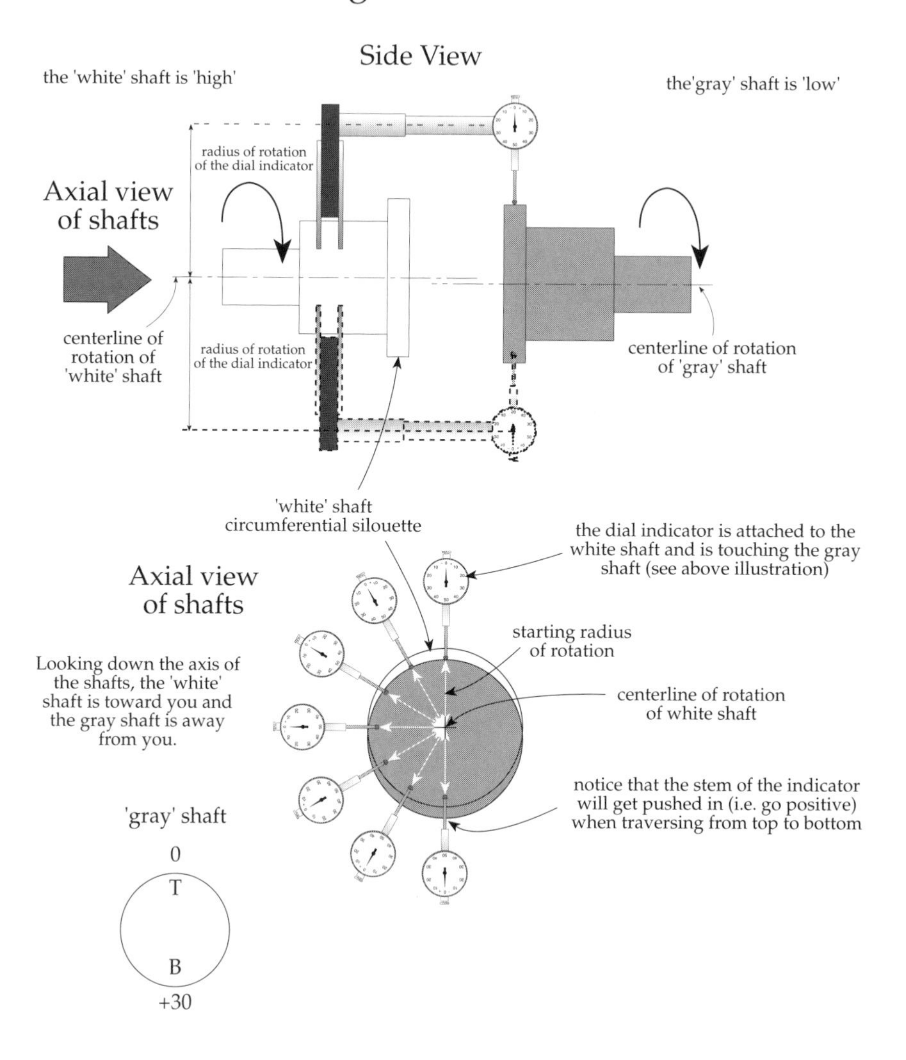

Figure 7-51. What happens to the indicator stem when taking rim readings.

vertical and lateral moves for the machinery and applies to all of the dial indicator measurement methods discussed previously except the face - face technique. It does apply to the rim readings taken in the face - peripheral method, but the face readings are taken at 'face' value (i.e. they are not twice the amount). Chapter 8 will discuss this in further detail.

Dial Indicator Shaft Alignment Measurement Systems

Dial indicator based shaft alignment measurement systems have been in existence since the 1930's although patents didn't start appearing until the mid-1940's (see Chapter 2). These measurement tools and fixtures are rugged and many are adaptable to meet the needs of virtually every alignment task. In many cases, alignment brackets / fixtures are custom made to fit a particular machine such as illustrated by the X-mas tree bracket shown in figures 7-32 and 7-45 , or a general purpose bracket can be made from pieces of angle iron or machined out of steel.

There are several companies who manufacture the general purpose alignment bracket systems that are shown in the following figures. The price range for these systems ranges from $600.00 up to $6000.00. If you are unable to manufacture your own shaft alignment bracket, it is recommended that you consider purchasing one of the systems shown for use in measuring off-line shaft positions.

Advantages of dial indicator measurement systems -

- relatively inexpensive
- some systems can perform all five dial indicator measurement techniques
- accuracies of +/- 1 mil
- dial indicators are needed to perform many of the preliminary steps (see Chapter 6)
- some systems allow for the coupling to be disconnected

when capturing readings (see Chapter 11)

Disadvantages of dial indicator measurement systems -
- bracket sag must be measured and compensated for
- user must know how to read a dial indicator
- user must know how to graph / model the shaft positions from the readings or be able to calculate required machinery moves (some systems include a calculator or computer software, see following pages and see Chapter 8)

A questionnaire was sent to all of the manufacturers of dial indicator based shaft alignment measurement systems. The results of the questionnaires are tabulated and appear in the following pages. Many of the answers were quite lengthy and were condensed for space saving purposes. If you are interested in one or more of these systems, it is recommended that you contact the manufacturers directly for additional information.

Dial Indicator Shaft Alignment System Questionnaire

1. What are the minimum and maximum shaft diameters that your alignment brackets can be clamped to? If you have different models that can be clamped to different ranges of shaft diameters, please indicate what range applies to each model.
2. What are the minimum and maximum distances your alignment brackets can span from shaft to shaft (or from bracket to bracket)?
3. What is the distance from the point where the bracket touches the shaft to the center of the span bar? If your bracket system can change the 'height' of the span bar what is the minimum and maximum distance from the point where the bracket is touching the shaft to the center of the span bar?
4. Can your brackets be attached to the 'face' of coupling hubs rather than clamp around a shaft diameter?
5. Can your aedlignment bracket system be used to perform the reverse indicator technique?
6. Can your alignment bracket system be used to perform the face-

peripheral (i.e. face-rim) technique?

7. Can your alignment bracket system be used to perform the shaft to coupling spool technique?
8. Can your alignment bracket system be used to perform the double radial technique?
9. Can your alignment bracket system be used to perform the face-face technique?
10. Do the shafts have to be rotated together while capturing readings? If so, what amount of rotational 'backlash' between the two shafts can be tolerated before measurement accuracy is sacrificed?
11. Do you supply dial indicators with your system? If so, are they standard types of dial indicators that can be purchased directly from the dial indicator manufacturers (e.g. Starrett, Mitutoyo, Central, etc.) in the event that the user breaks or loses a dial indicator?
12. Can other measuring devices such as LVDTs, proximity probes, lasers, or CCDs be used with or on your brackets?
13. How do you measure the rotational position of the brackets/indicators/sensors when capturing readings?
14. If a user damages or loses a component of the system, can the parts be purchased separately?
15. If you supply a calculator or computer software with your system, please describe the operation of the calculator or computer software.
16. Is the user required to name one machine element stationary and the other movable?
17. How do you enter information on how machinery will move from off-line to running conditions in the vertical and lateral planes?
18. Can the system be used on vertically oriented rotating machinery?
19. Can the system be configured to measure English or metric units?
20. If your system is patented, please send a copy of the patent (in English). If you have patents in other countries, please indicate where the patents are held and their corresponding patent number or identification.
21. What is your warranty period?
22. What is the recommended calibration interval, does the unit have to be sent back to the factory for calibration, and what is the charge for recalibration? Is the calibration you offer traceable back to US or International Standards?
23. Do you offer equipment for the user to check and adjust calibration? If so, what is the cost of the test equipment?
24. If you have introduced new shaft alignment system models, do the people who own your original models have the option to 'trade in' their

older models to upgrade to a newer model? If so, what is the cost of the upgrade?

25. What is the overall weight of the entire system?
26. What is the price (or price range) of the system(s)?

Figure 7-52. Acculign COACH system. Photo courtesy Acculign Inc., Willis, TX.

Figure 7-53. A-Line shaft alignment system. Photo courtesy A-Line Mfg. Co., LaPorte, TX.

Dial Indicator Shaft Alignment Systems	A-Line	A-Line
System Price	$450.00	$650.00
System Name	A-500	A-750
System Weight (lbs.)	5	7
Brackets Supplied?	yes	yes
Dial Indicators Supplied?	yes	yes
Use on horizontally mounted machines?	yes	yes
Use on vertically mounted machines?	yes	yes
Custom calculator supplied?	optional	optional
Standard computer supplied?	no	no
Alignment software included?	optional	optional
Warranty period (months)	12	12
Shaft Bracket Fixtures		
Minimum Shaft Diameter (in.)	0.50	0.50
Maximum Shaft Diameter (in.)	1.63	2.88
Minimum Shaft to Shaft Span (in.)	1.50	1.50
Maximum Shaft to Shaft Span (in.)	9.00	9.00
Minimum Span Bar Height (in.)	2.50	2.20
Maximum Span Bar Height (in.)	2.50	2.20
Brackets attach around shaft circumference?	yes	yes
Brackets attach to face of coupling?	yes	yes
Shafts must be rotated together?	yes	yes
How are rotational positions measured?	twin spirit level	twin spirit level
Shaft Alignment Technique Capabilities		
Face - Rim Method?	yes	yes
Reverse Indicator Method?	yes	yes
Shaft to Coupling Spool Method?	yes	yes
Double Radial Method?	no	no
Face - Face Method?	yes	yes
Software Information		
Operates on special computer?	yes	yes
Operates on DOS based computers?	no	no
Operates on Macintosh™ computers?	no	no
Number of machines in drive train	2	2
One machine is stationary?	yes	yes
Number of bolting planes per machine	2	2
Select different alignment methods?	no	no
Enter off-line to running movement?	yes	yes
Number of movement solutions	1	1
Graphical display of shaft positions?	no	no
Print results?	yes	yes
Save alignment data?	no	no
Edit alignment data?	no	no
Alignment accuracy display?	no	no

Dial Indicator Shaft Alignment Systems	A-Line	A-Line
System Price	$850.00	$1400.00
System Name	A-1000	A-3000
System Weight (lbs.)	17	30
Brackets Supplied?	yes	yes
Dial Indicators Supplied?	yes	yes
Use on horizontally mounted machines?	yes	yes
Use on vertically mounted machines?	yes	yes
Custom calculator supplied?	optional	optional
Standard computer supplied?	no	no
Alignment software included?	optional	optional
Warranty period (months)	12	12
Shaft Bracket Fixtures		
Minimum Shaft Diameter (in.)	1.00	1.00
Maximum Shaft Diameter (in.)	5.00	10.50
Minimum Shaft to Shaft Span (in.)	1.50	1.50
Maximum Shaft to Shaft Span (in.)	18.00	18.00
Minimum Span Bar Height (in.)	4.20	3.20
Maximum Span Bar Height (in.)	4.20	3.20
Brackets attach around shaft circumference?	yes	yes
Brackets attach to face of coupling?	yes	yes
Shafts must be rotated together?	yes	yes
How are rotational positions measured?	twin spirit level	twin spirit level
Shaft Alignment Technique Capabilities		
Face - Rim Method?	yes	yes
Reverse Indicator Method?	yes	yes
Shaft to Coupling Spool Method?	yes	yes
Double Radial Method?	no	no
Face - Face Method?	yes	yes
Software Information		
Operates on special computer?	yes	yes
Operates on DOS based computers?	no	no
Operates on Macintosh™ computers?	no	no
Number of machines in drive train	2	2
One machine is stationary?	yes	yes
Number of bolting planes per machine	2	2
Select different alignment methods?	no	no
Enter off-line to running movement?	yes	yes
Number of movement solutions	1	1
Graphical display of shaft positions?	no	no
Print results?	yes	yes
Save alignment data?	no	no
Edit alignment data?	no	no
Alignment accuracy display?	no	no

Dial Indicator Shaft Alignment Systems	CSi	CSi
System Price	$4995.00	$6995.00
System Name	8900-P3-15	8900-P1 Ultra-Spec
System Weight (lbs.)	36	36
Brackets Supplied?	yes	yes
Dial Indicators Supplied?	yes	yes
Use on horizontally mounted machines?	yes	yes
Use on vertically mounted machines?	yes	yes
Custom calculator supplied?	no	yes
Standard computer supplied?	no	no
Alignment software included?	yes	yes
Warranty period (months)	12	12
Shaft Bracket Fixtures		
Minimum Shaft Diameter (in.)	0.63	0.63
Maximum Shaft Diameter (in.)	26.00	26.00
Minimum Shaft to Shaft Span (in.)	3.00	3.00
Maximum Shaft to Shaft Span (in.)	34.00	34.00
Minimum Span Bar Height (in.)	4.50	4.50
Maximum Span Bar Height (in.)	18.00	18.00
Brackets attach around shaft circumference?	yes	yes
Brackets attach to face of coupling?	no	no
Shafts must be rotated together?	yes	yes
How are rotational positions measured?	twin spirit level	twin spirit level
Shaft Alignment Technique Capabilities		
Face - Rim Method?	no	no
Reverse Indicator Method?	yes	yes
Shaft to Coupling Spool Method?	yes	yes
Double Radial Method?	yes	yes
Face - Face Method?	no	no
Software Information		
Operates on special computer?	yes	yes
Operates on DOS based computers?	no	no
Operates on Macintosh™ computers?	no	no
Number of machines in drive train	2	2
One machine is stationary?	no	no
Number of bolting planes per machine	2	2
Select different alignment methods?	no	no
Enter off-line to running movement?	yes	yes
Number of movement solutions	6	6
Graphical display of shaft positions?	no	no
Print results?	yes	yes
Save alignment data?	yes	yes
Edit alignment data?	yes	yes
Alignment accuracy display?	yes	yes

Figure 7-54. Computational Systems Inc. 8100 system. Photo courtesy Computational Systems Inc., Knoxville, TN.

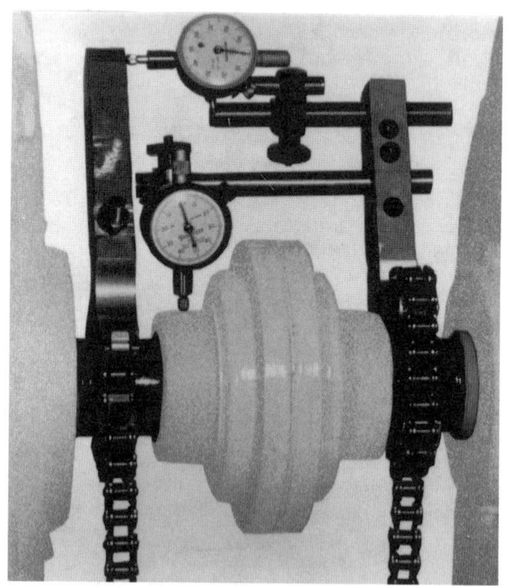

Figure 7-55. Peterson alignment tool system. Photo courtesy Peterson Alignment Tools Co., Plainfield, IL.

Figure 7-56. Murray & Garig Tool Works system. Photo courtesy Murray & Garig Tool Works, Baytown, TX.

Figure 7-57. Murray & Garig Tool Works system. Photo courtesy Murray & Garig Tool Works, Baytown, TX.

Dial Indicator Shaft Alignment Systems	CSi	MMS Inc.
System Price	$7995.00	$1375.00
System Name	8110 Ultra-Spec	REACT
System Weight (lbs.)	36	32
Brackets Supplied?	yes	yes
Dial Indicators Supplied?	yes (electronic)	yes
Use on horizontally mounted machines?	yes	yes
Use on vertically mounted machines?	yes	yes
Custom calculator supplied?	yes	no
Standard computer supplied?	no	no
Alignment software included?	yes	no
Warranty period (months)	12	12
Shaft Bracket Fixtures		
Minimum Shaft Diameter (in.)	0.63	0.50
Maximum Shaft Diameter (in.)	26.00	36.00
Minimum Shaft to Shaft Span (in.)	3.00	3.00
Maximum Shaft to Shaft Span (in.)	34.00	24.00
Minimum Span Bar Height (in.)	4.50	5.00
Maximum Span Bar Height (in.)	18.00	unlimited
Brackets attach around shaft circumference?	yes	yes
Brackets attach to face of coupling?	no	no
Shafts must be rotated together?	yes	yes
How are rotational positions measured?	twin spirit level	eyeball
Shaft Alignment Technique Capabilities		
Face - Rim Method?	no	yes
Reverse Indicator Method?	yes	yes
Shaft to Coupling Spool Method?	yes	yes
Double Radial Method?	yes	yes
Face - Face Method?	no	yes
Software Information		
Operates on special computer?	yes	n/a
Operates on DOS based computers?	no	n/a
Operates on Macintosh™ computers?	no	n/a
Number of machines in drive train	2	n/a
One machine is stationary?	no	n/a
Number of bolting planes per machine	2	n/a
Select different alignment methods?	no	n/a
Enter off-line to running movement?	yes	n/a
Number of movement solutions	6	n/a
Graphical display of shaft positions?	no	n/a
Print results?	yes	n/a
Save alignment data?	yes	n/a
Edit alignment data?	yes	n/a
Alignment accuracy display?	yes	n/a

Dial Indicator Shaft Alignment Systems	Murray & Garig
System Price	$1923.00
System Name	Murray & Garig
System Weight (lbs.)	24
Brackets Supplied?	yes
Dial Indicators Supplied?	yes
Use on horizontally mounted machines?	yes
Use on vertically mounted machines?	yes
Custom calculator supplied?	no
Standard computer supplied?	no
Alignment software included?	no
Warranty period (months)	12
Shaft Bracket Fixtures	
Minimum Shaft Diameter (in.)	1.50
Maximum Shaft Diameter (in.)	48.00
Minimum Shaft to Shaft Span (in.)	not answered
Maximum Shaft to Shaft Span (in.)	23.00
Minimum Span Bar Height (in.)	1.75
Maximum Span Bar Height (in.)	7.00
Brackets attach around shaft circumference?	yes
Brackets attach to face of coupling?	no
Shafts must be rotated together?	no
How are rotational positions measured?	automatic protractor
Shaft Alignment Technique Capabilities	
Face - Rim Method?	yes
Reverse Indicator Method?	yes
Shaft to Coupling Spool Method?	yes
Double Radial Method?	yes
Face - Face Method?	yes
Software Information	
Operates on special computer?	n/a
Operates on DOS based computers?	n/a
Operates on Macintosh™ computers?	n/a
Number of machines in drive train	n/a
One machine is stationary?	n/a
Number of bolting planes per machine	n/a
Select different alignment methods?	n/a
Enter off-line to running movement?	n/a
Number of movement solutions	n/a
Graphical display of shaft positions?	n/a
Print results?	n/a
Save alignment data?	n/a
Edit alignment data?	n/a
Alignment accuracy display?	n/a

Dial Indicator Shaft Alignment Systems	MMS Inc.	Peterson
System Price	$3350.00	$695.00
System Name	REACT	20RA
System Weight (lbs.)	32	9
Brackets Supplied?	yes	yes
Dial Indicators Supplied?	yes	yes
Use on horizontally mounted machines?	yes	yes
Use on vertically mounted machines?	yes	yes
Custom calculator supplied?	yes	optional
Standard computer supplied?	no	no
Alignment software included?	yes	optional
Warranty period (months)	12	12
Shaft Bracket Fixtures		
Minimum Shaft Diameter (in.)	0.50	0.75
Maximum Shaft Diameter (in.)	36.00	4.50
Minimum Shaft to Shaft Span (in.)	3.00	5.00
Maximum Shaft to Shaft Span (in.)	24.00	9.00
Minimum Span Bar Height (in.)	5.00	2.35
Maximum Span Bar Height (in.)	unlimited	2.35
Brackets attach around shaft circumference?	yes	yes
Brackets attach to face of coupling?	no	yes
Shafts must be rotated together?	yes	yes
How are rotational positions measured?	eyeball	eyeball
Shaft Alignment Technique Capabilities		
Face - Rim Method?	yes	yes
Reverse Indicator Method?	yes	yes
Shaft to Coupling Spool Method?	yes	no
Double Radial Method?	yes	no
Face - Face Method?	yes	no
Software Information		
Operates on special computer?	yes	yes
Operates on DOS based computers?	no	no
Operates on Macintosh™ computers?	no	no
Number of machines in drive train	2	2
One machine is stationary?	yes	yes
Number of bolting planes per machine	2	2
Select different alignment methods?	Face-Rim & Rev Indctr	no
Enter off-line to running movement?	yes	no
Number of movement solutions	1	1
Graphical display of shaft positions?	no	no
Print results?	yes	no
Save alignment data?	no	no
Edit alignment data?	no	no
Alignment accuracy display?	no	no

Dial Indicator Shaft Alignment Systems	Peterson	SPM Instruments
System Price	$895.00	$3500.00
System Name	30RA	SPM
System Weight (lbs.)	12	14
Brackets Supplied?	yes	yes
Dial Indicators Supplied?	yes	yes (electronic)
Use on horizontally mounted machines?	yes	yes
Use on vertically mounted machines?	yes	yes
Custom calculator supplied?	optional	yes
Standard computer supplied?	no	no
Alignment software included?	optional	yes
Warranty period (months)	12	12
Shaft Bracket Fixtures		
Minimum Shaft Diameter (in.)	1.00	0.63
Maximum Shaft Diameter (in.)	8.50	15.75
Minimum Shaft to Shaft Span (in.)	3.50	5.00
Maximum Shaft to Shaft Span (in.)	15.00	20.00
Minimum Span Bar Height (in.)	2.13	4.50
Maximum Span Bar Height (in.)	5.06	9.00
Brackets attach around shaft circumference?	yes	yes
Brackets attach to face of coupling?	yes	yes
Shafts must be rotated together?	yes	yes
How are rotational positions measured?	eyeball	built in Hg switch
Shaft Alignment Technique Capabilities		
Face - Rim Method?	yes	yes
Reverse Indicator Method?	yes	yes
Shaft to Coupling Spool Method?	no	no
Double Radial Method?	no	no
Face - Face Method?	no	no
Software Information		
Operates on special computer?	yes	yes
Operates on DOS based computers?	no	no
Operates on Macintosh™ computers?	no	no
Number of machines in drive train	2	2
One machine is stationary?	yes	yes
Number of bolting planes per machine	2	2
Select different alignment methods?	no	no
Enter off-line to running movement?	no	yes
Number of movement solutions	1	1
Graphical display of shaft positions?	no	no
Print results?	no	no
Save alignment data?	no	no
Edit alignment data?	no	no
Alignment accuracy display?	no	no

Figure 7-58. SPM MAC-5 system. Photo courtesy SPM Instruments Inc., Marlborough, CT.

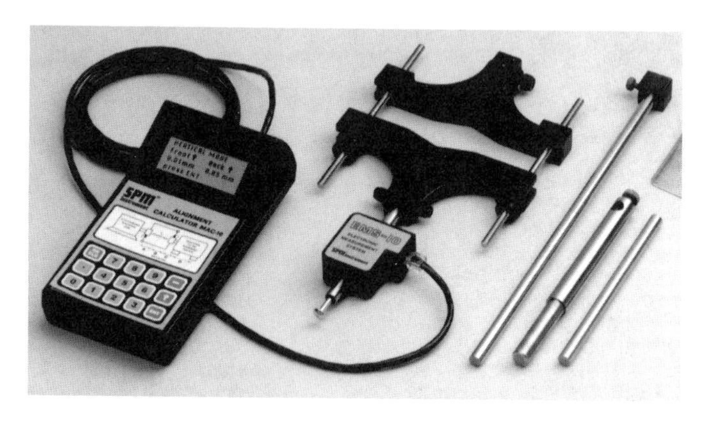

Figure 7-59. SPM MAC-10 system. Photo courtesy SPM Instruments Inc., Marlborough, CT.

Dial Indicator Shaft Alignment Systems	Turvac Inc.	Turvac Inc.
System Price	$995.00	$2695.00
System Name	Standard	Journeyman
System Weight (lbs.)	8	20
Brackets Supplied?	yes	yes
Dial Indicators Supplied?	yes	yes
Use on horizontally mounted machines?	yes	yes
Use on vertically mounted machines?	yes	yes
Custom calculator supplied?	no	no
Standard computer supplied?	no	no
Alignment software included?	no	yes
Warranty period (months)	12	12
Shaft Bracket Fixtures		
Minimum Shaft Diameter (in.)	0.50	0.50
Maximum Shaft Diameter (in.)	48.00	48.00
Minimum Shaft to Shaft Span (in.)	0.50	0.50
Maximum Shaft to Shaft Span (in.)	33.00	33.00
Minimum Span Bar Height (in.)	-0.50	-0.50
Maximum Span Bar Height (in.)	11.00	11.00
Brackets attach around shaft circumference?	yes	yes
Brackets attach to face of coupling?	yes	yes
Shafts must be rotated together?	no	no
How are rotational positions measured?	twin spirit level	twin spirit level
Shaft Alignment Technique Capabilities		
Face - Rim Method?	yes	yes
Reverse Indicator Method?	yes	yes
Shaft to Coupling Spool Method?	yes	yes
Double Radial Method?	yes	yes
Face - Face Method?	yes	yes
Software Information		
Operates on special computer?	n/a	no
Operates on DOS based computers?	n/a	yes (user choice)
Operates on Macintosh™ computers?	n/a	yes (user choice)
Number of machines in drive train	n/a	2
One machine is stationary?	n/a	no
Number of bolting planes per machine	n/a	2
Select different alignment methods?	n/a	yes
Enter off-line to running movement?	n/a	yes
Number of movement solutions	n/a	unlimited
Graphical display of shaft positions?	n/a	yes
Print results?	n/a	yes
Save alignment data?	n/a	yes
Edit alignment data?	n/a	yes
Alignment accuracy display?	n/a	yes

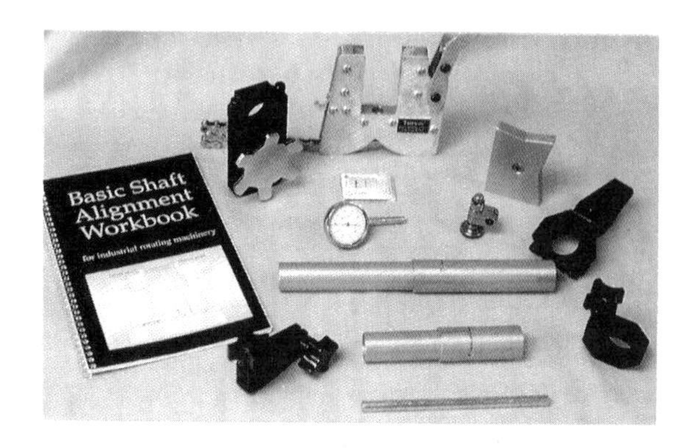

Figure 7-60. Turvac Standard system. Photo courtesy Turvac Inc., Cincinnati, OH.

Figure 7-61. Turvac Journeyman system. Photo courtesy Turvac Inc., Cincinnati, OH.

Dial Indicator Shaft Alignment Systems	Turvac Inc.	Turvac Inc.
System Price	$4995.00	$5695.00
System Name	Master	Professional
System Weight (lbs.)	25	25
Brackets Supplied?	yes	yes
Dial Indicators Supplied?	yes	yes
Use on horizontally mounted machines?	yes	yes
Use on vertically mounted machines?	yes	yes
Custom calculator supplied?	no	no
Standard computer supplied?	DOS	Apple© Macintosh™
Alignment software included?	yes	yes
Warranty period (months)	12	12
Shaft Bracket Fixtures		
Minimum Shaft Diameter (in.)	0.50	0.50
Maximum Shaft Diameter (in.)	48.00	48.00
Minimum Shaft to Shaft Span (in.)	0.50	0.50
Maximum Shaft to Shaft Span (in.)	33.00	33.00
Minimum Span Bar Height (in.)	-0.50	-0.50
Maximum Span Bar Height (in.)	11.00	11.00
Brackets attach around shaft circumference?	yes	yes
Brackets attach to face of coupling?	yes	yes
Shafts must be rotated together?	no	no
How are rotational positions measured?	twin spirit level	twin spirit level
Shaft Alignment Technique Capabilities		
Face - Rim Method?	yes	yes
Reverse Indicator Method?	yes	yes
Shaft to Coupling Spool Method?	yes	yes
Double Radial Method?	yes	yes
Face - Face Method?	yes	yes
Software Information		
Operates on special computer?	no	no
Operates on DOS based computers?	yes	no
Operates on Macintosh™ computers?	no	yes
Number of machines in drive train	2	2
One machine is stationary?	no	no
Number of bolting planes per machine	2	2
Select different alignment methods?	yes	yes
Enter off-line to running movement?	yes	yes
Number of movement solutions	unlimited	unlimited
Graphical display of shaft positions?	yes	yes
Print results?	yes	yes
Save alignment data?	yes	yes
Edit alignment data?	yes	yes
Alignment accuracy display?	yes	yes

Figure 7-62. Turvac Journeyman system. Photo courtesy Turvac Inc., Cincinnati, OH.

Figure 7-63. Turvac miniature bracket. Photo courtesy Turvac Inc., Cincinnati, OH.

Figure 7-64. Update alignment system. Photo courtesy Update International Inc., Denver, CO.

Dial Indicator Shaft Alignment Systems	Update Int'l.
System Price	$2490.00
System Name	Proaction
System Weight (lbs.)	35
Brackets Supplied?	yes
Dial Indicators Supplied?	yes
Use on horizontally mounted machines?	yes
Use on vertically mounted machines?	yes
Custom calculator supplied?	yes
Standard computer supplied?	no
Alignment software included?	yes
Warranty period (months)	12
Shaft Bracket Fixtures	
Minimum Shaft Diameter (in.)	0.625
Maximum Shaft Diameter (in.)	12
Minimum Shaft to Shaft Span (in.)	2.5
Maximum Shaft to Shaft Span (in.)	23
Minimum Span Bar Height (in.)	3
Maximum Span Bar Height (in.)	12
Brackets attach around shaft circumference?	yes
Brackets attach to face of coupling?	no
Shafts must be rotated together?	yes
How are rotational positions measured?	spiral level
Shaft Alignment Technique Capabilities	
Face - Rim Method?	yes (addtn'l parts req'd)
Reverse Indicator Method?	yes
Shaft to Coupling Spool Method?	yes (addtn'l parts req'd)
Double Radial Method?	yes (addtn'l parts req'd)
Face - Face Method?	yes (addtn'l parts req'd)
Software Information	
Operates on special computer?	no
Operates on DOS based computers?	yes
Operates on Macintosh™ computers?	no
Number of machines in drive train	2
One machine is stationary?	yes
Number of bolting planes per machine	2
Select different alignment methods?	yes (RevInd, Fc-Rm, Lsr)
Enter off-line to running movement?	yes
Number of movement solutions	2
Graphical display of shaft positions?	no
Print results?	yes
Save alignment data?	yes
Edit alignment data?	no
Alignment accuracy display?	yes

Laser Shaft Alignment Systems

The basics of semiconductor laser and detector operation were discussed earlier in this chapter. Since some of the manufacturers have taken slightly different approaches to using lasers and detectors for measuring the shaft positions. Figure 7-65, shows the three basic measurement techniques used by all of the manufacturers.

The laser/detector, roof prism system is used in most of the Prüftechnik systems. The dual laser / dual detector system is used by AlignX, FixtureLaser AB, and Computational Systems Inc. The laser / beam splitter / dual detector system is used by Hamar, MMS, and some Prüftechnik systems.

Advantages of laser measurement systems -

- bracket sag does not occur with laser beams but the user must insure the brackets that hold the lasers and detectors are firmly attached to the shafts
- accuracies of +/- 3 microns
- some systems allow for the coupling to be disconnected when capturing readings (see Chapter 11)
- most systems include a operator keypad / display module that prompts the user through the measurement steps and calculates the moves for one of the machines (some systems can solve for a variety of movement solutions see Chapter 8 and the tabulated questionairre results)

Disadvantages of laser measurement systems -

- relatively expensive
- range of measurement somewhat limited since most manufacturers use 10mm x 10mm detectors
- incapable of measuring runout conditions (it is possible to align bent shafts and not know it)
- many systems determine moves for one of the two ma-

chines (see 'stationary' - 'movable' concept in Chapter 8)
- most systems require that the coupling be bolted in place when capturing readings (see Chapter 11)
- both shafts have to be rotated
- sometimes difficult to capture readings in bright sunlight or well lit areas
- accuracy reduced in presence of excessive steam or heat

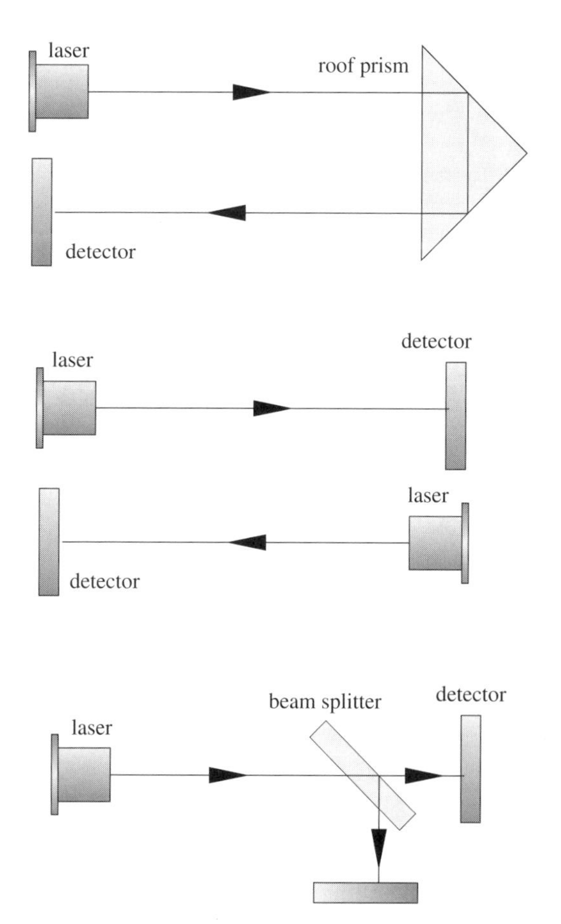

Figure 7-65. Basic principles of laser - detector shaft alignment measurement systems.

Imagine being able to look down the axis of two misaligned shafts. The gray shaft is away from us and the white shaft is toward us. The laser beam is attached to the white shaft and therefore rotates around its centerline. The detector target is attached to the gray shaft and therefore rotates around its centerline.

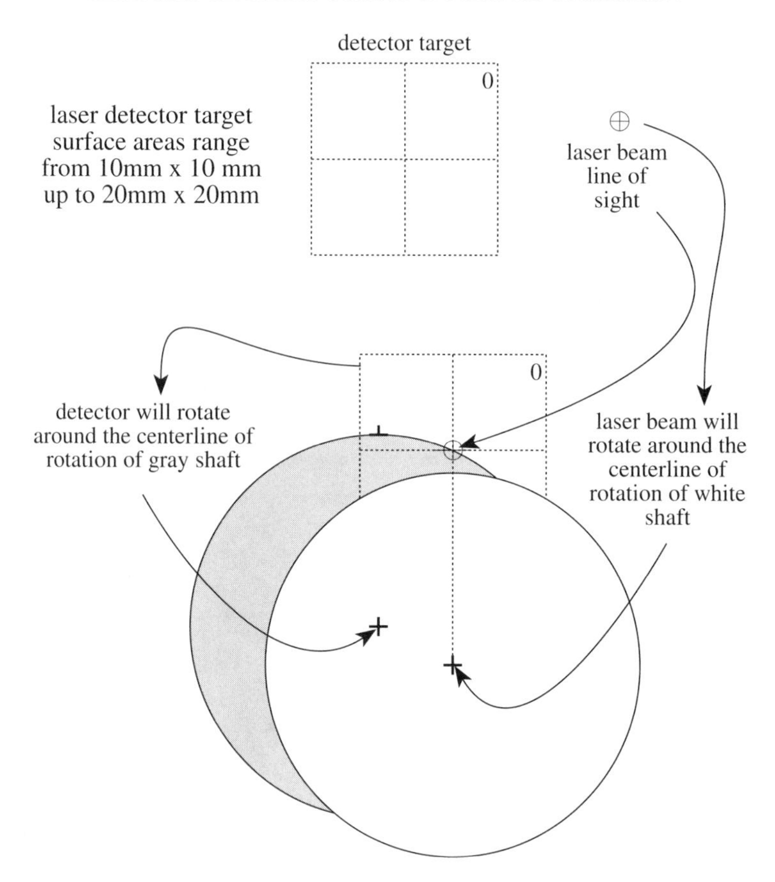

Figure 7-66. What a laser detector 'sees' looking down the axis of the shafts.

If we took a multiple exposure photograph when the shafts were rotated through 30 degree increments, this is what we would see through 360 degrees of rotation.

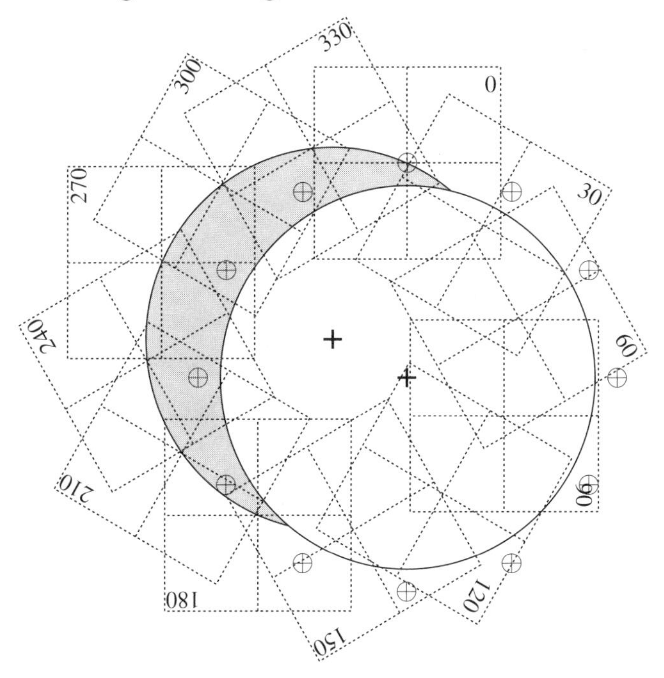

Figure 7-67. Where the laser beam strikes the detector through 360 degrees of rotation.

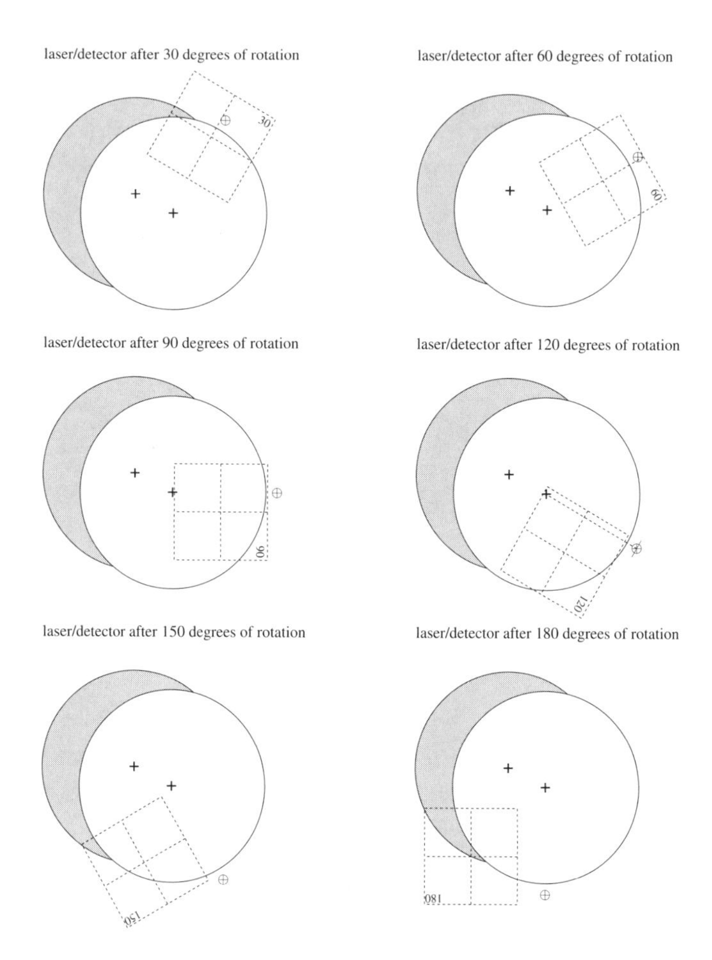

Figure 7-68. Where the laser beam strikes the detector from 30 through 180 degrees of rotation.

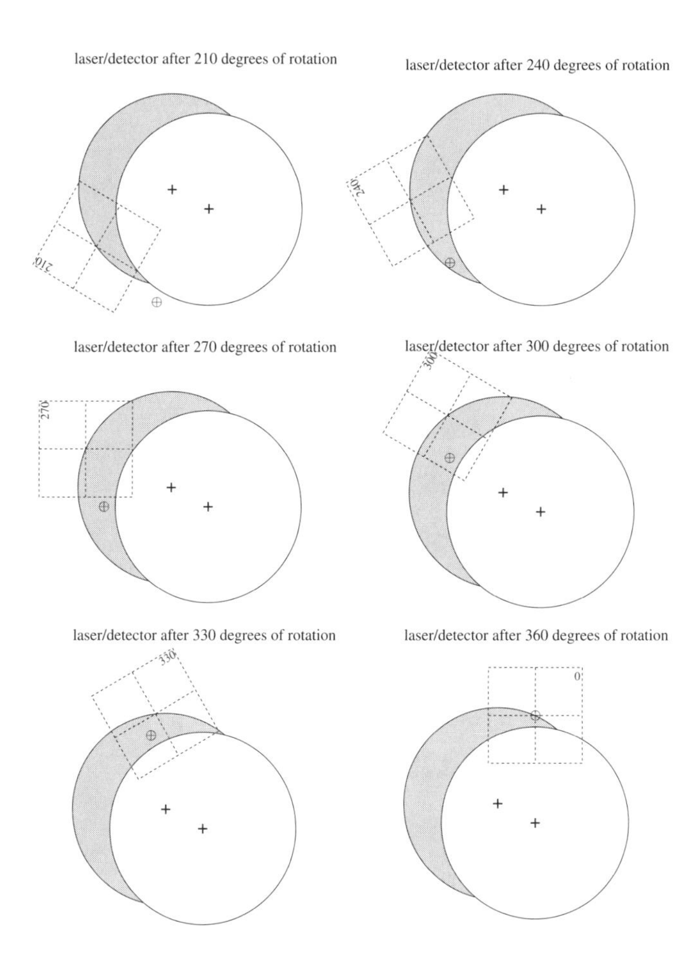

Figure 7-69. Where the laser beam strikes the detector from 210 through 360 degrees of rotation.

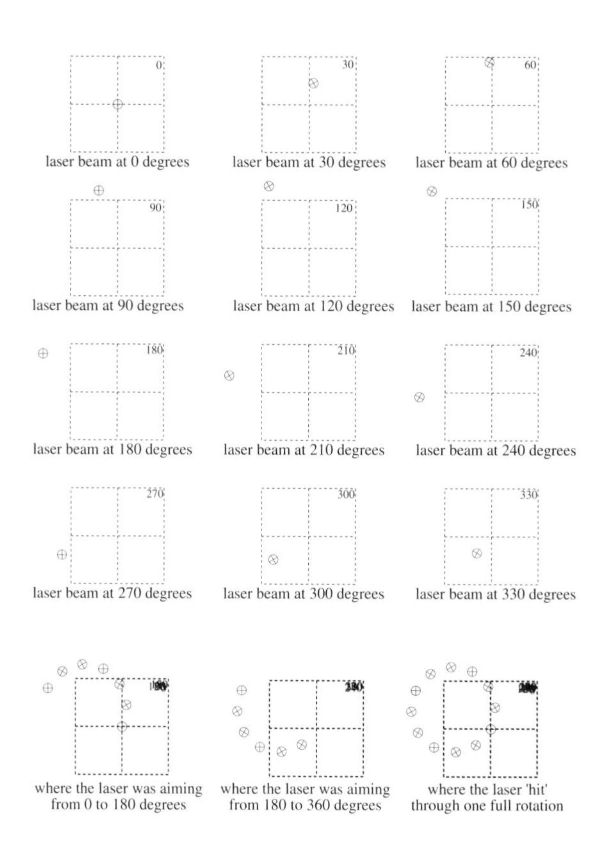

Figure 7-70. What we would see if we were standing next to the detector as the shafts were rotated.

A questionnaire was sent to all of the manufacturers of laser based shaft alignment measurement systems. The results of the questionnaires that were returned to the author are tabulated and appear in the following pages. Many of the answers were quite lengthy and were condensed for space saving purposes. If you are interested in one or more of these systems, it is recommended that you contact the manufacturers directly for additional information. The questionaire sent to the manufacturers was as follows ...

Laser Shaft Alignment System Questionnaire

1. Describe the basic operation of your laser measurement system. Use drawings or diagrams to enhance your explanation if possible.
2. What is the overall weight of the entire system?
3. What is the price (or price range) of the system(s)?
4. How is the unit powered?
5. If batteries are used, what size are required and how many are needed?
6. What battery life can be expected for continuous use on battery power?
7. Can the unit be plugged into AC electric supplies? Can it work on either 50 or 60 Hz circuits and voltage ranges from 90-220 VAC?
8. What type of laser is used in your system and how does it work (beam frequency in nm, continuous or pulse?, pulse duration, beam diameter at collimator exit, beam diameter @ 1, 3, 10 meters, power output, safety precautions, etc.)?
9. If you are using semiconductor junction diode lasers, which typically produce a low-quality beam that is divergent, elliptical, and astigmatic, what optical correction is incorporated in your system to correct for these deficiencies?
10. What effect does dirt, grease, oil, fingerprints, moisture, etc. have on the laser collimating lens or protective optics? If detrimental to the operation of the system, how should one clean the laser optics?
11. What type of detector is used in your system and how does it work (photodiode, CCD, single or dual axis?, detector target area, resolution, linearity, accuracy / repeatability, environmental limits, optical filtering, etc.)?
12. What effect does dirt, grease, oil, fingerprints, moisture, etc. have on the detector or protective optics? If detrimental to the operation of the

system, how should one clean the detector?

13. Can the output from the detector be connected directly to a computer (specifically computer using Intel or Motorola processors)?
14. How is the data from the detectors transmitted to the computer / operator interface? Cables or wireless transmission? If cables are used, how long are they? Do the cables have quick disconnect type fittings? If cables are used, what happens if an operator can not stop a shaft(s) from rotating once they get it started?
15. What type of shock protection is incorporated in your system to prevent damage to the laser / detector? What happens if an operator drops the system from 6 feet?
16. Can the user store a shaft alignment job in the computer / operator interface? Can they store more than one alignment job? If, so how many? Can the data that is stored in the alignment system be transferred to a personal computer? If so, how is this done?
17. If an operator mistakenly pressed the 'OFF' button in the middle of an alignment job, what happens?
18. What type of thermal protection is incorporated in your system to prevent measurement distortion?
19. If the unit is operated near strong electromagnetic fields (e.g. motor windings, magnetic bases) is there any effect on the electronics that would affect the accuracy of the instrument?
20. Do you offer an intrinsically safe and/or explosion proof models?
21. What type of environmental protection for the laser / detector is incorporated? Does it comply with US or International environmental protection standards?
22. If your system includes an operator interface / keypad entry device, describe the basic function of the device.
23. If a customer has more than one of your measurement systems in their possession, can the operator interface modules and laser/detectors be interchanged or are the detectors matched to a specific operator interface? If they can be interchanged, explain why.
24. Do the shafts have to be rotated together while capturing readings? If so, what amount of rotational 'backlash' between the two shafts can be tolerated before measurement accuracy is sacrificed?
25. How do you measure the rotational position of the sensors when capturing readings?
26. What is the minimum and maximum shaft diameters the brackets can be clamped to?
27. What is the minimum and maximum height range that the laser/

detector can be placed from the point of contact on the shaft to the position of the laser/detector?

28. Please describe the operation of the software that interfaces with your system.
29. Is the user required to name one machine element stationary and the other movable?
30. How do you enter information on how machinery will move from off-line to running conditions in the vertical and lateral planes?
31. Can the sensors be reconfigured to capture off-line to running machinery movement? If so, how is this accomplished?
32. Can the system be used on vertically oriented rotating machinery?
33. Can the system be configured to enter English or metric units?
34. If your system is patented, please send a copy of the patent (in English). If you have patents in other countries, please indicate where the patents are held and their corresponding patent number or identification.
35. What is your warranty period?
36. What are the recommended calibration intervals, does the unit have to be sent back to the factory for calibration, and what is the charge for recalibration? Is the calibration you offer traceable back to US or International Standards?
37. Do you offer equipment for the user to check and adjust calibration? If so, what is the cost of the test equipment?
38. If you have introduced new shaft alignment system models, do the people who own your original models have the option to 'trade in' their older models to upgrade to a newer model? If so, what is the cost of the upgrade?

Figure 7-71. Optalign™ laser system. Photo courtesy Prüftechnik Dieter Busch + Partner GmbH, Ismaning, Germany.

Laser Shaft Alignment Systems	Align-X
System Price Range	$15000.00
System Information	
Total Unit Weight	not answered
Alignment Operating Principle	reverse indicator
Intrinsically safe / explosion proof models?	no
Shock Protection	yes
Operating Temperature Range	not answered
Electromagnetic Field Protection	no effect
Environmental protection	sealed detector & laser
Computer / Software	
Alignment computer & software included?	yes
Interface system with standard computers?	yes
Software installation	floppy disk
Req'd to name one machine stationary?	yes
Number of possible movement solutions	1
Maximum number of machines in drive train	2
alignment jobs can be stored for future recall	unlimited
battery type / number req's.	12 VDC NiCad / 1
battery life @ full charge	8-10 hrs.
optional AC power & requirements	90-220 VAC / 50-60 Hz
enter off-line to running movement?	yes
Laser Emitter (Source)	
laser type	not answered
wavelength	670 nm
beam shape	elliptical (1mm x 1.5 mm)
beam diameter at exit	1mm x 1.5 mm
beam diameter @ 3 ft.	not answered
beam diameter @ 10 ft.	not answered
beam diameter @ 30 ft.	not answered
max. beam power output	0.9 mW
Laser Safety Class	not answered
max. recommended distance	not answered
Continuous or pulsed Beam?	not answered
beam pulse rate	not answered
laser power supplied by ...	cable
cable length	not answered
Laser Detector (Receiver)	
Type	Dual Axis photodiode
detector measurement planes	2
detector measurement area	10mm x 10 mm (20x20 mm optional)
number of detectors used	2
linearity	99.9% over 90% of active area
accuracy	0.0005"
optical filtering?	not answered
beam position data transfer via ...	cable
decector power supplied by ...	cable
battery life @ full charge	8 hours

Operational Info	
Shafts must be rotated together?	no rotation required
How are rotational positioned measured?	n/a
Max allowable rotational backlash	n/a
Can be used on vertical shafts?	yes
Additional Features & Accessories	
shaft clamping brackets inluded?	yes
minimum shaft diameter	.0625"
maximum shaft diameter	unlimited with special brackets
minimum sensor height from shaft surface	not answered
maximum sensor height from shaft surface	not answered
adapted for measuring OL2R?	yes (special brackets req'd)
Calibration	
Recommended calibration interval	12 months
User can by calibration equipment?	yes
Cost of calibration equipment	included with system

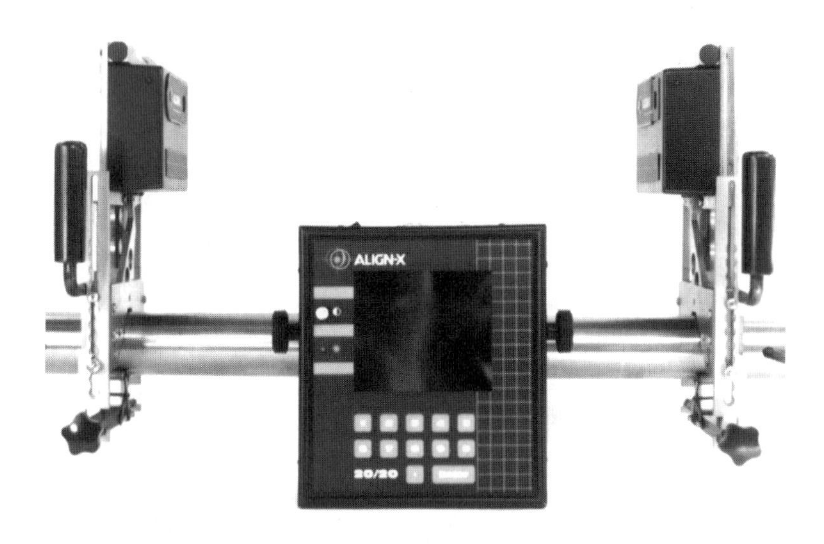

Figure 7-72. AlignX laser shaft alignment system. Photo courtesy AlignX Co. Milwuakee, WI.

Laser Shaft Alignment Systems	CSi UltraSpec 8200
System Price Range	$12495.00
System Information	
Total Unit Weight	30 lbs. (13.6 kg)
Alignment Operating Principle	reverse indicator
Intrinsically safe / explosion proof models?	no
Shock Protection	yes (foam protect in case)
Operating Temperature Range	15-120 deg F
Electromagnetic Field Protection	no effect
Environmental protection	sealed detector & laser
Computer / Software	
Alignment computer & software included?	yes
Interface system with standard computers?	no
Software installation	downloaded to Keypad / analyzer
Req'd to name one machine stationary?	no
Number of possible movement solutions	6
Maximum number of machines in drive train	2
alignment jobs can be stored for future recall	100
battery type / number req's.	custom / 3
battery life @ full charge	4 hours continuous / 8 hrs typical
optional AC power & requirements	90-220 VAC / 50-60 Hz (optional)
enter off-line to running movement?	yes
Laser Emitter (Source)	
laser type	In-Ga-Al-P semiconductor
wavelength	670 nm
beam shape	circular
beam diameter at exit	2 mm
beam diameter @ 3 ft.	2.5 mm
beam diameter @ 10 ft.	3.0 mm
beam diameter @ 30 ft.	5.0 mm
max. beam power output	< 1.0 mW
Laser Safety Class	Class II
max. recommended distance	30 ft.
Continuous or pulsed Beam?	pulsed
beam pulse rate	220 Hz
laser power supplied by ...	battery (internal to head)
cable length	n/a
Laser Detector (Receiver)	
Type	Dual Axis photodiode
detector measurement planes	2
detector measurement area	.4 x .4 sq. in. (10 x 10 sq. mm)
number of detectors used	2
linearity	99.9% over 90% of active area
accuracy	not answered
optical filtering?	yes
beam position data transfer via ...	cable or infrared beam (no cable)
decector power supplied by ...	battery (internal to head)
battery life @ full charge	4 hours

Operational Info	
Shafts must be rotated together?	no
How are rotational positioned measured?	internal angle sensors
Max allowable rotational backlash	4 deg.
Can be used on vertical shafts?	yes
Additional Features & Accessories	
shaft clamping brackets inluded?	yes
minimum shaft diameter	.0625"
maximum shaft diameter	26" (with additional chain)
minimum sensor height from shaft surface	4.375"
maximum sensor height from shaft surface	10.375"
adapted for measuring OL2R?	yes (special brackets req'd)
Calibration	
Recommended calibration interval	12 months
User can by calibration equipment?	yes
Cost of calibration equipment	$7000.00

Figure 7-73. Computational Systems Inc. 8200 laser shaft alignment system. Photo courtesy CSi, Knoxville, TN.

Laser Shaft Alignment Systems	FixtureLaserAB
System Price Range	$9995.00 to $16500.00
System Information	
Total Unit Weight	28 lbs.
Alignment Operating Principle	reverse indicator
Intrinsically safe / explosion proof models?	yes
Shock Protection	not answered
Operating Temperature Range	not answered
Electromagnetic Field Protection	not answered
Environmental protection	not answered
Computer / Software	
Alignment computer & software included?	yes
Interface system with standard computers?	no (has RS232 output of raw data)
Software installation	not answered
Req'd to name one machine stationary?	yes
Number of possible movement solutions	1
Maximum number of machines in drive train	2
alignment jobs can be stored for future recall	not answered
battery type / number req's.	not answered / 1
battery life @ full charge	20 hrs.
optional AC power & requirements	90-220 VAC / 50-60 Hz
enter off-line to running movement?	not answered
Laser Emitter (Source)	
laser type	diode type
wavelength	not answered
beam shape	not answered
beam diameter at exit	not answered
beam diameter @ 3 ft.	not answered
beam diameter @ 10 ft.	not answered
beam diameter @ 30 ft.	not answered
max. beam power output	not answered
Laser Safety Class	not answered
max. recommended distance	not answered
Continuous or pulsed Beam?	not answered
beam pulse rate	not answered
laser power supplied by ...	not answered
cable length	not answered
Laser Detector (Receiver)	
Type	single axis photodiode (two)
detector measurement planes	1
detector measurement area	20mm x 20mm
number of detectors used	2
linearity	+/- 1%
accuracy	0.0005"
optical filtering?	not answered
beam position data transfer via ...	not answered
decector power supplied by ...	not answered
battery life @ full charge	n/a

Operational Info	
Shafts must be rotated together?	no
How are rotational positioned measured?	not answered
Max allowable rotational backlash	no effect
Can be used on vertical shafts?	yes
Additional Features & Accessories	
shaft clamping brackets inluded?	yes
minimum shaft diameter	.375"
maximum shaft diameter	17"
minimum sensor height from shaft surface	not answered
maximum sensor height from shaft surface	not answered
adapted for measuring OL2R?	not answered
Calibration	
Recommended calibration interval	none recommended
User can by calibration equipment?	not answered
Cost of calibration equipment	use shim stock to check

Figure 7-74. Fixturelaser AB laser shaft alignment system. Photo courtesy Fixturelaser AB, Sweden and Vibralign Inc., Richmond, VA.

Laser Shaft Alignment Systems	Hamar
System Price Range	$10900.00
System Information	
Total Unit Weight	48 lbs.
Alignment Operating Principle	modified double radial technique
Intrinsically safe / explosion proof models?	yes
Shock Protection	yes
Operating Temperature Range	not answered
Electromagnetic Field Protection	no effect
Environmental protection	yes
Computer / Software	
Alignment computer & software included?	yes
Interface system with standard computers?	yes
Software installation	floppy disk
Req'd to name one machine stationary?	yes
Number of possible movement solutions	1
Maximum number of machines in drive train	2
alignment jobs can be stored for future recall	700
battery type / number req's.	depends on computer supplied
battery life @ full charge	2 hours
optional AC power & requirements	depends on computer supplied
enter off-line to running movement?	yes
Laser Emitter (Source)	
laser type	not answered
wavelength	670 nm
beam shape	circular
beam diameter at exit	not answered
beam diameter @ 3 ft.	not answered
beam diameter @ 10 ft.	not answered
beam diameter @ 30 ft.	not answered
max. beam power output	not answered
Laser Safety Class	not answered
max. recommended distance	not answered
Continuous or pulsed Beam?	continuous
beam pulse rate	n/a
laser power supplied by ...	battery or cable
cable length	10 ft.
Laser Detector (Receiver)	
Type	4 axis photodiode
detector measurement planes	2
detector measurement area	10mm x 10mm
number of detectors used	2
linearity	+/- 2.5% over target area
accuracy	not answered
optical filtering?	not answered
beam position data transfer via ...	cable
decector power supplied by ...	cable
battery life @ full charge	n/a

Operational Info	
Shafts must be rotated together?	yes
How are rotational positioned measured?	switch & inclinometer
Max allowable rotational backlash	not answered
Can be used on vertical shafts?	yes
Additional Features & Accessories	
shaft clamping brackets inluded?	yes
minimum shaft diameter	.5"
maximum shaft diameter	unlimited
minimum sensor height from shaft surface	not answered
maximum sensor height from shaft surface	not answered
adapted for measuring OL2R?	yes
Calibration	
Recommended calibration interval	12 months
User can by calibration equipment?	yes
Cost of calibration equipment	$2700.00

Figure 7-75. Hamar Coupling1 laser shaft alignment system. Photo courtesy Hamar Laser, Danbury, CT.

Laser Shaft Alignment Systems	MMS
System Price Range	$11950.00
System Information	
Total Unit Weight	42 lbs.
Alignment Operating Principle	modified double radial technique
Intrinsically safe / explosion proof models?	no
Shock Protection	yes
Operating Temperature Range	32-185 deg F
Electromagnetic Field Protection	no effect
Environmental protection	sealed
Computer / Software	
Alignment computer & software included?	yes
Interface system with standard computers?	no
Software installation	not answered
Req'd to name one machine stationary?	yes
Number of possible movement solutions	1
Maximum number of machines in drive train	2
alignment jobs can be stored for future recall	not answered
battery type / number req's.	'C' size / 6
battery life @ full charge	6-8 hrs.
optional AC power & requirements	90-220 VAC / 50-60 Hz
enter off-line to running movement?	
Laser Emitter (Source)	
laser type	not answered
wavelength	830 nm
beam shape	not answered
beam diameter at exit	not answered
beam diameter @ 3 ft.	not answered
beam diameter @ 10 ft.	not answered
beam diameter @ 30 ft.	not answered
max. beam power output	< 1 mW
Laser Safety Class	Class IIIb
max. recommended distance	20 ft.
Continuous or pulsed Beam?	not answered
beam pulse rate	not answered
laser power supplied by ...	9 VDC alkaline battery or cable (AC)
cable length	not answered
Laser Detector (Receiver)	
Type	not answered
detector measurement planes	not answered
detector measurement area	not answered
number of detectors used	not answered
linearity	not answered
accuracy	.0001"
optical filtering?	not answered
beam position data transfer via ...	cable
decector power supplied by ...	cable
battery life @ full charge	n/a

Operational Info	
Shafts must be rotated together?	yes
How are rotational positioned measured?	inclinometer
Max allowable rotational backlash	0 deg.
Can be used on vertical shafts?	yes
Additional Features & Accessories	
shaft clamping brackets inluded?	yes
minimum shaft diameter	.5"
maximum shaft diameter	36" (optional equip.)
minimum sensor height from shaft surface	4.5"
maximum sensor height from shaft surface	8"
adapted for measuring OL2R?	yes (special brackets req'd)
Calibration	
Recommended calibration interval	none
User can by calibration equipment?	n/a
Cost of calibration equipment	n/a

Figure 7-76. MMS laser shaft alignment system. Photo courtesy MMS, Lombard, IL.

Bibliography

King, W. F., Peterman, J. E., "Align Shafts, Not Couplings!", Allis Chalmers Electrical Review, 2nd Quarter, 1951, pgs. 26-29.

Samzelius, J. W., "Check Points for Proper Coupling Alignment", Plant Engineering, June 1952, pgs. 92-95.

Brotherton M. , **Masers and Lasers**, McGraw-Hill Book Co., 1964, LCCCN 63-23249.

Yarbrough, C. T., "Shaft Alignment Analysis Prevents Shaft and Bearing Failures", Westinghouse Engineer, May 1966, pgs. 78-81.

Nelson, Carl A., "Orderly Steps Simplify Coupling Alignment", Plant Engineering, June 1967, pgs. 176-178.

Beckwith, T. G., Buck, N. Lewis, **Mechanical Measurements**, Addison-Wesley Publishing Co., 1969, LCC# 70-85380.

Blubaugh, R. L., Watts, H. J., "Aligning Rotating Equipment", Chemical Engineering Progress, Vol. 65, No. 4, April 1969, pgs. 44-46.

Dreymala, James, **Factors Affecting and Procedures of Shaft Alignment**, Technical and Vocational Dept., Lee College, Baytown , Texas, 1970.

Dodd, V.R., **Total Alignment**, The Petroleum Publishing Co., Tulsa, Okla., 1975.

Doeblin, Ernest, **Measurement Systems : Application and Design**, McGraw Hill Book Co., 1975, ISBN # 0-07-017336-2.

"Two Step Dial Indicator Method", bulletin no. MT-SS-04-001, Rexnord, Thomas Flexible Coupling Div., Warren, Pa., 1979.

Durkin, Tom, "Aligning Shafts, Part I - Measuring Misalignment", Plant Engineering, Jan. 11, 1979.

Murray, M.G., "Choosing and Alignment Measurement Setup", Murray and Garig Tool Works, Baytown, Texas, personal correspondence, Oct. 12, 1979.

Piotrowski, John D., "The Graphical Alignment Calculator", Machinery Vibration Monitoring and Analysis, Vibration Institute, Clarendon Hills, Ill., 1980.

Beynon, J. D., Lamb, David Robert, **Charge Couple Devices and Their Applications**, 1980, McGraw-Hill Book Co., ISBN # 0-07-084522-0.

Piotrowski, John D.,"Alignment Techniques", Proceedings Machinery Vibration Monitoring and Analysis Meeting, New Orleans, LA., June 26-28, 1984, Vibration Institute, Clarendon Hills, Ill.

Murray, M. G., "OPTALIGN - Laser - Optic Machinery Alignment System - Report Following Four Month Test", Murray and Garig Tool Works, Baytown, Texas, April 2, 1985.

Malak, Stephen P., Solomon, Donald W., "An Analysis of Non-Rotated 3-Dimensional Shaft Alignment Instrumentation Compared with Rotated 2-Dimensional Shaft Alignment Measurement Methods Obtained Currently by Dial Indicator Techniques", SMI-LI-TR1.1, August, 1985, Lineax Instruments, SMI Spring Mornne, Inc.

Murray, M. G., **Alignment Manual for Horizontal, Flexibly Coupled Rotating Machines**, Murray and Garig Tool Works - Third Edition, Baytown, Texas, April 21, 1987.

66 Centuries of Measurement, Sheffield Corp., 1987, LCCC# 530.81.

Franklin, Douglas E., "Active Alignment", presented at the 10th Biennial Machinery Dynamics Seminar, The National Research Council Canada, September 1988.

Carman, David B., "Determining Small Planar Rotations of Cylinders Using Computer Vision", doctoral thesis, 1988.

Carman, David B., "Measurement of Shaft Alignment Using Computer Vision", doctoral thesis, 1989.

"Boiler and Machinery Engineering Report - Shaft Alignment for Rotating Machinery", Section 4.0, # 4.26, October, 1989, American Insurance Services Group, Inc., New York, NY.

Evans, Galen, Casanova, Pedro, Azcarate, Ana Maria, **The Optalign Training Book**, Ludeca Inc. Miami, FL, 1990, Catalog # 01-705-01.

Techniques for Digitizing Rotary and Linear Motion - Encoder Division, 4th printing, Dynamics Research Corp., Wilmington, MA, 1992.

Hecht, Jeff, **The Laser Guidebook**, McGraw-Hill Book Co., 1992, ISBN # 0-07-027737-0.

Mims, Forrest M. III, **Getting Started in Electronics**, 1993, Tandy Corp. Catalog # 276-5003.

Brook, Ken, personal correspondence for partial arc mathematics, June, 1994.

Heid, Jim, "Photography Without Film", MacWorld, September 1994, pgs. 140-147.

Teskey, W. F., "Dynamic Alignment Project", University of Calgary, November 18, 1994.

Laser Diode User's Manual, Ref. No. HT519D, Sharp Electronics Corporation, 22-22, Nagaike-Cho, Abeno-Ku, Osaka, 545, Japan.

Optoelectronics Components Catalog, UDT Sensors Inc., 12525 Chadron Ave., Hawthorne, CA 90250.

8
Graphing / Modeling Techniques to Determine Proper Machinery Movement

Even for people who align rotating machinery on a regular basis, it is very difficult to visualize exactly where the centerlines of rotation are by just looking at dial indicator, laser, or optical encoder measurement data. Your goal is to position each machine so that both shafts run in the same axis of rotation and you invariably begin to wonder ... is one shaft higher or lower than the other one, is it to the west or is it to the east, and if so how much?

Still today, many people who align rotating equipment will do 'trial and error' realignment. They install some shim stock under the feet and move the machinery sideways a little bit, take another set of readings, and see if the measurements get any better. This sophisticated technique is called 'trial and error alignment' and will eventually produce frustration, anxiety, and anger if continued for long periods of time. There happens to be a much better way to determine how to accurately position the machinery instead of guessing.

There is a geometric relationship that exists between the size of the machinery (i.e. where the foot bolts are located), where the shaft position measurements are taken, and the shaft measurements themselves. Once all of the measurements have been taken, the movement solutions can be mathematically or graphically solved.

This chapter will explain how to take the measurement data collected when using these techniques to ascertain exactly where the shaft centerlines are and show you how to calculate the moves needed to correct a misalignment condition on the machinery. Many people consider this to be the most important part of the overall alignment process, but it in fact is just one of many important parts. There is nothing worse than trying to align machinery with bad measurement data, or aligning machines with bent shafts or uncorrected 'soft foot' (and on and on).

Mathematical relationship in machinery alignment

Figure 8-1 shows the mathematical relationship between the machinery dimensions and the dial indicator readings captured using the face-rim and reverse indicator technique. The equations will solve for the moves that need to be made to correct the misalignment condition (ie. bring the shafts into a collinear relationship when off-line) on one or the other machine case. It's an either or condition ... if you decide to keep the driver stationary, you solve for the moves on the driven machine or vice versa. This is often referred to as the 'Stationary - Movable' alignment concept and is not recommended for reasons explained later on in the chapter. However, the vast majority of machinery alignment systems shown in Chapter 7 use this limited technology despite the fact that the graphing / modeling methods and the 'overlay line' concept have been in use since the late 1970's.

Face - Rim Method Mathematics

where :
A, B, C, D, E = distances shown (in.)
H = diameter of face readings (in.)
F = face reading difference (from top to bottom or side to side in mils)
Y = rim reading difference (from top to bottom or side to side in mils)

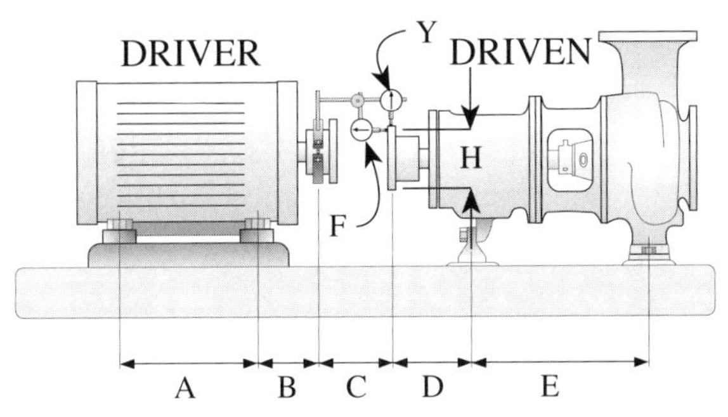

$$\text{inboard feet of DRIVER} = \frac{F\,(B+C)}{\sqrt{H^2 + F^2}} - (Y) \qquad \text{inboard feet of DRIVEN} = \frac{F\,D}{\sqrt{H^2 + F^2}} + (Y)$$

$$\text{outboard feet of DRIVER} = \frac{F\,(A+B+C)}{\sqrt{H^2 + F^2}} - (Y) \qquad \text{outboard feet of DRIVEN} = \frac{F\,(D+E)}{\sqrt{H^2 + F^2}} + (Y)$$

Reverse Indicator Method Mathematics

where :
A, B, C, D, E = distances shown (in.)
X = DRIVER rim reading difference (from top to bottom or side to side in mils)
Y = DRIVEN rim reading difference (from top to bottom or side to side in mils)

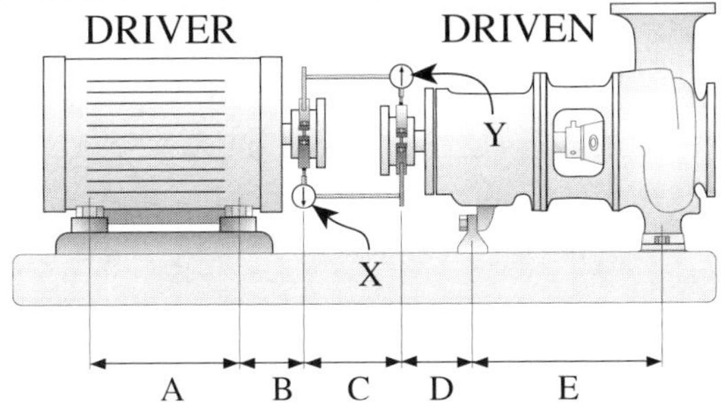

$$\text{inboard feet of DRIVER} = \frac{(B+C)\,(X+Y)}{C} - (Y) \qquad \text{inboard feet of DRIVEN} = \frac{(C+D)\,(X+Y)}{C} - (X)$$

$$\text{outboard feet of DRIVER} = \frac{(A+B+C)\,(X+Y)}{C} - (Y) \qquad \text{outboard feet of DRIVEN} = \frac{(C+D+E)\,(X+Y)}{C} - (X)$$

Figure 8-1. Equations for determining moves on one or the other machine case for face-rim and reverse indicator readings.

Graphing / modeling alignment techniques

Regardless of the device used to measure the positions of the centerlines of rotation (be it dial indicators, optical encoders, lasers, and the like), virtually every alignment measurement system utilizes one (or a slight variation) of the following measurement principles:

Reverse Indicator Method
Face and Rim Method
Double Radial Method
Shaft to Coupling Spool Method
Face - Face Method

In order to understand how each of these techniques work, dial indicator readings will be used to illustrate how each method can be graphed or modeled to determine the relative positions of each shaft. All of these techniques can be graphed or modeled by hand. Typically, all you need is some graph paper (20 divisions per inch is a good choice), a straightedge, and a pencil (with an eraser just in case!). You don't even really need graph paper; all that is required is a scaled grid or some sort of measurement device like a ruler.

The graphical shaft alignment modeling techniques use two different scaling factors. One scaling factor proportions the overall dimensions of the machinery drive 'train' to fit within the boundaries of the graph paper and another different scaling factor is used to exaggerate the misalignment between the machinery shafts. If we limit our discussion to horizontally mounted rotating machinery drive trains for now, there will be two graphs that need to be drawn. As depicted in figure 8-2, one graph will show the exaggerated positions of each shaft in the SIDE VIEW, illustrating the up and down or vertical positions of the machinery. Another graph will be constructed in the TOP VIEW that will illustrate the side to

Combining the SIDE and TOP VIEWS

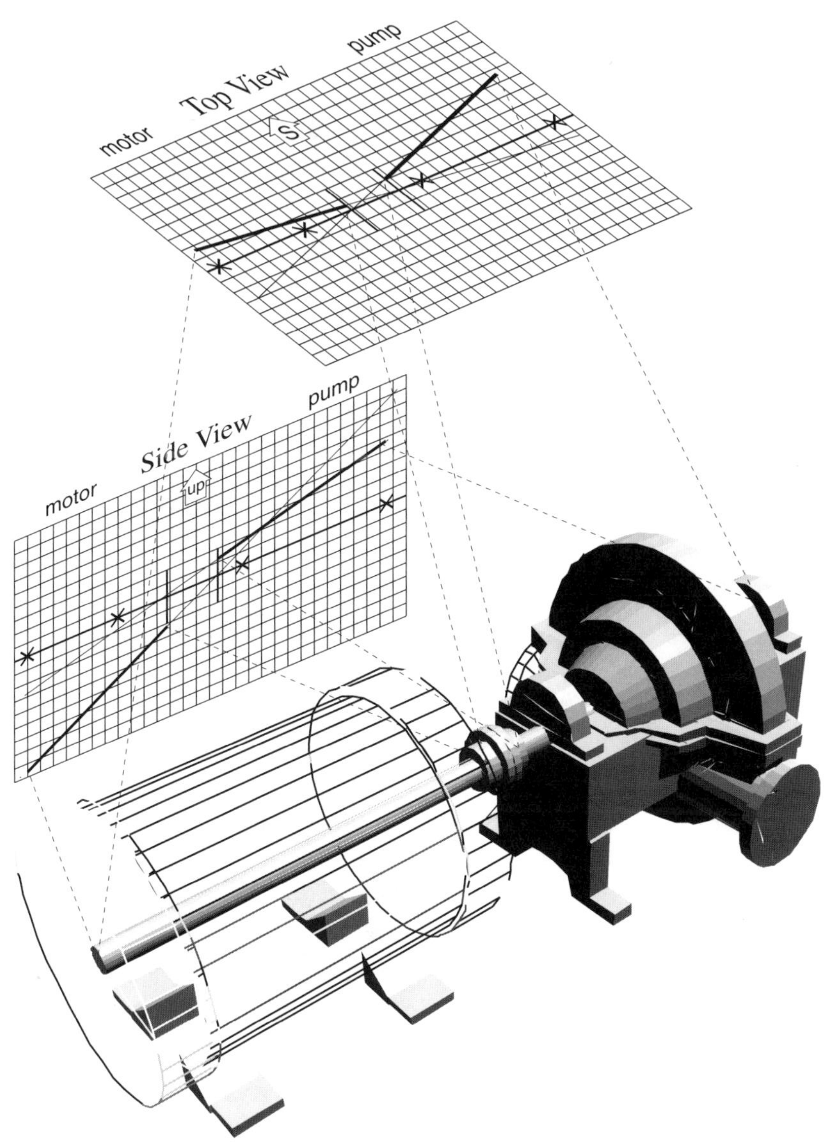

Figure 8-2. Splitting the misalignment condition on a drive train into two different views.

side or lateral positions of the machinery.

Once the relative positions of the machinery shafts are constructed on the graph, a wide variety of different solutions can be determined to bring the two centerlines of rotation in line with each other. The benefit of modeling rotating machinery is to visually represent an exaggerated, but accurately scaled picture of the misalignment condition, so you can easily ascertain what positions the machinery could be moved to that would make it easy to align the shafts within the boundary conditions imposed by the baseplate / foundation and the allowable lateral restrictions between the machinery casing bolts and the holes drilled in the machine cases (aka 'bolt bound' conditions).

Additionally, the modeling technique can include other measurement parameters such as improperly fitted piping, air gap clearances between stators and armatures, and fan rotor to shroud clearances. Finally, the graph is a permanent record of the alignment of the machinery and can be kept for future reference.

In summary, this chapter will review the following key steps in correcting the misalignment situation (refer to step 6 in Chapter 1):

1. determine the current positions of the centerlines of rotation of all the machinery
2. observe any movement restrictions on the machines at the control / adjustment points (usually the machinery feet / hold down bolts)
3. plot the restrictions on the graph / model
4. determine the moves for either or both of the machinery casings on the graph / model that will be feasible to perform.

The modeling techniques shown in the chapter are by far the most efficient and accurate methods to align rotating machinery and are among the most important aspects of mastering rotating machinery alignment principles.

We will first begin by illustrating how to construct the relative positions of the two centerlines of rotation and then show how you can determine the wide variety of movement options available to you when repositioning misaligned machinery.

Modeling Reverse Indicator Method Using the the 'Point to Point' Technique

Perhaps the easiest modeling technique to learn is the point to point reverse indicator modeling method and will, therefore, be the first one illustrated.

There are eight pieces of information that you need to properly construct the shaft positions using this technique:

1. The distance from the outboard to inboard feet (bolting planes) of the first machine.
2. The distance from the inboard bolting plane of the first machine to the point on the shaft where the bracket is being held in place.

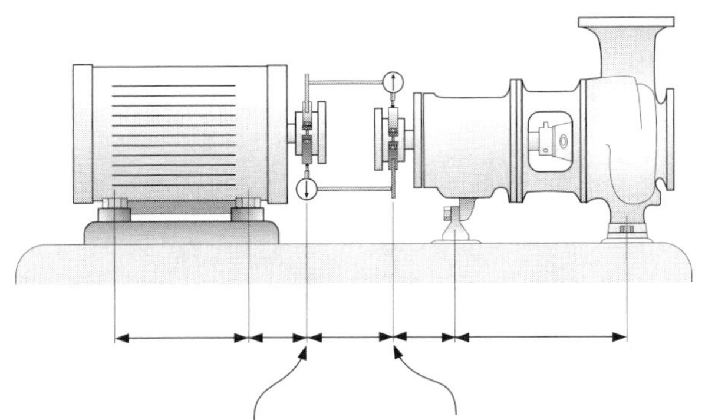

Figure 8-3. Measuring distances along the length of the machinery for the reverse indicator graphing technique.

3. The distance from where the bracket is being held in place to the point where the dial indicator is capturing the rim readings on the first machine. Note that this distance could be zero if you are using a symmetrical arrangement where you are clamping and reading at the same points on each shaft.
4. The distance from where the dial indicator is capturing the rim readings on the first machine to the point where the dial indicators are capturing the rim readings on the second machine.
5. The distance from where the dial indicator is capturing the rim readings on the second machine to the point where the bracket is being held in place. Note that this distance could be zero if you are using a symmetrical arrangement where you are clamping and reading at the same points on each shaft.
6. The distance from where the bracket is being held in place to the inboard bolting plane of the second machine.
7. The distance from the inboard to outboard feet (bolting planes) of the second machine.
8. The eight dial indicator readings taken at the top, bottom, and both sides on both shafts after compensating for sag (i.e. what a perfect, 'no sag' bracket system would have measured).

From the standpoint of geometry, the reverse indicator method measures shaft centerline deviations at two 'slices' in space at a known distance apart.

Accurately scale the distances along the length of the drive train onto the graph centerline as shown in figure 8-4 and prepare two graph sheets, one for the side view and one for the top view.

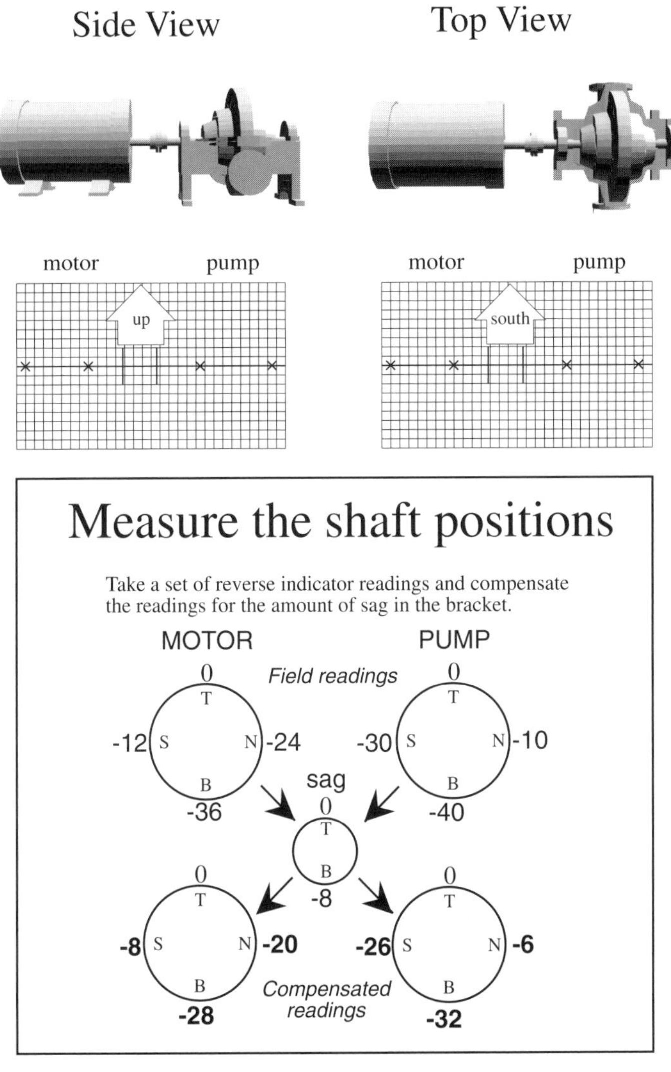

Figure 8-4. Prepare two graphs to show the side and top views of the centerlines of rotation.

Rim Readings are always twice the offset amount

Anytime a rim or circumferential reading is taken the amount measured from one side to the other side of the shaft (180 degrees of rotation) is twice the amount of the actual distance between the centerlines of rotation at that point. Figure 8-5 shows why that happens (also see figures 7-49 and 7-50).

Procedure for plotting the point to point reverse indicator technique:

1. Start with the top to bottom or side to side dial indicator readings on the shaft where the largest top to bottom or side to side reading occurred (this will help you pick the best scaling factor for the entire graph).

Why the dial indicator reads twice the actual amount the centerlines are really offset.

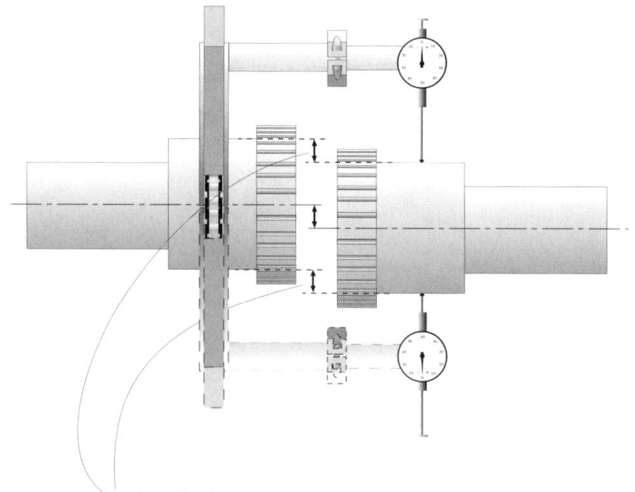

The dial indicator 'sees' both of these distances when it traverses from one side to the other.

Figure 8-5. Rim / peripheral / circumferential readings are always twice the amount of actual shaft centerline offset.

3. At the intersection of the graph centerline and the point where the dial indicator captured the larger of the two bottom (or side) readings, plot a point above or below this intersection one-half of the top to bottom or side to side dial indicator reading. If the bottom (or side) reading was negative, place a point half of the bottom (or side) reading

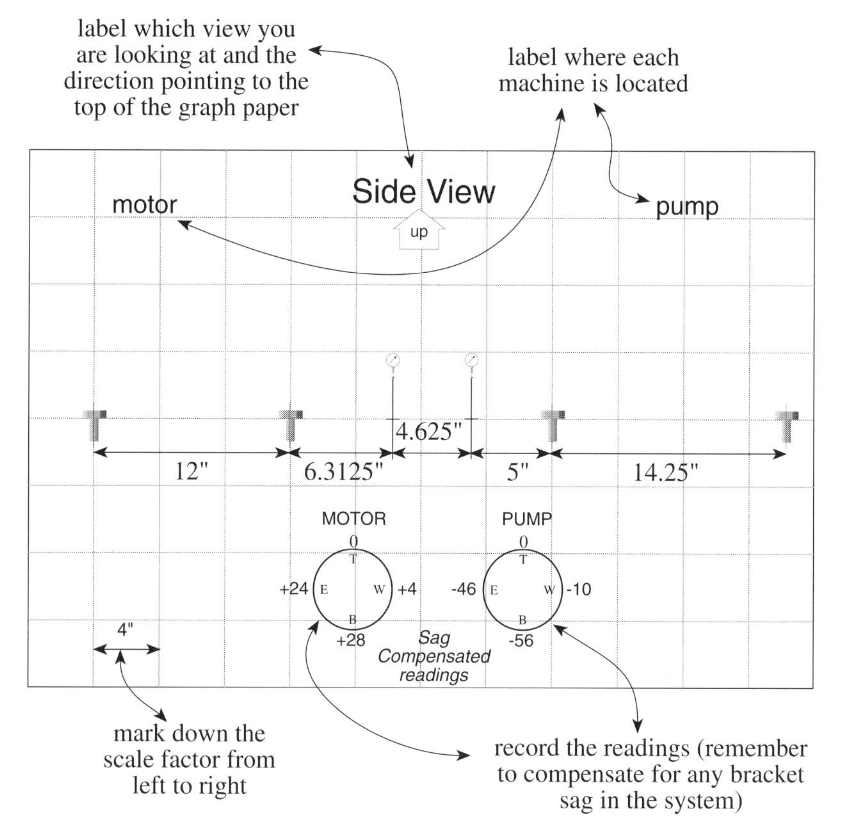

Figure 8-6. Setting up the graph / model.

from the graph centerline toward the top of the graph. If the bottom (or side) reading was positive, place a point half of the bottom (or side) reading from the graph centerline toward the bottom of the graph (the same as in the point to point modeling techniques). Lay a straightedge from the point on the graph centerline where the bracket was held through the point on the graph where the dial indicator captured the reading. Draw a line from the point on the graph where the dial indicator captured the reading to the outboard end of that shaft. Remember, whatever shaft the dial indicator has captured the readings on, that's the shaft that will be drawn on the graph paper.

4. Next, at the intersection of the graph centerline and the point where the dial indicator captured the smallest reading, plot a point above or below this intersection one-half of the top to bottom or side to side dial indicator reading. If the bottom (or side) reading was negative, place a point half of the bottom (or side) reading from the graph centerline toward the top of the graph. If the bottom (or side) reading was positive, place a point half of the bottom (or side) readings from the graph centerline toward the bottom of the graph (the same as in the point to point modeling techniques). Lay a straightedge from the point on the graph centerline where the bracket was held through the point on the graph where the dial indicator captured the reading. Draw a line from the point on the graph where the dial indicator captured the reading to the outboard end of that shaft.

Figures 8-7, 8-8, and 8-9 illustrate an example of this modeling method.

Notice that there is a consistency to this plotting technique. If the top to bottom or side to side dial indicator reading is negative, plot half of the reading toward the top of the graph paper, for either shaft. If the top to bottom or side to side dial indicator reading is positive, plot half of the reading toward the bottom of the graph paper, for either shaft.

Step 2 - To select the appropriate up and down scale factor, start with the shaft that had the larger of the two bottom readings. In this example, it is the pump shaft. When you pick an up and down scale factor that will insure that the entire pump shaft fits within the boundaries of the graph paper ... the motor shaft (which has the smaller bottom reading) will also typically fit within the boundaries of the graph paper.

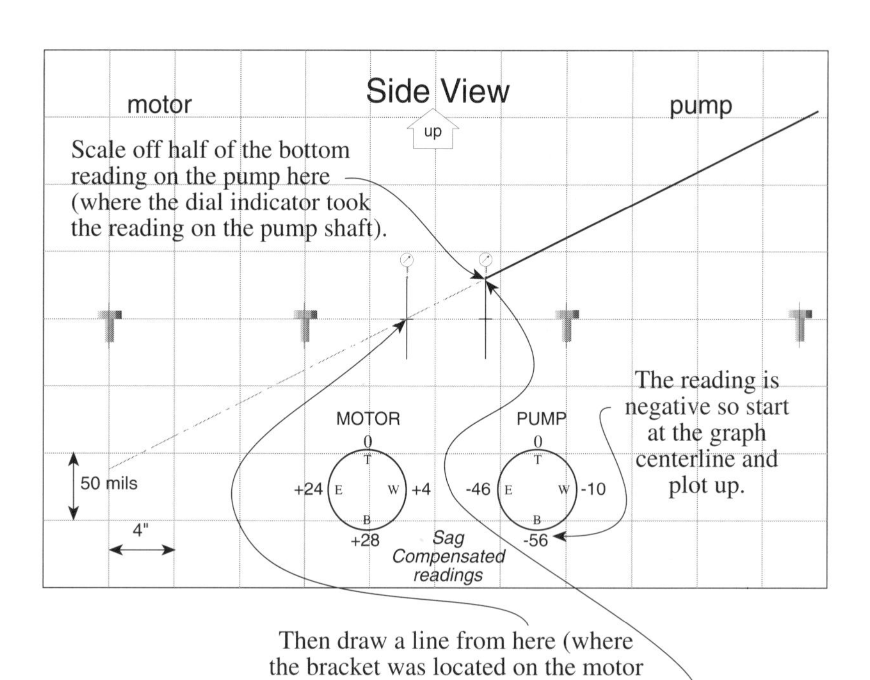

Figure 8-7. Drawing the first shaft on the graph.

Step 3 - Construct the position of the motor shaft based on the bottom reading captured by the dial indicator as shown.

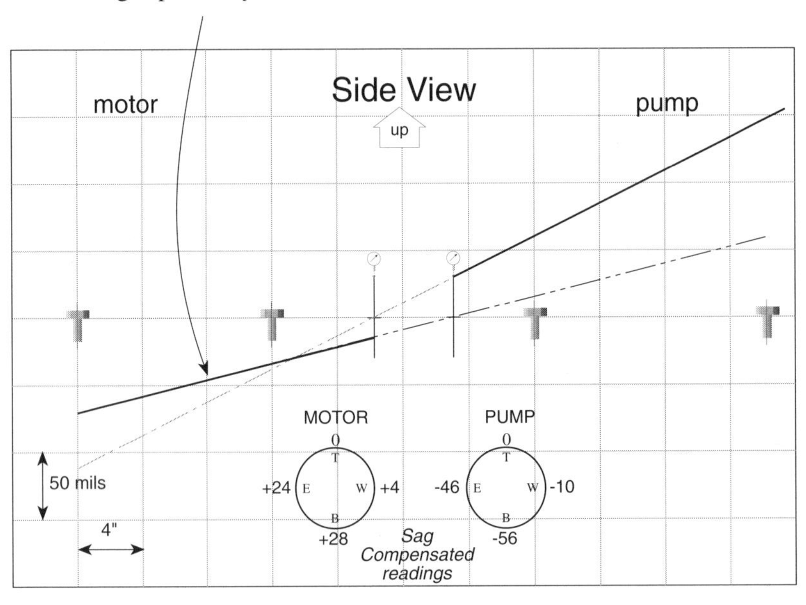

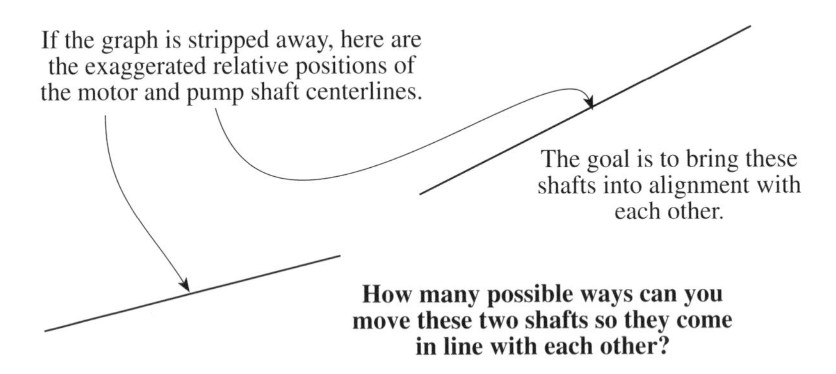

Answer : There are an *infinite number of ways* to align two shafts when you consider that they are both movable!

Figure 8-8. Drawing the second shaft on the graph.

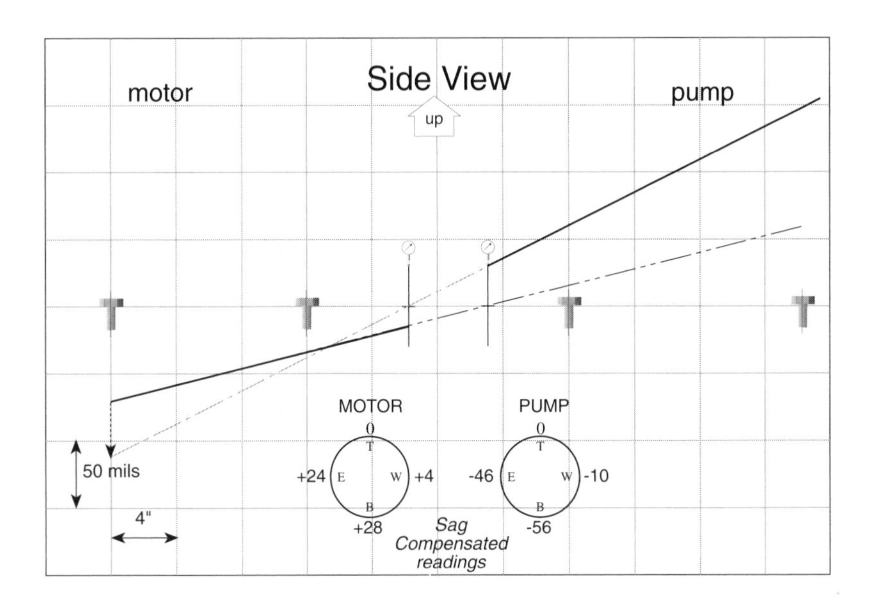

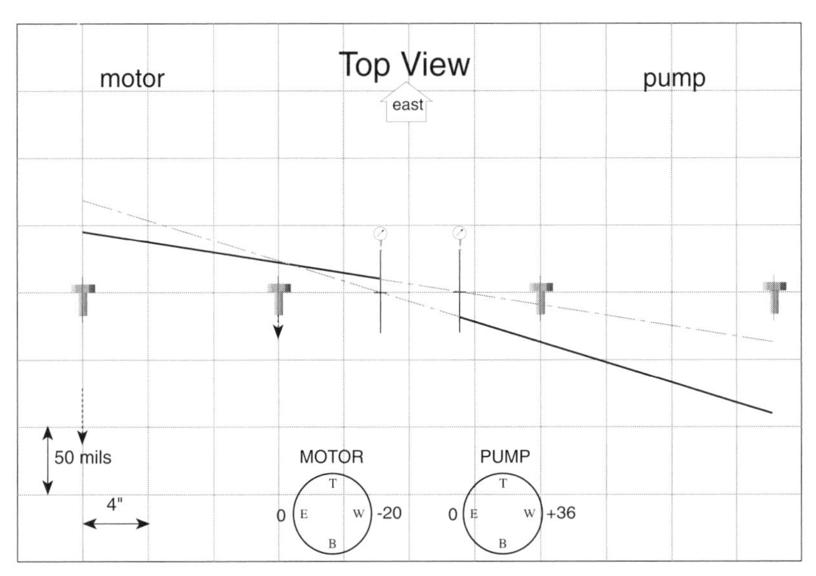

Figure 8-9. Reverse indicator point to point modeling technique example viewing the exaggerated shaft centerline positions in the side and top views.

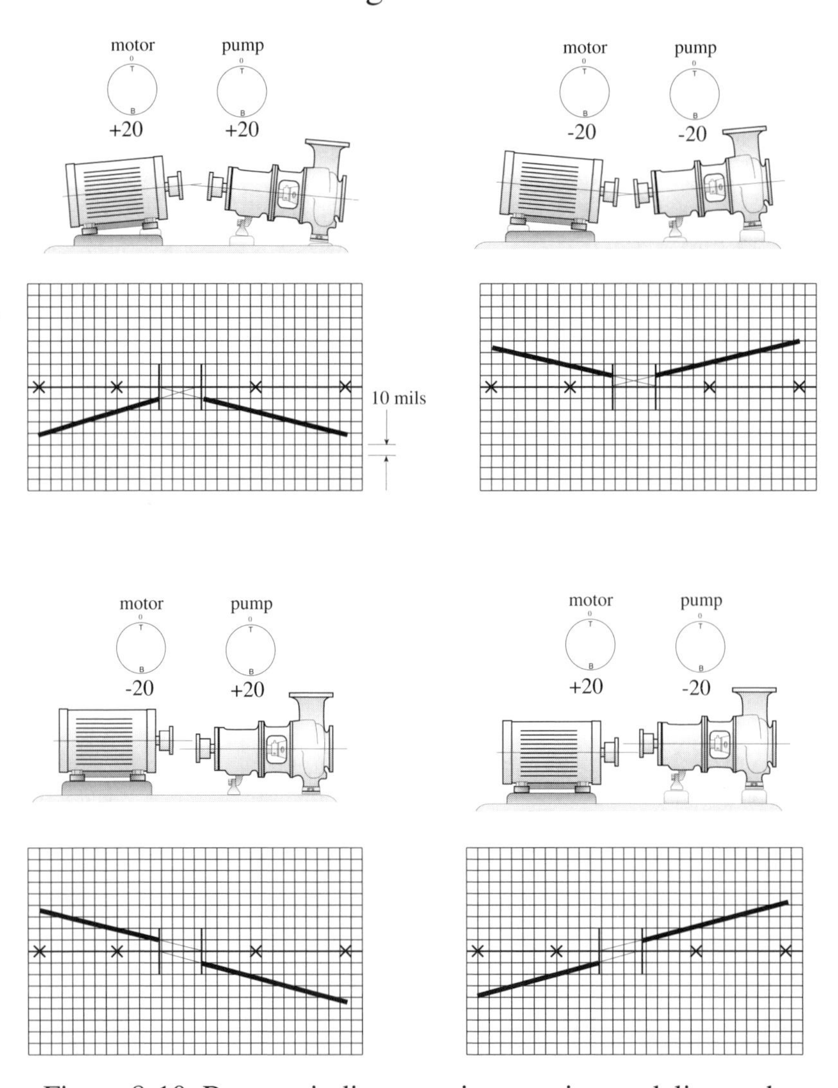

Figure 8-10. Reverse indicator point to point modeling technique examples showing the positions of the shafts based on the sign (+/-) value of the bottom readings.

Alter the graph setup for asymetric bracket and indicator positions

The ideal bracket and indicator placement has the indicators in the same plane as the bracket clamping position.

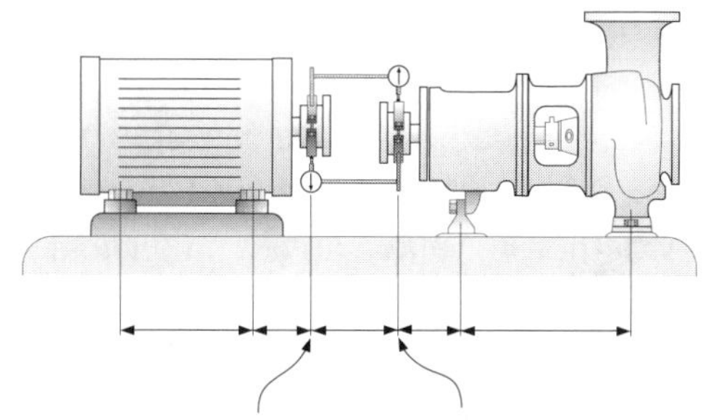

An asymetric bracket and indicator setup requires that you position the centerlines based on the points from where you are clamped on one shaft (where the bracket is attached) to the point where the dial indicators are taking their measurements on the other shaft.

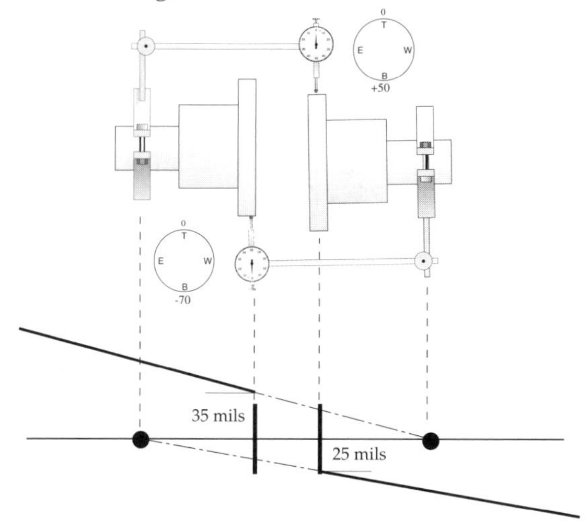

Figure 8-11. When performing the reverse indicator point to point modeling technique, the bracket positions should be plotted correctly for increased graphing accuracy.

Modeling Reverse Indicator Method Using the Line to Points Technique

There is an alternative method to model reverse indicator readings. There are two advantages to this technique as opposed to the point to point method:

• It is somewhat easier to model multiple element drive trains where reverse indicator readings were captured at two or more flexible couplings (this will be covered in Chapter 12).

• Regardless of whether you have an asymmetrical or symmetrical bracket arrangement, the points where the brackets are being clamped to the shaft are not relevant; only the points where the dial indicator readings are being captured are required.

There are six pieces of information that you need to properly construct the shaft positions using this technique:

1. The distance from the outboard to inboard feet (bolting planes) of the first machine.
2. The distance from the inboard bolting plane of the first machine to the point on the shaft where the dial indicator is capturing the rim readings on the first machine.
3. The distance from where the dial indicator is capturing the rim readings on the first machine to the point where the dial indicators are capturing the rim readings on the second machine.
4. The distance from where dial indicator is capturing the rim readings on the second machine to the inboard bolting plane of the second machine.
5. The distance from the inboard to outboard feet (bolting planes) of the second machine.
6. The eight dial indicator readings taken at the top, bottom, and both sides of both shafts after compensating for sag (i.e. what a perfect, 'no sag' bracket system would have measured).

Accurately scale the distances along the length of the drive train onto the graph centerline as shown in figures 8-11 and 8-12.

Procedure for plotting the line to point reverse indicator technique:

1. Select one of the two machinery shafts and draw one of those shafts on top of the graph centerline.

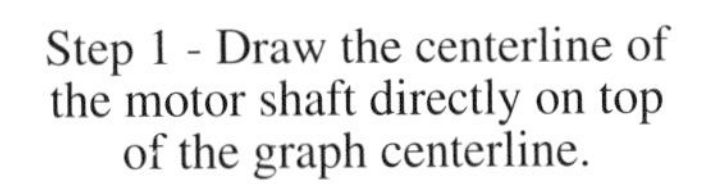

Step 2 - Mark a scaled point one-half the bottom reading above the graph centerline (since the reading was negative ... it looks high) where the dial indicator was taking readings on the pump.

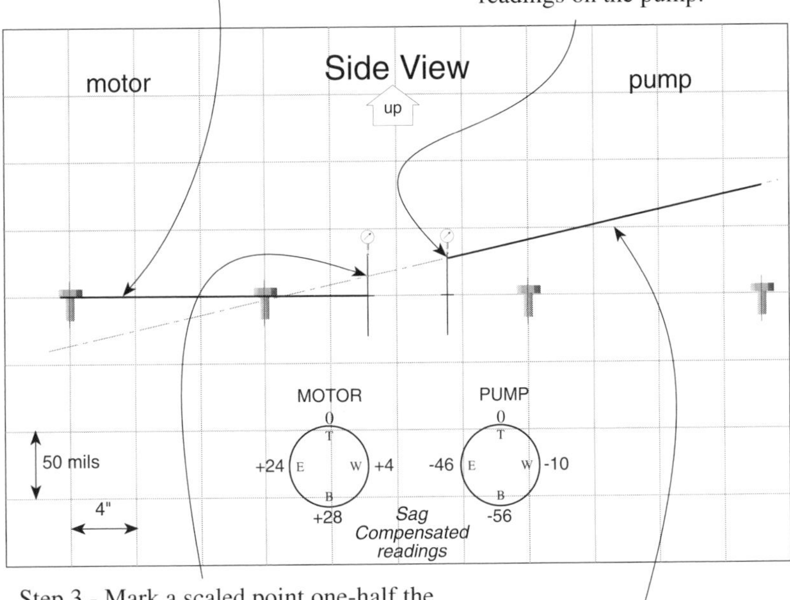

Step 3 - Mark a scaled point one-half the bottom reading above the graph centerline (since the reading was positive ... it looks low from the line of sight of the pump) where the dial indicator was taking readings on the motor.

Step 4 - Draw a line through the two marked points. This is the line of sight of the pump shaft.

Figure 8-12. Reverse indicator line to point modeling technique example where the motor shaft is placed directly on top of the graph centerline.

2. Start with the top to bottom or side to side dial indicator readings on the other shaft (i.e. the one you did not draw on the graph centerline).
3. Plot the other shaft centerline position by starting at the intersection of the graph centerline and the point where the

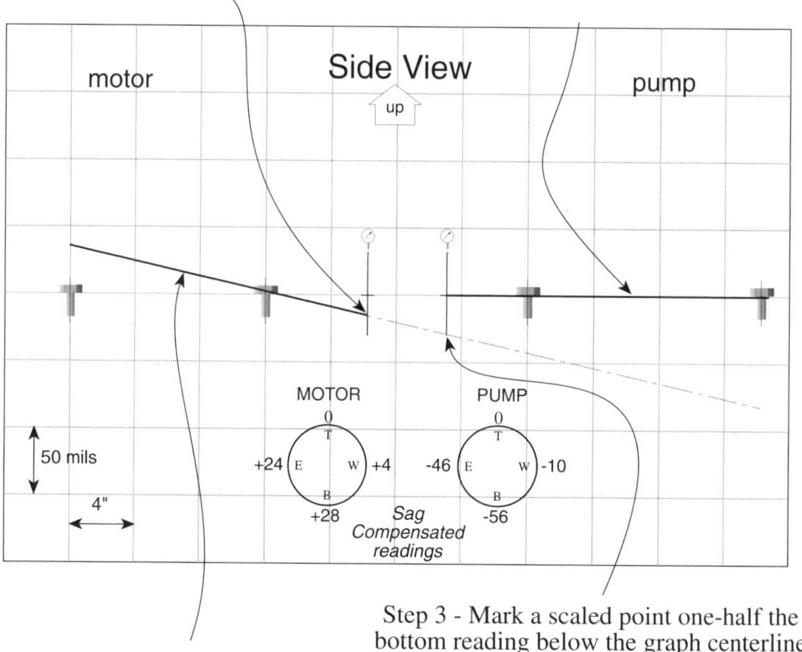

Figure 8-13. Reverse indicator line to point modeling technique example where the pump shaft is placed directly on top of the graph centerline.

dial indicator was capturing the readings on the other shaft. If the bottom (or side) reading was negative, place a point half of the bottom (or side) reading from the graph centerline toward the top of the graph. If the bottom (or side) reading was positive, place a point half of the bottom (or side) reading from the graph centerline toward the bottom of the graph (the same as in the point to point modeling techniques). Do not draw any lines yet!

4. Next, start at the intersection of the graph centerline and the point where the dial indicator was capturing the readings on the shaft that was drawn on top of the graph centerline. If the bottom (or side) reading was negative, place a point half of the bottom (or side) reading from the graph centerline toward the bottom of the graph. If the bottom (or side) reading was positive, place a point half of the bottom (or side) reading from the graph centerline toward the top of the graph (opposite of the point to point modeling technique).
5. These two points marked on the graph at the dial indicator reading points define the line of sight (i.e. the centerline of rotation) of the other shaft. Draw a straight line through these two points from the coupling end to the outboard end of the other shaft.

Modeling the Face and Rim Method

The Face and Rim method measures an offset and an angle of another shaft's centerline of rotation with respect to the line of sight of a 'reference shaft'.

To graph the face-peripheral method you need to have a clear piece of plastic with a 'T' inked onto the plastic similar to what is shown in figure 8-14. The 'T' bar overlay will represent the shaft where the dial indicators are capturing the readings.

There are nine pieces of information that you need to properly construct the shaft positions using this technique:

1. Which shaft will the bracket be attached to and on which shaft will the dial indicators be taking readings?
2. The distance from the outboard to inboard feet (bolting planes) of the machine where the bracket is attached.
3. The distance from the inboard bolting plane of the machine where the bracket is attached to the point on the shaft where the bracket is being held in place.
4. The distance from where the bracket is being held in place to the point on the other shaft where the dial indicators are capturing the face and rim readings.
5. The distance from where the dial indicators are capturing the face and rim readings to the inboard bolting plane of that machine.
6. The distance from the inboard to outboard feet (bolting planes) of the machine where the dial indicators are capturing the readings.
7. The diameter the face readings are being taken on.
8. Whether the face readings are being taken on the 'front' or 'back' side of the coupling hub or face measurement surface.
9. The eight dial indicator readings taken at the top, bottom, and both sides of the rim and face measurement points.

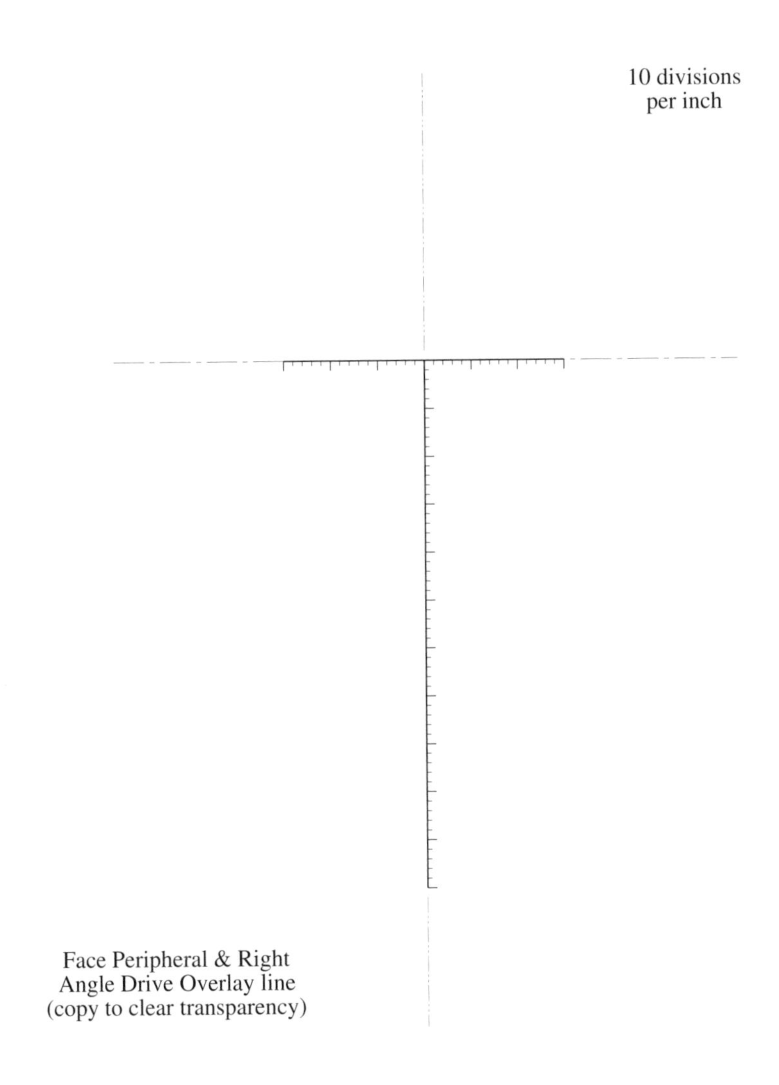

Figure 8-14. The 'T' bar overlay. This figure can be photocopied and scaled 200% to be used for face-rim graphing / modeling.

Scale the distances onto a piece of graph paper and scale the diameter of the face reading onto the 'T' shaped clear overlay. Be sure that you use the same scale for all these dimensions. The top part of the 'T' represents the face of the shaft you are taking readings on and the 'base' of the 'T' represents the centerline of rotation of the shaft.

Procedure for plotting the Face-Rim Technique :

1. Draw the shaft where the alignment bracket is attached directly on top of the graph centerline.
2. Next, position the clear 'T' bar overlay to reflect the readings captured on the rim or perimeter of the other shaft. If the bottom (or side) rim reading was negative, slide the 'T' bar toward the top of the graph paper so that the base of the 'T' is one-half of the rim reading from the graph centerline. If the bottom (or side) rim reading was positive, slide the 'T' bar toward the bottom of the graph paper so that the base of the 'T' is one-half of the rim reading

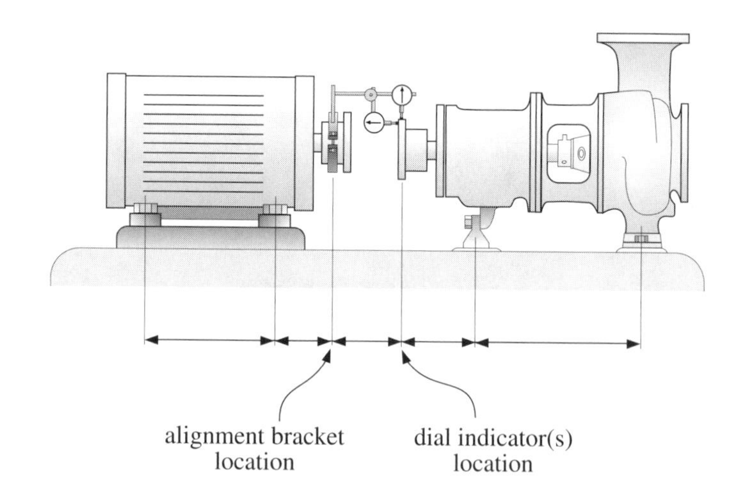

Figure 8-15. Dimensions needed for face-rim graph setup.

from the graph centerline.

3. Pivot the ‘T’ bar overlay to reflect the face readings captured. There are several ways to accomplish this. You can pivot or rotate the ‘T’ bar from the upper point on the top of the ‘T’ bar where the dial indictor was zeroed and move the bottom point.

After you have properly positioned the ‘T’ overlay, determine the best realignment move to bring the two centerlines into a collinear relationship based on your knowledge of the restrictions at the machinery feet.

The 'T' Bar Overlay

The 'T' bar overlay represents the shaft where the dial indicators are capturing readings. In the examples shown here, the 'T' bar represents the pump shaft since the dial indicators are taking readings on the pump with the bracket attached to the motor. In this arrangement, you are trying to 'see' where the pump shaft is with respect to a point on the motor shaft centerline.

Remember where the dial indicators are taking their readings. There would be nothing wrong with mounting the bracket to the pump shaft and capturing readings on the motor shaft. If the bracket was attached to the pump shaft, then the 'T' bar would have to represent the motor shaft position.

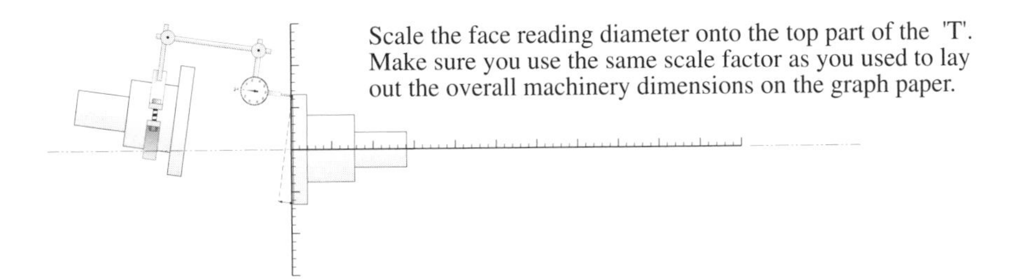

Figure 8-16. Scale the diameter the face readings were captured on to the top of the 'T' on the T-bar overlay.

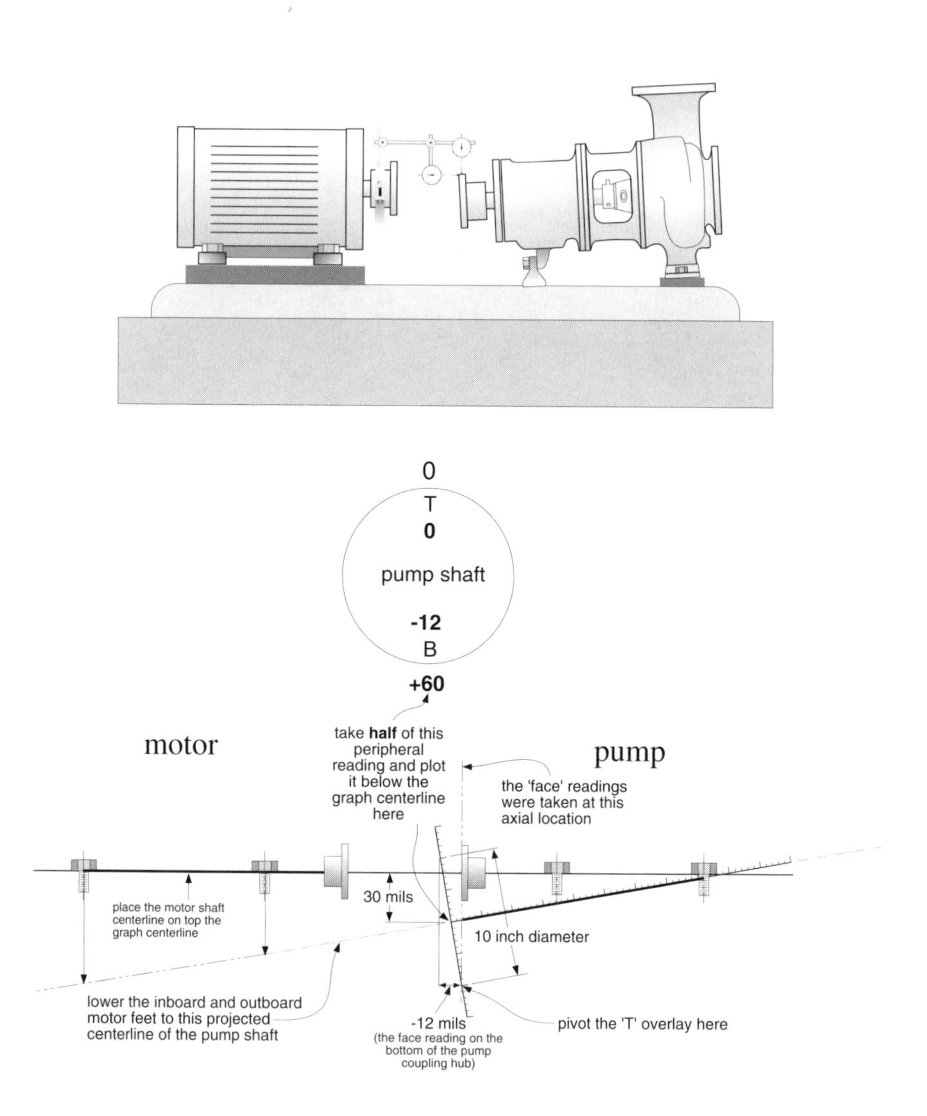

Figure 8-17. Face-rim example in the side view.

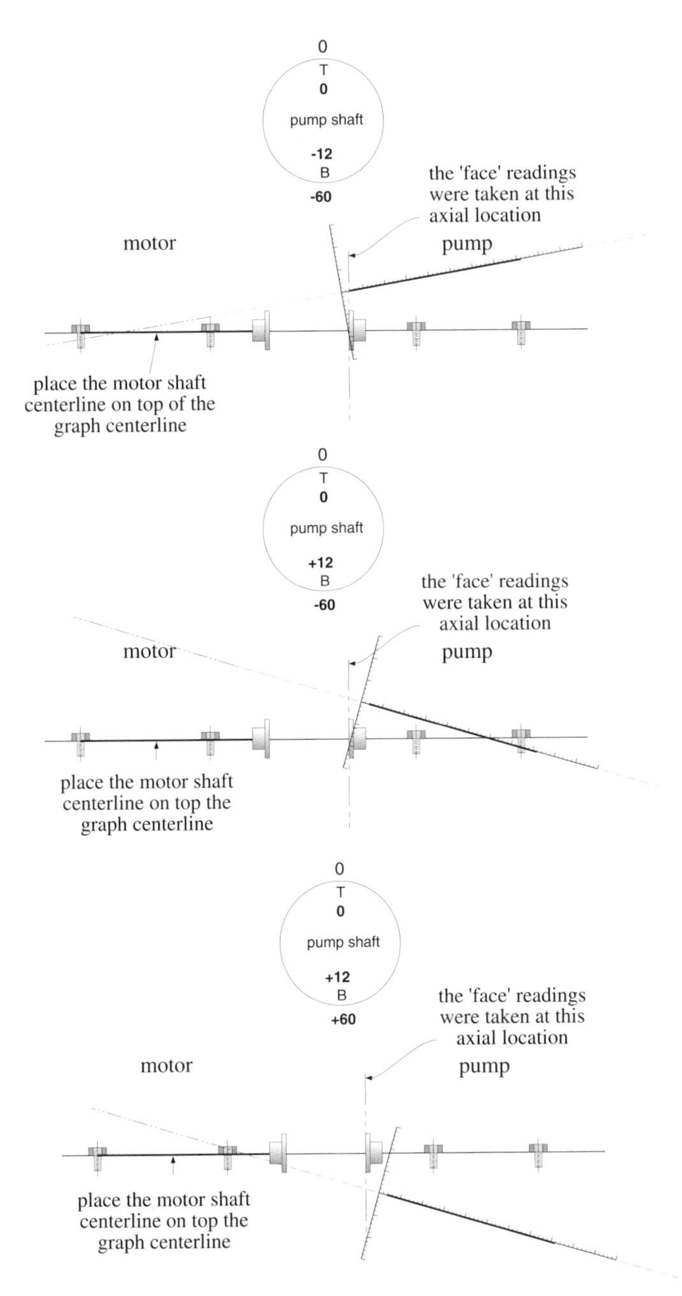

Figure 8-18. More face-rim examples in the side view.

Modeling the Double Radial Method

The basic principle of the double radial technique is to capture two (or more if desired) circumferential readings at different points along the length of a shaft.

There are six pieces of information that you need to properly construct the shaft positions using this technique:

1. The distance from the outboard to inboard feet (bolting planes) of the first machine.
2. The distance from the inboard bolting plane of the first machine to the point on the shaft where the bracket is located on the first machine.
3. The distance from where the 'near' dial indicator is capturing the rim readings on the second machine to the point where the the 'far' dial indicator is capturing the rim readings on the second machine.
4. The distance from where the 'far' dial indicator is capturing the rim readings on the second machine to the inboard bolting plane of the second machine.
5. The distance from the inboard to outboard feet (bolting planes) of the second machine.
6. The eight dial indicator readings taken at the top, bottom, and both sides on both shafts after compensating for sag (i.e. what a perfect, 'no sag' bracket system would have measured). Be aware of the fact that there will probably be two different 'sag' amounts at each of the dial indicator locations.

Accurately scale the distances along the length of the drive train onto the graph centerline as shown in figure 8-19.

Procedure for plotting the double radial technique:

1. Draw the shaft where the bracket is clamped on top of the graph centerline.
2. Start with the top to bottom or side to side dial indicator

readings on the other shaft (i.e. the one you did not draw on the graph centerline).

3. Plot the other shaft centerline position by starting at the intersection of the graph centerline and the point where the 'near' dial indicator was capturing the readings on the other shaft. If the bottom (or side) reading was negative, place a point half of the bottom (or side) reading from the graph centerline toward the top of the graph. If the bottom (or

The alignment bracket was attached to the motor shaft and the indicators were positioned at two different locations along the length of the fan shaft capturing circumferential readings. The bottom reading at the 'NEAR' indicator showed the fan shaft high by 10 mils (-20) so a point was plotted above the graph centerline 10 mils. The bottom reading at the 'FAR' indicator showed the fan shaft low by 6 mils (+12) so a point was plotted below the graph centerline 6 mils. These two points define the line of sight (i.e. centerline) of the fan shaft.

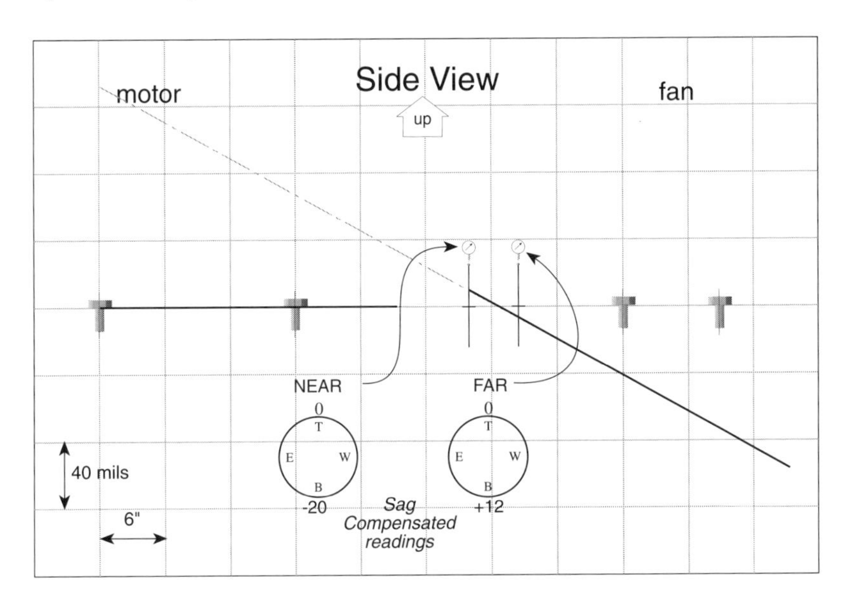

Figure 8-19. Double radial graphing example in the side view.

side) reading was positive, place a point half of the bottom (or side) reading from the graph centerline toward the bottom of the graph (the same as in the point to point modeling techniques). Do not draw any lines yet!

4. Next, start at the intersection of the graph centerline and the point where the 'far' dial indicator was capturing the readings on the shaft. If the bottom (or side) reading was negative, place a point half of the bottom (or side) reading from the graph centerline toward the bottom of the graph. If the bottom (or side) reading was positive, place a point half of the bottom (or side) reading from the graph centerline toward the top of the graph (opposite of the point to point modeling technique).
5. These two points marked on the graph at the dial indicator reading points define the line of sight (i.e. the centerline of rotation) of the other shaft. Draw a straight line through these two points from the coupling end to the outboard end of the other shaft.

Modeling the Shaft to Coupling Spool Method

The basic measurement principle of the shaft to coupling spool technique lies in the ability to measure the angle between each shaft centerline and the centerline of the coupling spool / jackshaft. Since there is only one flex point at the end of each shaft, near perfect angular alignment exists between each shaft and the coupling spool. The coupling spool remains intact (i.e. connected to the shafts) during this procedure.

There are eight pieces of information that you need to properly construct the shaft positions using this technique:

1. The distance from the outboard to inboard feet (bolting planes) of the first machine.
2. The distance from the inboard bolting plane of the first machine to the flexing point between the shaft and the

coupling spool on the first machine. Note that the point where the bracket is being clamped on the shaft is not relevant. A good distance to span past the flex point is anywhere from 3 to 24 inches. The greater this distance is, the more accurate this technique becomes.

3. The distance from the flexing point between the shaft and the coupling spool on the first machine and the point where the dial indicator is capturing the rim readings on the coupling spool.
4. The distance from where the dial indicator is capturing the rim readings on the coupling spool near the first machine to the point where the dial indicator is capturing the rim readings on the coupling spool near the second machine.
5. The distance from where the dial indicator is capturing the rim readings on the coupling spool near the second machine to the flexing point between the shaft and the coupling spool on the second machine. Note that this distance does not have to be the same distance as measured in no. 2 above.

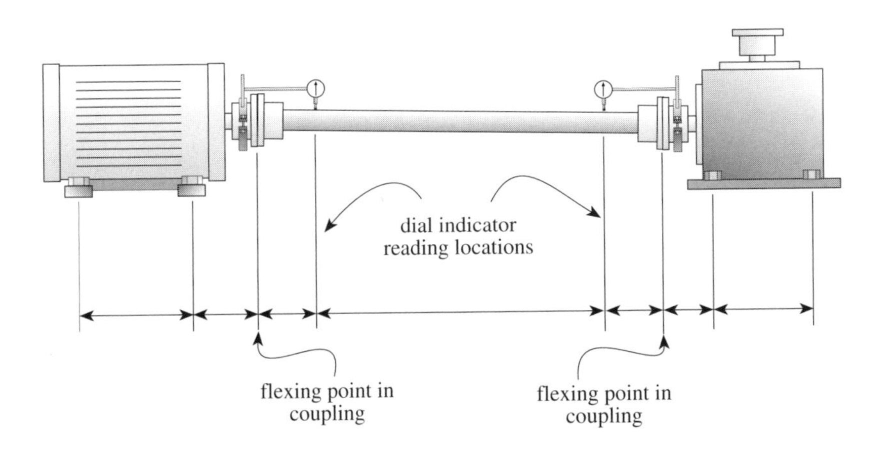

Figure 8-20. Shaft to coupling spool dimensions needed for graphing set-up.

Setting up the graph for Shaft to Coupling Spool Modeling

Measure the distances between the inboard and outboard feet on both machines, the distance from the inboard feet to the points where the flexing points are located at both ends of the coupling spool / jackshaft, the distance from where the flexing points are to where the dial indicator stems are touching the coupling spool, and the distance between the dial indicators.

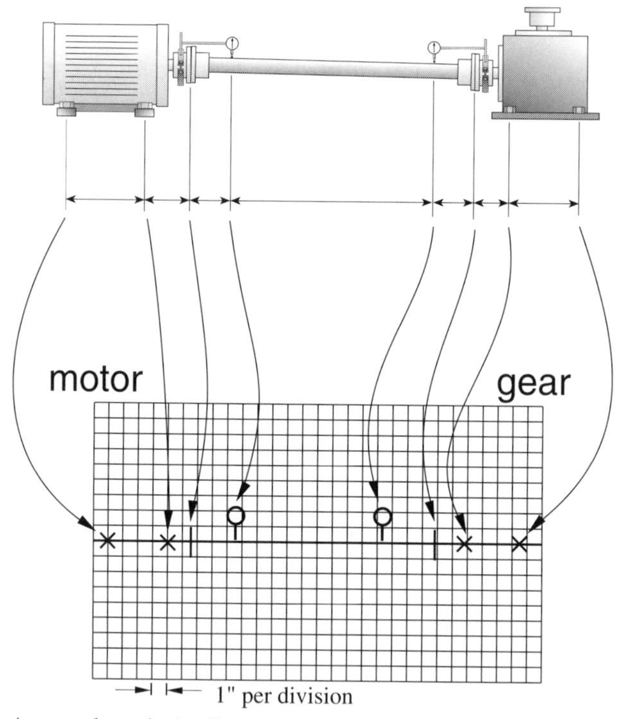

Accurately scale the distances measured above onto the graph paper.
*Note: The scale can be 1", 2", 3", 10" per division. Select the smallest scale factor that fits the entire drive train onto the graph paper.

Figure 8-21. Graph / model set up for shaft to coupling spool technique.

6. The distance from the flexing point between the shaft and the coupling spool on the second machine to the inboard bolting plane of the second machine.
7. The distance from the outboard to inboard feet (bolting planes) of the second machine.
8. The eight dial indicator readings taken at the top, bottom, and both sides on the coupling spool after compensating for sag (i.e. what a perfect, 'no sag' bracket system would have measured). Be aware of the fact that there will probably be two different 'sag' amounts at each of the dial indicator locations if the distances were not the same from the flex points to where the dial indicator readings were taken.

Accurately scale the distances along the length of the drive train onto the graph centerline as shown in figure 8-21.

Procedure for plotting the shaft to coupling spool technique:

1. Draw the coupling spool on top of the graph centerline.
2. Start with the top to bottom dial indicator readings (or side to side dial indicator readings if you want to plot the Top View) taken from the first machine to the spool piece.
3. At the intersection of the graph centerline and the point where the dial indicator was capturing the readings on the spool piece, place a point half of the bottom (or side to side) reading above or below the graph centerline. If the bottom (or side) reading was negative, place a point half of the bottom (or side) reading from the graph centerline toward the bottom of the graph. If the bottom (or side) reading was positive, place a point half of the bottom (or side) reading from the graph centerline toward the top of the graph. Draw a line through this point and the point where the flexing point and the graph centerline intersect. These two points define the line of sight (i.e. the centerline of rotation) of the first machine.
4. Next, start at the intersection of the graph centerline and

the point where the dial indicator was capturing the readings on the spool piece from the bracket attached to the second machine shaft. Place a point half of the bottom (or side to side) reading above or below the graph centerline. If the bottom (or side) reading was negative, place a point half of the bottom (or side) reading from the graph centerline toward the bottom of the graph. If the bottom (or

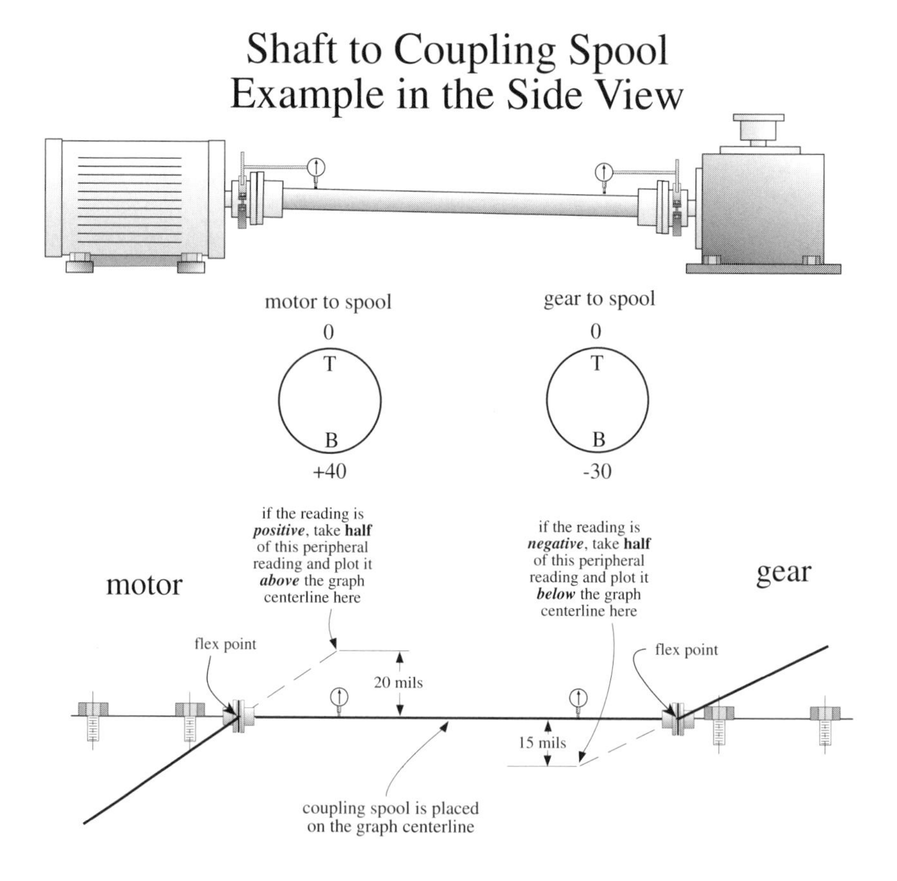

Figure 8-22. Shaft to coupling spool graphing example in the side view.

side) reading was positive, place a point half of the bottom (or side) reading from the graph centerline toward the top of the graph. Draw a line through this point and the point where the flexing point and the graph centerline intersect. These two points define the line of sight (i.e. the centerline of rotation) of the second machine.

Modeling the Face - Face Method

The face-face coupling spool technique also measures the angle between each shaft centerline and the centerline of the coupling spool / jackshaft, except here the angles are measured across the faces at each flexing point. The coupling spool remains intact (i.e. connected to the shafts) during this procedure.

It is interesting to note that the bracket can be clamped on the shaft or on the coupling spool. Although we are showing use of brackets and dial indicators to capture the face measurements, traditionally these measurements have been taken with feeler gauges at four points around the faces (see the 16 point method in Chapter 7). The same principles applied to the face-rim technique apply here in that the larger the diameter the face readings are taken on the more accurate the technique becomes.

To plot the position of the two shafts and the spool piece, the 'T' bar overlay will be used twice (once at each end of the spool piece). The graph is 'dual scaled' similar to the face-rim technique.

There are eight pieces of information that you need to properly construct the shaft positions using this technique:

1. The distance from the outboard to inboard feet (bolting planes) of the first machine.
2. The distance from the inboard bolting plane of the first machine to the flexing point between the shaft and the coupling spool on the first machine. Note that the point where the bracket is being clamped on the shaft (or on the

coupling spool) is not relevant. Again, the larger the diameter the face readings are taken on, the more accurate this technique becomes.

3. The diameter the face readings were taken on at the flexing point between the first machine and the spool piece.
4. The distance from the flexing point between the shaft and the coupling spool on the first machine to the flexing point between the shaft and the coupling spool on the second machine.
5. The diameter the face readings were taken on at the flexing point between the first machine and the spool piece.
6. The distance from the flexing point between the shaft and the coupling spool on the second machine to the inboard bolting plane of the second machine.
7. The distance from the outboard to inboard feet (bolting planes) of the second machine.
8. The eight 'face' dial indicator readings taken at the top, bottom, and both sides on the coupling spool after compen-

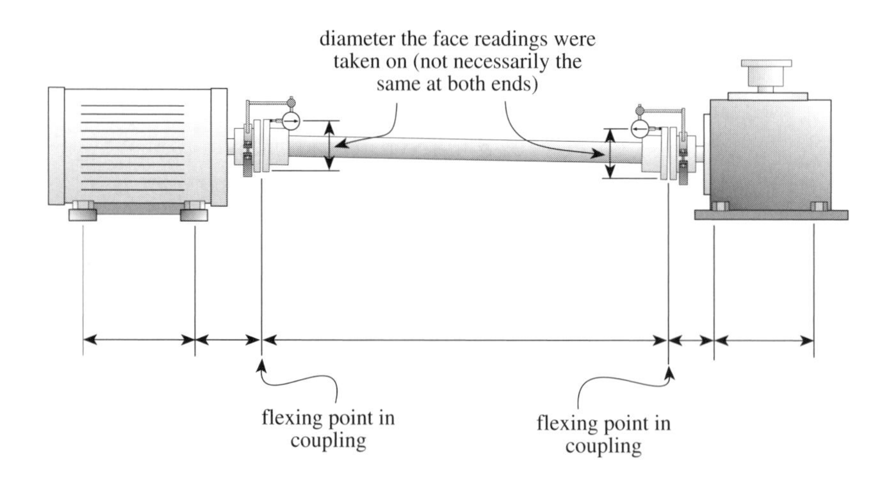

Figure 8-23. Face-face dimensions needed for graphing set-up.

sating for sag (i.e. what a perfect, 'no sag' bracket system would have measured). Be aware of the fact that there will probably be two different 'sag' amounts at each of the dial indicator locations if the bracket span distances were not the same from the flex points to where the face dial indicator readings were taken.

Accurately scale the distances along the length of the drive train onto the graph centerline and the diameters the face readings were taken on at each flexing point as shown in figure 8-23.

Procedure for plotting the shaft to coupling spool technique:

1. Draw the coupling spool on top of the graph centerline.
2. Start with the top to bottom face dial indicator readings (or side to side face dial indicator readings if you want to plot the Top View) taken across the flex point on the first machine to the spool piece.
3. At the intersection of the graph centerline and the point where the flexing occurs between the first machine shaft and the spool piece, pitch / rotate the 'T' bar overlay to reflect the difference in gap between the top and bottom (or side to side) readings. If the bottom (or side) reading was negative, pitch the 'T' bar clockwise the full amount of the bottom (or side) readings across the diameter the face readings were captured on. If the bottom (or side) reading was positive, pitch the 'T' bar counterclockwise the full amount of the bottom (or side) readings across the diameter the face readings were captured on. The base of the 'T' represents the centerline of rotation of the first machine. If you are going to use the 'T' bar again at the other flexing point, draw in the position of the shaft so it lines up with the base of the 'T'.
4. Next, start at the intersection of the graph centerline and the point where the flexing occurs between the second machine shaft and the spool piece, pitch / rotate the 'T' bar

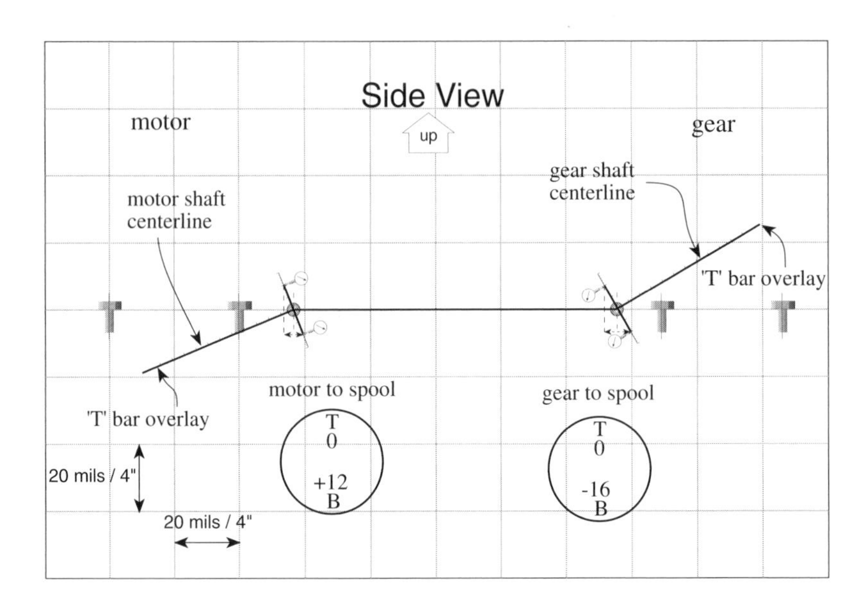

Figure 8-24. Face-face graphing example.

overlay to reflect the difference in gap between the top and bottom (or side to side) readings. If the bottom (or side) reading was negative, pitch the 'T' bar clockwise the full amount of the bottom (or side) reading across the diameter the face readings were captured on. If the bottom (or side) reading was positive, pitch the 'T' bar counter-clockwise the full amount of the bottom (or side) reading across the diameter the face readings were captured on. The base of the 'T' represents the centerline of rotation of the second machine. Again, draw in the position of the second machine shaft so it lines up with the base of the 'T'.

Checking the misalignment tolerance

The modeling techniques just reviewed have enabled us to visually represent the relative positions of the centerlines of rotation of the two shafts. Once the positions of the shafts have been determined, the first step is to determine whether the amount of misalignment is within tolerance. Refer back to the definition of shaft misalignment in Chapter 5 and study figures 5-2 and 5-3.

At any point in time, the machinery shafts are somewhat misaligned side to side and misaligned up and down (or any other coordinates they happen to lie in). The key is to find the largest of the four deviations at the points of power transmission (the flexing points) and divide it by the distance between points of power transmission (the flexing points). Two of these deviations occur in the TOP VIEW which will show the amount of lateral (side to side) misalignment and two more deviations occur in the SIDE VIEW which show the amount of vertical (up/down) misalignment as illustrated in figure 8-25.

The maximum misalignment deviation can be determined directly from the graph. Note that the flexing points don't necessarily occur at the same points where the readings were taken. However, since the misalignment is expressed in mils per inch, whether the deviations are measured where the readings were taken, or at the ends of the shafts, or at the flexing points themselves, the misalignment deviation (remember it's in mils/inch) will always be the same amount (i.e. the deviation might be different at each of the points around the coupling area, but so is the distance between them).

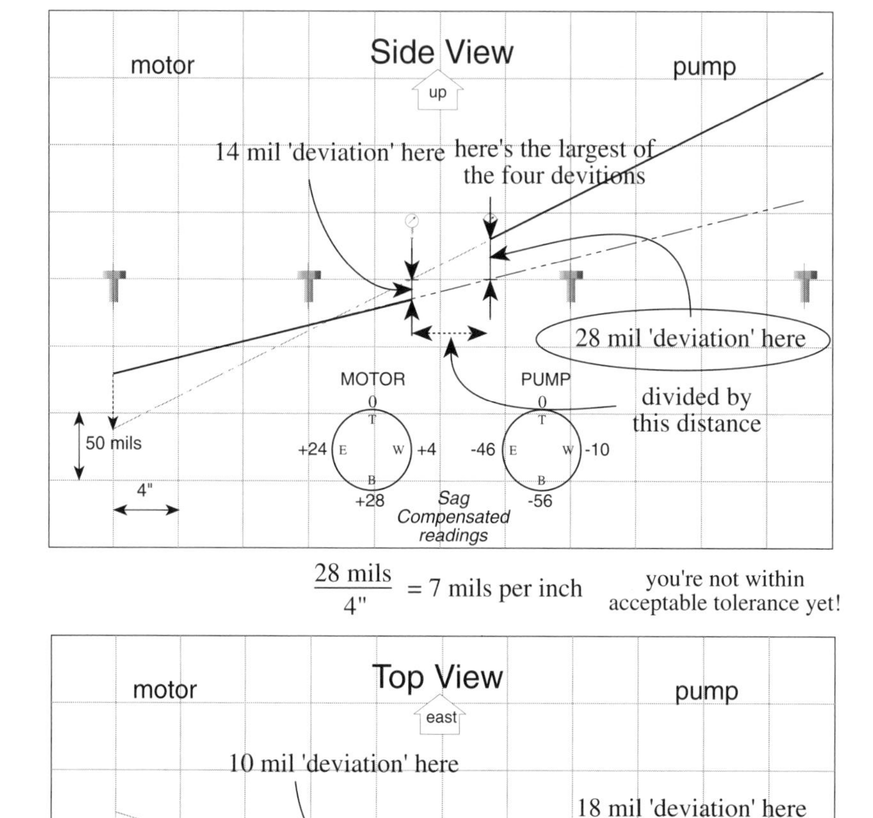

Figure 8-25. Determining the current misalignment accuracy / tolerance.

Movement restrictions and the allowable movement envelope

The various graphing / modeling techniques have shown the relative positions of the machinery centerlines of rotation, and the misalignment deviations that have been extracted from the side and top view graphs have shown that the amount of misalignment is excessive. Now what do we do?

The next step is to determine the movement restrictions imposed on the machine cases at the control / adjustment points. Movement restrictions define the allowable amount of 'easy' movement on the machinery.

When viewing the machinery in the up / down direction (SIDE VIEW), the movement restrictions are defined by the amount of movement the machinery can be adjusted in the up and down directions.

How far can machinery casings be moved upward? There is virtually an unlimited amount of movement in the up direction (within reason - see Chapter 11). Machine cases are typically moved upward by installing shims (i.e. sheet metal of various thicknesses) between the undersides of the machinery feet and the baseplate (refer to Chapter 6 and Chapter 11 for more information).

Shim stock / shim typically refer to sheet metal thicknesses ranging from 1 mil (0.001") to 125 mils (0.125"). There are several companies that manufacture 'pre-cut' shim stock in 4 standard sizes and 17 standard thicknesses (see Tables 11-1 and 11-2). Once shim thickness gets over 125 mils, they are typically referred to as 'spacers' or 'plates' and are custom-made from plate steel.

Equally important is ... How far can machinery casings be moved downward? Machine cases are typically moved downward by removing shims between the undersides of the machinery feet and the baseplate.

So, if you want to move a machinery downward and there

are no shims under the machinery feet, that is defined as a vertical movement restriction. If however, there are 10, 20, or maybe 50 mils under the machinery feet that can be removed, that is referred to as a movement envelope, or in this particular case, the 'basement floor'.

In addition to aligning machinery in the up / down direction, it is also imperative that the machinery be aligned properly side to side. Machinery is aligned side to side by translating the machine case laterally. This sideways movement is typically monitored by setting up dial indicators along the side of the machine case at the inboard and outboard hold down bolts, anchoring the indicators to the frame or baseplate, zeroing the indicators, and then moving the inboard and outboard ends the prescribed amounts. Here is where realignment typically becomes extremely frustrating since there is a limited amount of room between the shanks of the hold down bolts and the holes drilled in the machine case feet as illustrated in figure 8-26.

If, for example, you wanted to move the outboard end of a machine 120 mils to the south, began moving the outboard end monitoring the move with a dial indicator, and the machine case stopped moving after 50 mils of translation, this would be considered a movement restriction commonly referred to as a 'bolt bound' condition. The problem when moving machinery laterally is that there is a limited amount of allowable movement in both sideways directions. The total amount of side to side movement at each end of the machine case is referred to as the lateral movement envelope. To find the allowable lateral movement envelope, remove a bolt from each end of the machine case, look down the hole, and see how much room exists between the shank of the bolt and the hole drilled in the machine case at that foot. If necessary, thread the bolt into the hole a couple of turns, and measure the gaps between the bolt shank and the sides of the hole with feeler or wire gauges.

It is very important for one to recognize that trouble free

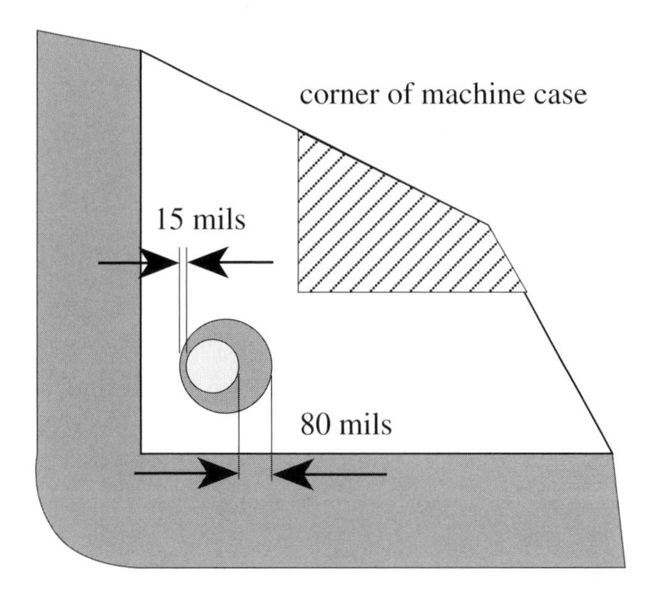

lateral movement envelope

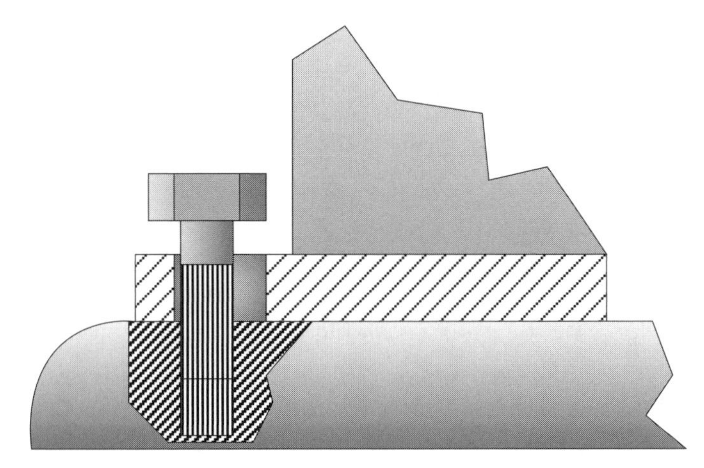

Figure 8-26. Finding the lateral movement restrictions and defining the allowable movement envelope at a machinery foot.

alignment corrections can only be achieved when the allowable movement envelope is known. Perhaps one of the most important statements that will be made in this book is:

When you consider that both machine cases are movable, there are an infinite number of possible ways to align the shafts, some of which fall within the allowable movement envelope.

It seems ridiculous, but many people have ground baseplates or the undersides of machinery feet away because they felt that a machine had to be lowered. When machinery becomes 'bolt bound' when trying to move it sideways, people frequently cut down the shanks of the bolts or grind holes open more when there is typically an easier solution. Disappointingly, the vast majority of alignment measurement systems shown in Chapter 7 force the user to name one machine case stationary and the other one movable which will invariably cause repositioning problems when the machine case has to be moved outside its allowable movement envelope.

Please, for your own sake, follow these four basic steps to prevent you from wasting hours or days of your time correcting a misalignment condition:

1. Find the positions of every shaft in the drive train by the graphing / modeling techniques shown in this chapter.
2. Determine the total allowable movement envelope of all the machine cases in both directions.
3. Plot the restrictions on the graph / model.
4. Select a 'final desired alignment line' that fits within the envelope (hopefully) and move the machinery to that line.

If you are involved with aligning machinery, by following the three steps above, it is guaranteed that you will save countless hours of wasted time trying to move a machine where it doesn't really want to go. This has been repeated several times so it might appear to be somewhat important!

Where did the Stationary - Movable alignment concept come from?

I don't know. Every piece of rotating machinery in existence has, at one time or another, been placed there. Mother earth never gave birth to a machine. They are not part of the earth's mantle nor are they firmly imbedded in bedrock. Every machine is movable, it is just a matter of effort (pain) to reposition it. So why have the vast majority of people who align machinery called one machine stationary and the other machine movable?

The only viable reason that I can come up with is this ... in virtually every industry there is an electric motor driving a pump. When you first approach a motor pump arrangement, you immediately notice that the pump has piping attached to it and the only appendage attached to the motor is a conduit (usually a flexible conduit). From your limited vantage point at this time it would appear to be easier to move the motor because there is no piping attached to it like the pump. You would prefer to just move the motor because it looks easier to move than the pump (and so would I). The assumption is made that the pump will not be moved, no matter what position you find the motor shaft in with respect to the pump shaft. But what do you do when you have to align a steam turbine driving a pump? Uh-oh, they're both piped! Which machine do you call the stationary machine ... the pump or the turbine? No matter what your answer is, you are going to have to move one of them and they both have piping attached to their casings.

Piping is no excuse not to move a piece of machinery, particularly in light of what most of us know about how piping is really attached to machinery. For some people, they are afraid to loosen the bolts holding machinery with piping attached to it because the piping strain is so severe that they fear the machine will shift so far that it will never get back into alignment. So is the problem with the alignment process or the piping fit-up? Refer to Chapter 3 for information on this subject.

If you align enough machinery and insist that one machine will be stationary, eventually you will get exactly what you deserve for your shallow range of thinking.

Determining machinery moves to correct the misalignment

Once the centerlines of rotation have been determined and the allowable movement envelope illustrated on the graph, it

Here are the two different Stationary / Movable solutions for this arrangement of shafts ...
If you keep the motor stationary, the pump must be lowered 45 mils at the inboard feet and 87 mils at the outboard feet.
If you keep the pump stationary, the motor must be lowered 5 mils at the inboard feet and 41 mils at the outboard feet.

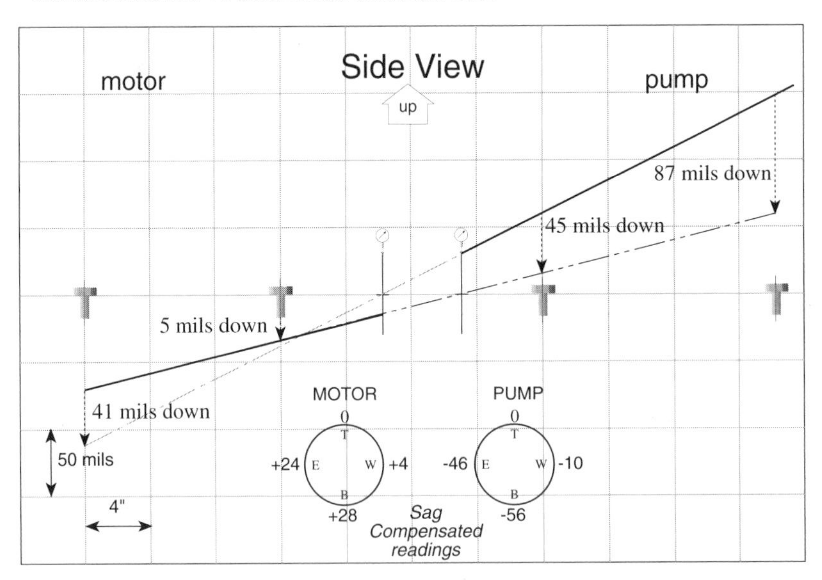

Figure 8-27. Stationary movable choices.

becomes very apparent what repositioning moves will work easily and which ones won't.

The final desired alignment line (also known as the overlay line) is a straight line drawn on top of the graph showing the desired position both shafts should be in to achieve colinearity of centerlines. It should be apparent that if one machine case is stationary, that machine centerline of rotation is the final desired alignment line.

Since adjustments are made at the inboard and outboard feet of the machinery, some logical alternative solutions would be to consider using one or more of these feet as pivot points. Both

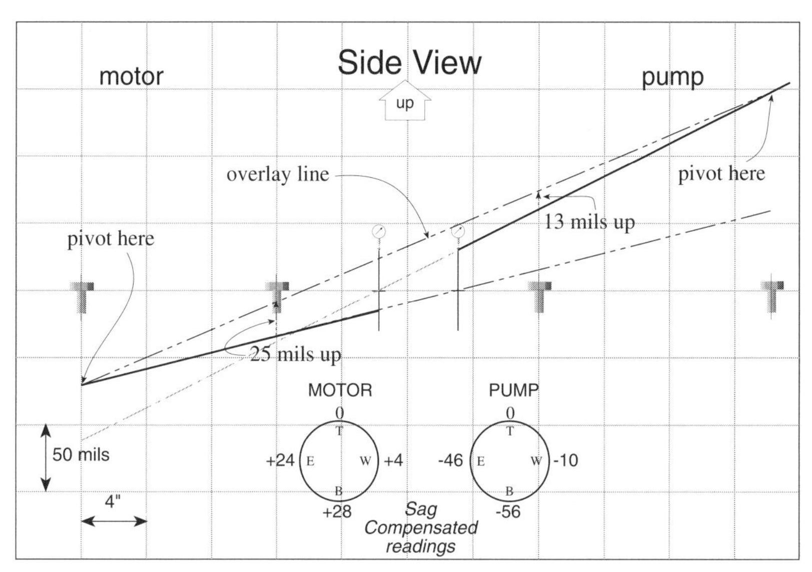

Figure 8-28. Using the two outboard feet as pivot points.

In this example, the positions of the steam turbine and boiler feedwater pump shafts have been determined and the existing shim stock under all of the hold down bolts has been plotted onto the graph (52 mils under the outboard and 20 mils under the inboard feet of the steam turbine and 40 mils under the inboard and 46 mils under the outboard feet of the BFW pump). The dashed line represents the lower restriction points at each of the hold down bolts. Notice that neither machine case can be kept 'stationary' without resorting to grinding baseplates or the undersides of the machinery feet.

In this case, one possible solution might be to pivot at the outboard feet of the pump and the inboard feet of the steam turbine lowering the other two feet the prescribed amounts shown on the graph.

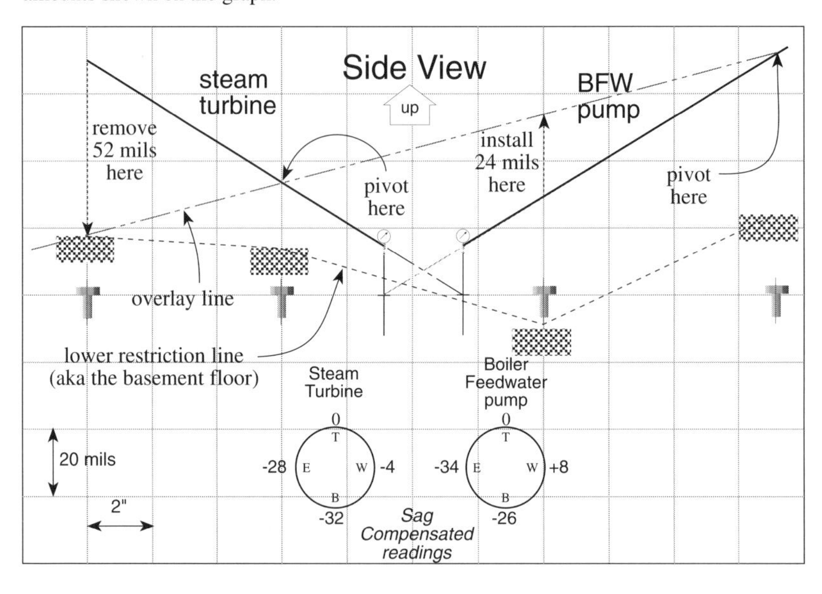

Figure 8-29. Plotting the lower restriction points on the graph.

Here the side to side (lateral) positions of the steam turbine and boiler feedwater pump shafts have been determined and one hold down bolt at each of the inboard and outboard ends of both machine cases has been removed and the amount of side to side clearance between the shank of the bolt and the holes drilled in the machine case feet has been plotted onto the graph. The dashed lines represent the east and west restriction points (sometimes referred to as the 'canyon walls'). Notice that neither machine case can be kept 'stationary' without resorting to undercutting bolt shanks, or grinding, or over-drilling the holes of the machinery feet.

In this case, one possible solution might be to pivot at the outboard feet of the steam turbine and the inboard feet of the pump, translating the other two feet the prescribed amounts and directions shown on the graph.

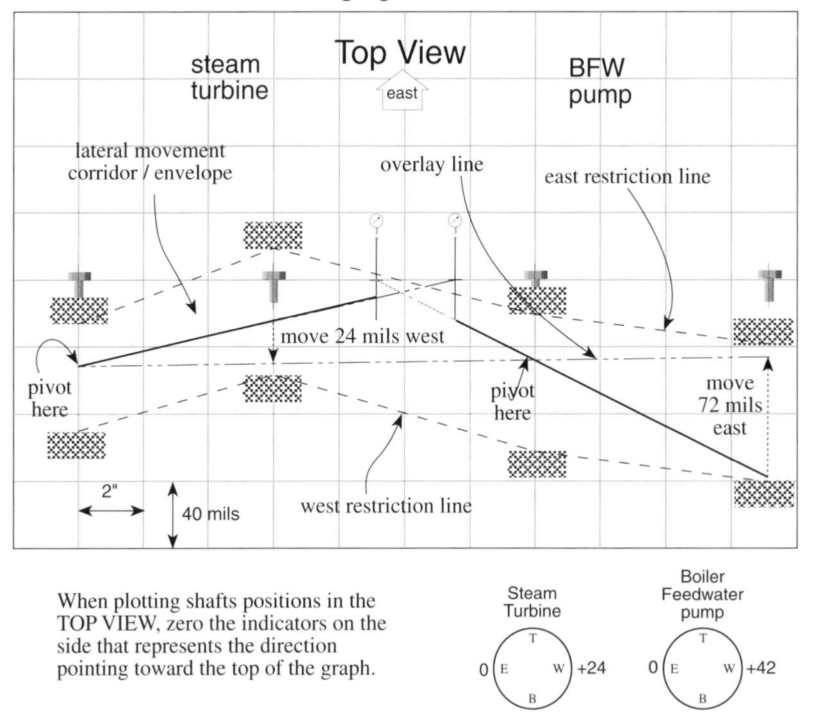

Figure 8-30. Plotting the lateral restrictions on the graph.

outboard feet or both inboard feet, or the outboard foot of one machine case and the inboard foot of the other machine case could be used as pivot points. By drawing the overlay line through these foot points, shaft alignment can usually be achieved with smaller moves. In real-life situations, you will typically have greater success aligning two machine cases a little bit rather than moving one machine case a lot.

In this example, if the gear was to be kept stationary, the motor would have to be moved over 3/8' downward. Instead, an overlay line is projected go go through the outboard feet of the gear and the inboard feet of the motor minimizing the move to just 17 and 12 mils.

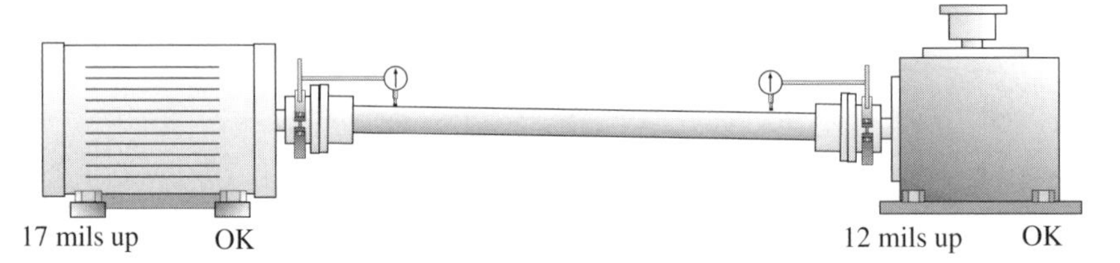

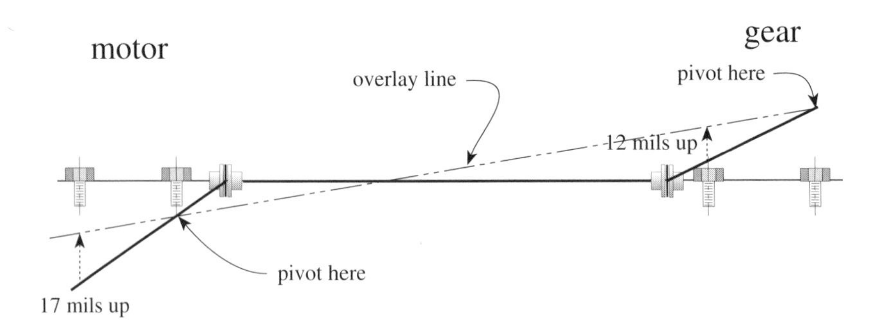

Figure 8-31. Using an inboard bolt plane and an outboard bolt plane as pivot points.

Solving piping fit-up problems and aligning the shafts

The suction pipe flange is not lining up with the pump suction flange due to the pipe being 175 mils too high. In addition, there is a 175 mil gap between the flange faces on the discharge piping. If the pump is not repositioned there will be too much piping strain.

In this case, the overlay line is projected to raise the outboard end of the pump 175 mils to relieve the piping strain and the other end of the overlay line crosses through the outboard feet of the motor which will be used as a pivot point. Shim changes are then calculated at the inboard feet of the motor and pump to bring the centerlines to the overlay line.

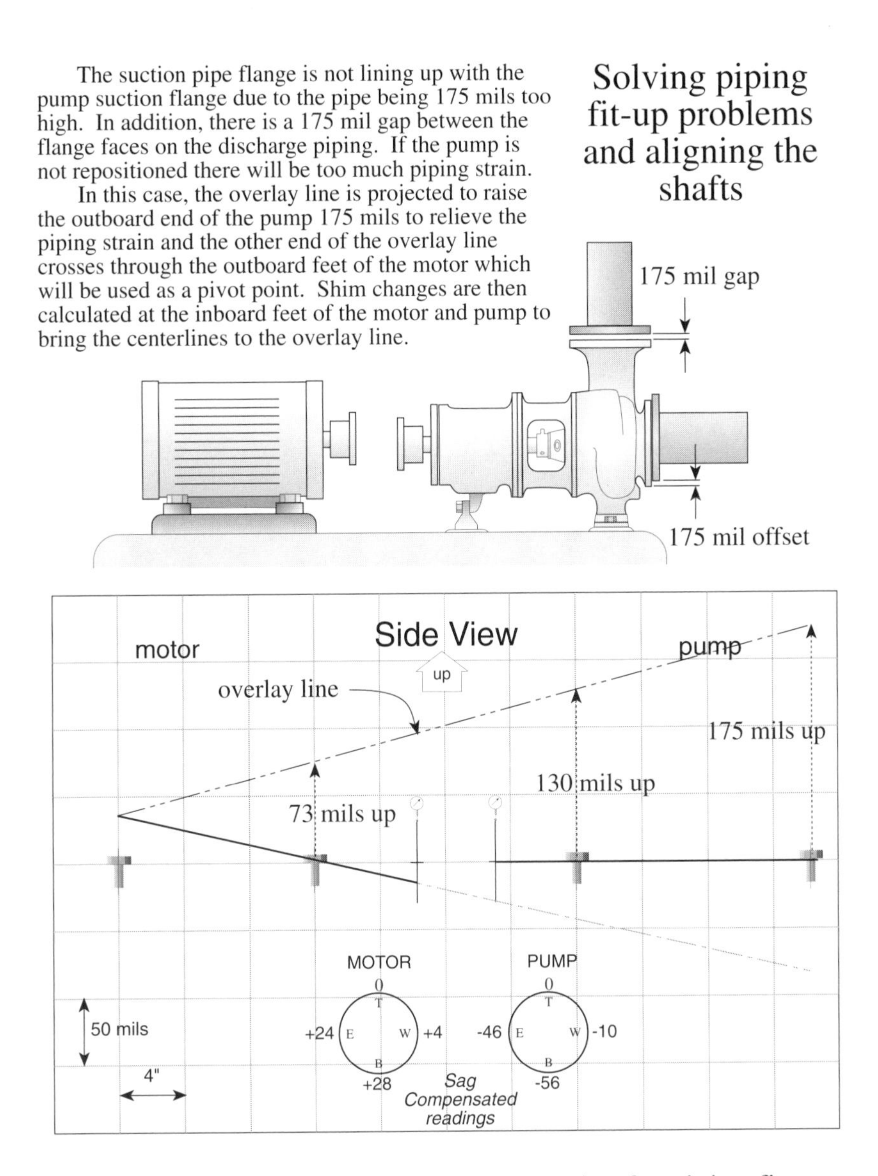

Figure 8-32. Using the overlay line to solve for piping fit-up problems and align shafts at one time.

Solving piping fit-up problems with the overlay line

Although we have been showing the overlay line (aka final desired alignment line) being drawn through foot bolt points, it is important to see that the overlay line could be drawn anywhere and the machinery shafts moved to that line.

This can be particularly beneficial if there are other considerations that have to be taken into account such as piping fit-up problems. The overlay line can be placed to relieve or eliminate piping stress problems as illustrated in figure 8-32.

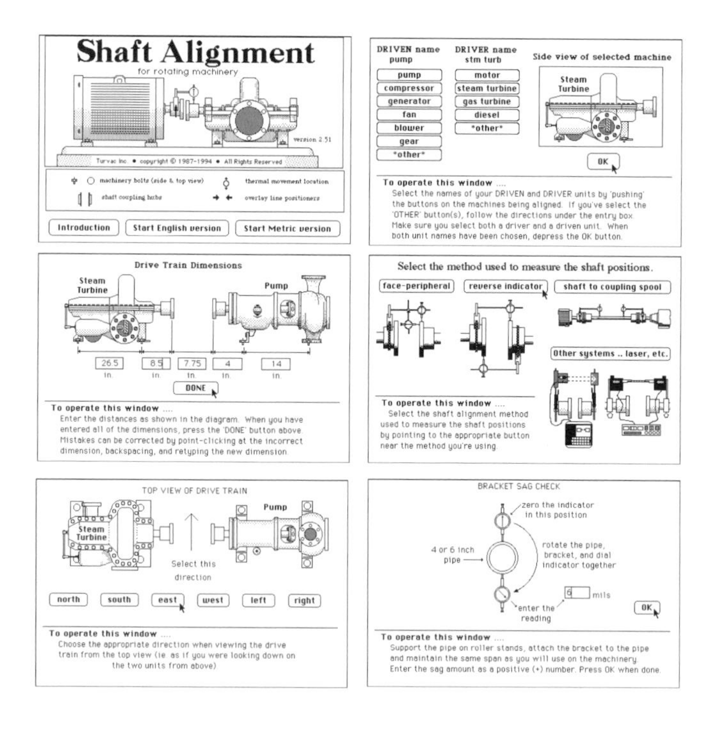

Figure 8-33. Shaft alignment software program screen shots. Courtesy Turvac Inc., Cincinnati, OH.

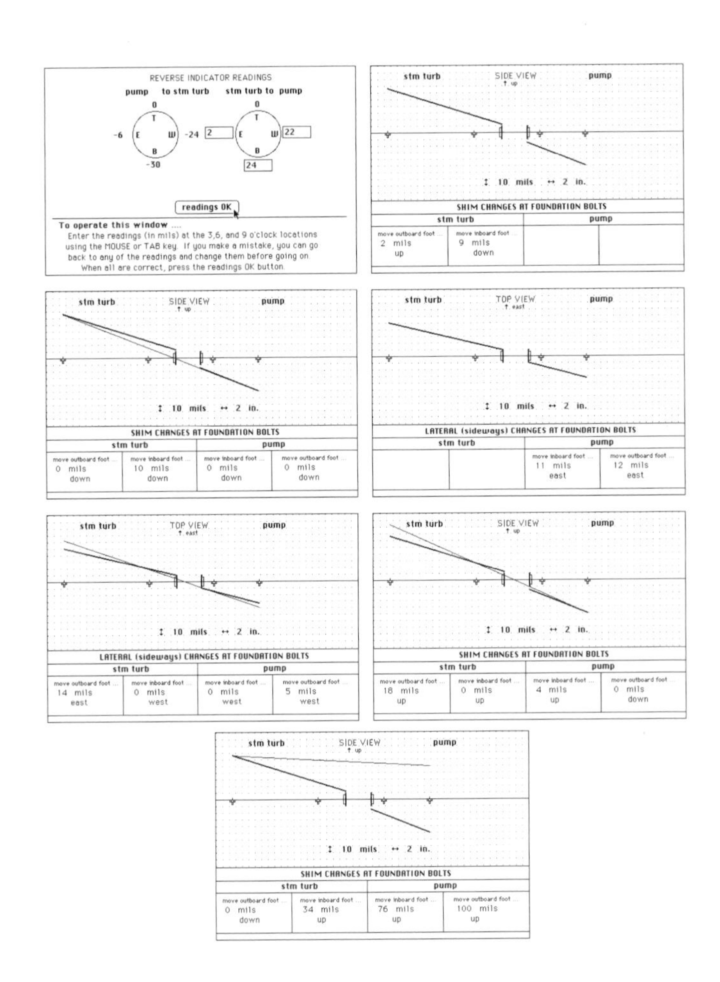

Figure 8-34. Shaft alignment software program screen shots. Courtesy Turvac Inc., Cincinnati, OH.

Bibliography

Dreymala, James, **Factors Affecting and Procedures of Shaft Alignment**, Technical and Vocational Dept., Lee College, Baytown, Texas, 1970.

Dodd, V.R., **Total Alignment**, The Petroleum Publishing Co., Tulsa, Okla., 1975.

Piotrowski, John, **Basic Shaft Alignment Workbook**, Turvac Inc., Cincinnati, OH.

9
Measuring and Compensating for Off-line to Running Machinery Movement

Virtually all rotating equipment will undergo a change in position during start-up and while running that affects the alignment of the shafts. In order for the shafts to run collinearly under normal operating conditions, it is desirable to know the amount and direction of this movement to properly position the machinery during what is commonly called the 'cold' alignment process (i.e. off-line or not running) to compensate for this change.

What type of machinery is likely to change its position when running?

The off-line to running movement characteristics of the vast majority of rotating machinery drive systems in existence have never been measured. On perhaps 60% of the drive systems in existence, the amount of movement is negligible and can pretty much be ignored. In the remaining cases, however, it can make all the difference between a smooth running drive system and one that is plagued with problems. It is important to know how much movement is occurring before you deem it insignificant and ignore it. The most baffling question is ... which rotating machinery do you have that is moving enough from off-line to running conditions that you need to measure and compensate for this movement?

Below is a broad list of machinery that is likely to change its position enough from off-line to running conditions to warrant measuring this occurrence.

- rotating machinery drive systems running at or above 200 hp and speeds of 1200 rpm or greater

- machinery that undergoes a change in casing temperature for example ...
  - electric motors and generators
  - steam turbines
  - gas turbines
  - internal combustion engines (diesels, etc.)

- speed changers (e.g. gear boxes)

- machinery that is pumping or compressing fluids or gases where the fluid or gas undergoes a change in temperature by 50 degrees or greater from intake to discharge (this could be either a rise or drop in temperature) for example ...
  - centrifugal or reciprocating compressors
  - centrifugal pumps
  - HVAC air moving equipment
  - furnace fans
- equipment with poorly supported piping attached to the

machine case where expansion or contraction of the piping induces forces into the machine case or where fluid flows can cause a reactionary moment in the piping

What causes this movement to occur?

There are a variety of factors that cause machinery to move once it is on-line and running. The most common cause is due to temperature changes in the machinery itself (as it compresses gases or heats the lubricant from friction in the bearings) and is therefore generally referred to as 'thermal' or 'cold' to 'hot' movement. The temperature change in rotating machinery is rarely uniform throughout the casing which causes most equipment to 'pitch' at some angle rather than grow (or shrink) straight up (or down). For compressors and pumps, thermal movement of the attached piping will also cause the equipment to shift.

Other sources of movement in machinery can be caused by: loose foot bolts, varying weather conditions for equipment located outdoors, heating or cooling of concrete pedestals, changes in the operating condition of equipment from unloaded to loaded postures, and/or casing and support counter-reactions to the centrifugal force of rotors as they are rotating.

Special considerations must be taken into account for equipment that is started and stopped frequently or where loads may vary considerably while running. In cases like these, a compromise has to be made that weighs factors such as: periods of time at certain conditions, total variation of machinery movement from maximum to minimum, coupling and alignment tolerances, etc. To properly observe and record these changes, periodic checks should be made of this change in movement to properly understand how to effectively position the equipment for optimum performance. Continuous shaft position monitoring systems are available and are explained later in this chapter.

It has been my experience, however, that the majority of

rotating equipment will typically maintain one specific position regardless of varying loads. What usually turns out to be a bigger problem is that some equipment may have to be offset aligned 'cold' by a considerable distance making startups very nerve-wracking. In the majority of cases, equipment will undergo the greatest rate of change of movement shortly after start-up. 'Shortly' can mean anywhere from 5 minutes to 1 hour for most types of equipment and may settle at some 'final' position several hours or even days later.

Going from running to off-line conditions virtually anything can occur. Some equipment may make a very rapid change immediately after shutdown. Other drive systems may flounder around and then slowly move back near their original position. To my knowledge, there has never been any published information where someone has observed several start-up / shut-down sequences to see if the machinery continues to attain the same positions in the off-line and in the running condition. Most people just assume and hope that this happens.

Conducting the Off-line to Running Machinery Movement Survey

All of this may seem quite complicated at first glance, but these measurements are nothing more than a comparison between the position of the centerlines of rotation when the machinery is off-line to the position of the centerlines of rotation when the machinery is running (frequently referred to as OL2R for off-line to running and conversely, R2OL for running to off-line). Therefore, data must be taken when the machinery is off-line (cold), and then again when the machinery is running (hot).

There are many inventive ways to measure shaft alignment positions from off-line to running (or vice versa) and this chapter will review the more commonly used techniques. The off-line to running machinery movement data that is typically measured is often quite surprising and there is a great tendency not to believe the results. Each method has its advantages and disadvantages and it is a good idea to compare the measurements from two or more methods just to see if the results are similar.

There are currently eleven different techniques that have been used to measure this movement that will be explained in greater detail in this chapter. One very popular practice is not recommended since there are a number of potential problems with it as explained herein.

Taking 'hot' alignment readings immediately after shutdown

For many years, people have attempted to take quick 'hot' alignment readings, that is, take shaft position readings immediately after a unit has been shut down. This turns out to be quite a difficult task to perform for a variety of reasons: mounting brackets and indicators or lasers quickly enough to get a set of readings, safety tagging driver units to prevent them from inadvertently starting up with the brackets mounted on the shafts, and getting an accurate set of readings while the shafts are still moving proves to be a real challenge for the personnel doing the work. If you plan to attempt this, it is wise to take three or four readings (say..... every five minutes) during the 'cool' down period to plot the movement and extrapolate them back to the instant when the unit was first shut down to determine the actual shaft positions when running. The data is usually non-linear and guessing the slope of the curve during the first five to ten minute period after the unit has been shut down is a hit or miss proposition. Therefore, this method is not recommended, but if it is all you are willing to do, then it is better than guessing or doing nothing at all.

Four general categories of OL2R measurements

There are four broad classifications of measurement techniques that are used to capture off-line to running machinery movement:

- Movement of the centerline of the machine cases with respect to its baseplate or frame
- Movement of the centerline of the machine cases with respect to some remote reference or observation point
- Movement of one machine case with respect to another machine case
- Movement of one shaft with respect to another shaft

Just as there are advantages and disadvantages of each of the shaft measurement techniques covered in Chapter 7, there are advantages and disadvantages of each of the OL2R techniques shown in this chapter. After conducting these surveys for many years on a wide variety of equipment, there is always a certain level of uncertainty in the measurement data or in the set-up of the equipment. Since the results of these surveys can be quite surprising, there is a tendency to disbelieve the first set of measurements. So, what do you do?

Often you repeat the test again and see if the results of the second set of measurements somewhat agree with the results of the first set. If there is a wide variation between both sets of results (e.g. the inboard end of the compressor looked like it moved 62 mils upward the first time and 38 mils downward the second time), you need to carefully review every aspect of the OL2R equipment set-up, the measurement sensors, and when and how you collected the data to determine how this occurred.

One of the reasons why several methods are shown in this chapter is to give you an opportunity to compare the results from two or more different techniques. It is not uncommon to repeat the

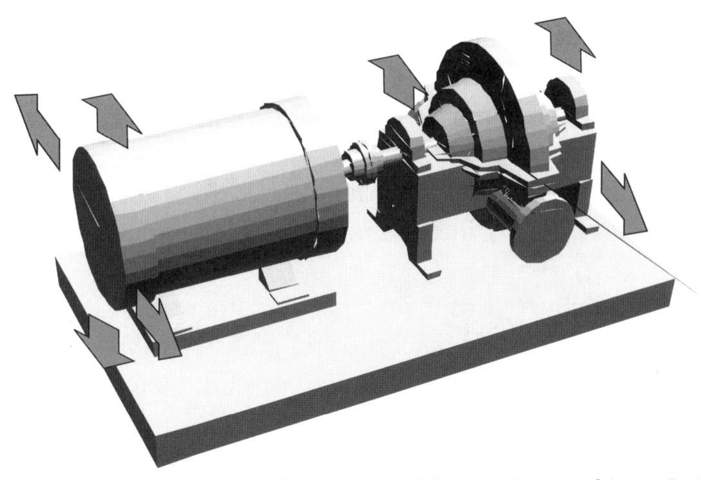

All of these techniques compare the position of the rotating machinery in the drive train when the equipment is off-line (not running) to the position of the machinery when it is on-line (running)

Movement of machine case with respect to a remote reference point

- Optical tilting levels and jig transits

- laser tooling

Movement of machine case with respect to another machine case

- alignment bars with proximity probes

- laser - detector systems

- Plug in - back zeroing - laser / target mounts

- Ball - Rod - Tubing connector system

Movement of machine case with respect to its baseplate or foundation

- casing thermal expansion using the strain equation

- inside micrometers and/or angle measurement devices - tooling balls

- proximity probes with water cooled stands

Movement of one shaft with respect to the other shaft

Instrumented coupling system

Vernier - Strobe system

Figure 9-1. Tools and techniques used to measure off-line to running conditions on rotating machinery.

OL2R survey 2 or 3 times getting very similar results each time, but the data might still be unbelievable in your and other peoples eyes. So, what do you do? Try a different measurement method to see if the results from that method agree with the first method attempted.

There are several important issues that need to be mentioned concerning off-line to running measurements:

- Less than 10% of the people who are responsible for alignment of rotating machinery in industry have ever actually conducted off-line to running machinery movement surveys (1 out of 5 may be aware of this phenomena but only half of them have actually tried to measure it).

- Many of the people who have conducted OL2R surveys or are aware of this phenomena, believe that machinery only moves in the up and down direction disregarding the possibility (and probability) of sideways movement.

- Compared to the amount of time usually taken to align rotating machinery when off-line (which takes 3-6 hours on the average), off-line to running machinery movement surveys can take days and even months to complete. These types of surveys are frequently conducted on the more critical pieces of rotating machinery in a plant and these drive systems usually can't be turned on and turned off at the convenience of the people performing the test. So be prepared to spend a lot more time planning, designing, fabricating, installing, measuring, and analyzing this data than you typically would for 'off-line / in-line' shaft alignment jobs.

- In several cases where this data has been collected, and it was verified that the machinery was not in the right alignment position when off-line, there was a great reluctance (and in some cases, pure refusal) to change the existing positions of the machinery to reflect the newly discovered information. The reasoning typically expressed was ... "The machinery has been in this position for several years, why change it now?" or "If we move the machines, we will void our warranty." or "We have to wait until the next shutdown to change the alignment (which seems to be forgotten at the designated shutdown time)."

It seems very silly to spend inordinate amounts of time, effort, and money on aligning machinery to extreme precision when the equipment is not running and ignore off-line to running machinery movement. What has been accomplished if the machinery has been aligned to 1/2 mil per inch or better in an off-line condition if the machinery moves 20 or 30 mils when it running?

Movement of the centerline of the machine cases with respect to its baseplate or frame

The principle of operation in this category is to measure points very near the centerline of rotation at the inboard and outboard ends of each machine case in the drive train when the equipment is not running, start the machinery up, periodically observe the measurement points during operation until they have stopped moving, and compare the running positions of the points to the off-line positions of the points. The 'reference' point is the machinery baseplate or foundation.

It is also feasible to reverse the sequence of data collection. In other words, take measurement readings when the machines are running, shut the machinery down and take readings again (i.e. running to off-line). In fact, it is recommended that both off-line to running (OL2R) and running to off-line (R2OL) measurements be taken just to see if the movement is consistent.

Tools and techniques that fall into this category are:

- calculating machine case thermal expansion using the strain equation
- inside micrometer / tooling ball / angle measurement devices
- proximity probes with water cooled stands

Movement of the centerline of the machine cases with respect to some remote reference or observation point

The principle of operation in this category is to observe a series of points very near the centerline of rotation at the inboard and outboard ends of each machine case from a position away from the drive train when the equipment is not running, start the machinery up, periodically observe the measurement points during operation until the measurement points have stopped moving, and compare the running positions of the points to the off-line positions of the points.
Tools and techniques that fall into this category are:

- optical alignment equipment
- laser - detector measurement systems
- videometry system (dual CCD's) & electronic theodolite system (see references University of Calgary)

Movement of one machine case with respect to another machine case

The objective of these techniques is to measure the positional change of one machine case with respect to another machine case.
Tools and techniques that fall into this category are:

- alignment bars or custom fixtures with proximity probes
- laser / detector systems with custom fabricated brackets or

special mounting systems
- Ball-Rod-Tubing connector systems

Movement of one shaft with respect to another shaft

Since we are trying to determine where the shaft centerlines are actually moving from off-line to running conditions, it would seem that the ultimate method would be to watch the movement of one shaft with respect to the other shaft directly. The obvious problem with this is that when the machinery is running, the shafts are rotating and placing sensors on moving shafts seems to be an impossible task, yet there are two methods that have been invented to do just that.
Tools and techniques that fall into this category are:
- vernier - strobe method
- instrumented coupling system

Calculating machine case thermal expansion using the strain equation (machine case to baseplate category)

At the atomic level in solid materials, the temperature and volume of the material is dictated by the vibration of the individual

Strain Equation

$$\Delta L = L(\alpha)(\Delta T)$$

where:
ΔL = change in dimension/length (inches)
L= length of the object (inches)
α = coefficient of thermal expansion/contraction (in/ in-^{o}F).
T = change in temperature (^{o}F).

Figure 9-2. Calculating off-line to running machinery movement using the strain equation.

molecules. In other words, the hotter a solid material gets, the more the molecules vibrate and the farther apart the molecules are spaced. This phenomena causes changes in dimensions (i.e. strain) that can be calculated by the equation shown in figure 9-2.

The coefficients of thermal expansion for the majority of materials used in machinery casings and foundations are shown in Table 9-1. These coefficients can be used for temperatures between 32 - 212 degrees F. There is a slight shift in the value of the coefficient for higher or lower temperatures due to the non-linearity of molecular vibration in materials.

Material	**Coefficient of Thermal expansion (α) (in./in. degF)**
Aluminum alloys	12.5×10^{-6}
Brass (70% Cu - 30% Zn)	11.0×10^{-6}
Carbon Steel (AISI 1040)	6.3×10^{-6}
Cast iron (grey)	5.9×10^{-6}
Concrete	7.2×10^{-6}
Invar	0.68×10^{-6}
Nickel steel	7.3×10^{-6}
Stainless Steel	9.8×10^{-6}
Vulcanized rubber	45.0×10^{-6}
Nylon	55.0×10^{-6}

Table 9-1. Coefficients of Thermal Expansion for Different Materials

What we are doing is measuring the temperature of a machine case or support leg using thermometers, surface temperature probes such as RTD's or thermocouples, measuring the distance from the point where the feet are touching at the base to the centerline of rotation of the shaft, determining what type of material the machine case or support leg is made of, and plugging this data into the strain equation.

Key considerations for capturing good readings:

- carefully observe how the machine case is attached to the baseplate to insure that you measure the support leg, machine case, sway bar, or frame support that is truly repositioning the shafts' centerline
- capture several temperature measurements particularly during running conditions in the event that there is an uneven temperature profile on the support mechanism
- measure the expansion / contraction at both ends of each machine case and realize that the amount of OL2R movement may not be the same at both ends
- capture a set of readings from off-line to running conditions and another set of readings from running to off-line conditions to determine if there is a consistent pattern of movement

Advantages -

- relatively inexpensive and easy to do

Disadvantages -

- results may be suspect based on when and where the temperature reading was taken on the machine case
- unable to calculate how machinery moves laterally (sideways)

Infrared thermographic techniques to determine thermal profiles of rotating equipment

Rotating machinery casings transfer heat to the environment when running which is generated by friction, compression of process gases, steam flow, or motor stator windings. Surface temperatures at various locations on a casing vary widely and virtually no rotating equipment maintains a constant thermal gradient across the entire casing. Gas turbines, for instance, may have inlet casing temperatures below the ambient air temperature and six feet away have a 1200 degree F casing temperature at the combustor section. Since our eyesight is limited only to the visible spectrum (see information on the electromagnetic spectrum in Chapter 6), we are unable to 'see' the temperature gradient profile of machine cases as they emit the longer, infrared radiation. Infrared radiation can be categorized into four general ranges:

- actinic range - incandescent objects such as bulb filaments having wavelengths near the visible red region
- hot object range - objects with temperatures around 400 degrees F
- calorific range - objects with temperatures around 250 degrees F
- warm range - objects having temperatures below 200 degrees F (approx. 9000 nm)

The infrared radiation emitted from an object can be observed by thermography equipment. The type of equipment used for thermographic studies is shown in figure 9-3. These instruments scan the object for the infrared radiation and amplify the

Figure 9-3. Thermographic scanning equipment.

converted electrical signals from a supercooled photodetector onto a cathode ray tube (CRT) where a photographic image of the object can be recorded. Figure 9-4 shows a three stage centrifugal compressor case and figure 9-5 illustrates the temperature profile

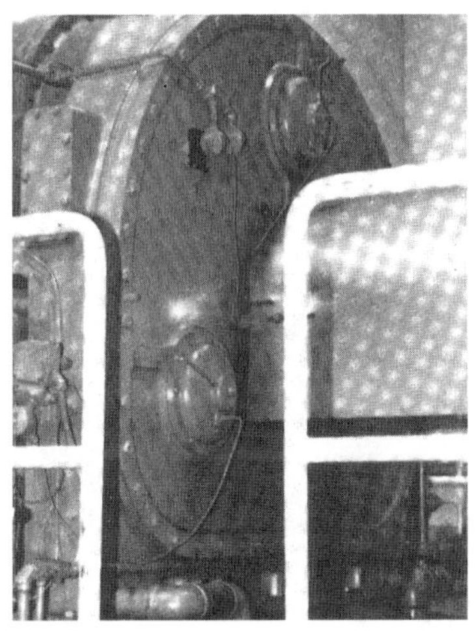

Figure 9-4. Compressor case.

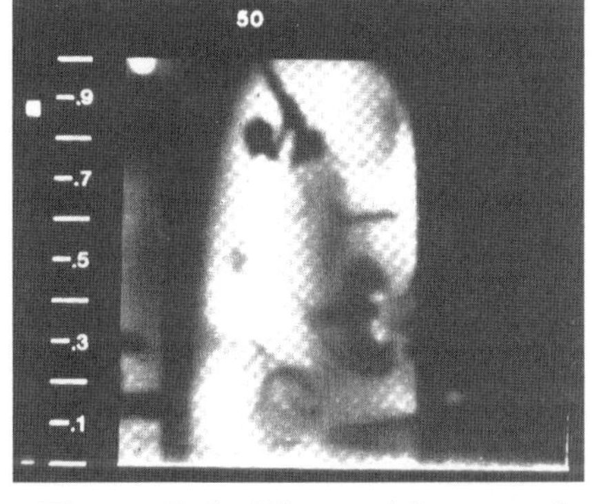

Figure 9-5. Thermal image of compressor case.

Figure 9-6. Axial compressor.

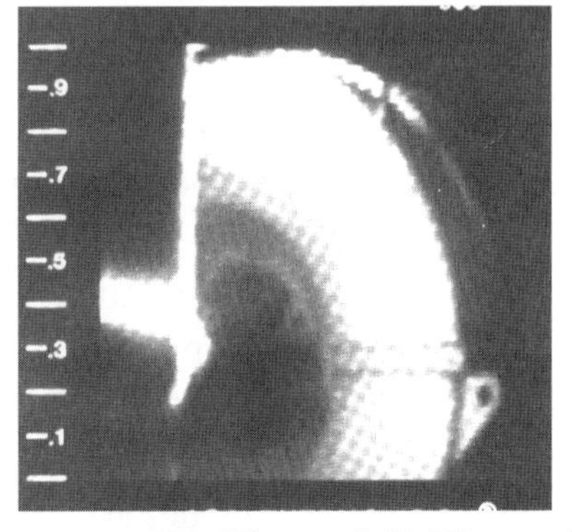

Figure 9-7. Thermal image of axial compressor.

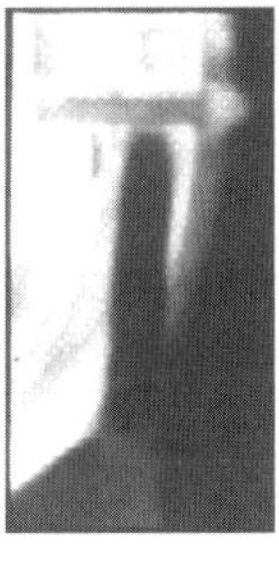

Figure 9-8. Thermal image of support leg.

when the compressor is running under full load. The white areas show where the infrared radiation (heat) is the greatest. The hottest areas in this image are approximately 135 degrees F.

Figure 9-6 shows an axial flow compressor with rigid supports at the inlet end and flexible supports at the discharge end. Figure 9-7 illustrates the thermal profile of the discharge end with the compressor running under load (note the hot spot at the 1 o'clock position). Figure 9-8 shows a closer view of the flexible support leg. The lifting eye is at the left side of the photograph and the flex leg is the black portion just to the right of the lifting eye. The photograph clearly shows that the support leg stays at an ambient temperature and does not expand thermally (as originally thought when the machinery was installed).

Although movement of rotating machinery casings does not occur solely from temperature changes in the supporting structures and the casings themselves, infrared thermographic studies can assist in understanding the nature of the thermal expansion taking place.

Inside micrometer / tooling ball / angle measurement devices (machine case to baseplate category)

One of the easiest and least expensive techniques for measuring casing movement in the field can be done with inside micrometers or with an inside micrometer and an inclinometer (angle measuring device). Tooling balls or similar reference point devices are rigidly attached to the foundation and to the inboard and outboard ends of the machine cases. Distances between the tooling balls (and angles if desired) are captured between each tooling ball when the machinery is at rest and then measured again when the equipment is running and has stabilized thermally. Three tooling balls can be set up in a triangular pattern or four tooling balls can be set up in a four sided 'pyramid' arrangement as illustrated in figure 9-9.

Tooling ball arrangements are placed at both ends of both machines. The tooling balls attached to the machinery case should be as close to the centerline of rotation as possible.

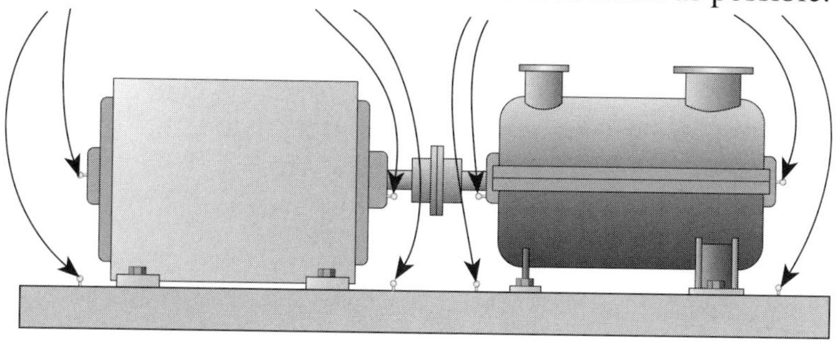

Tooling Ball Set-up

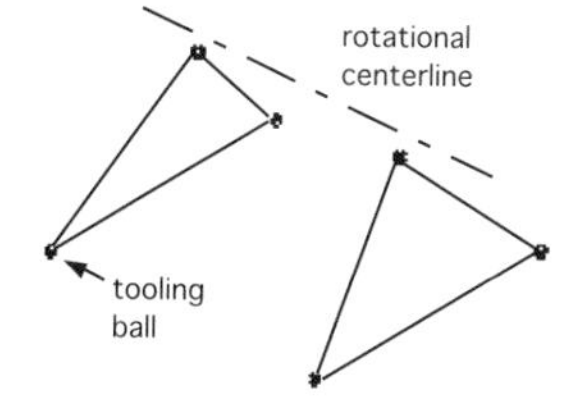

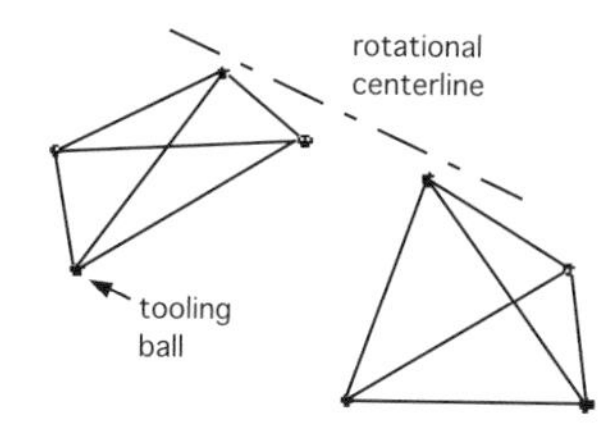

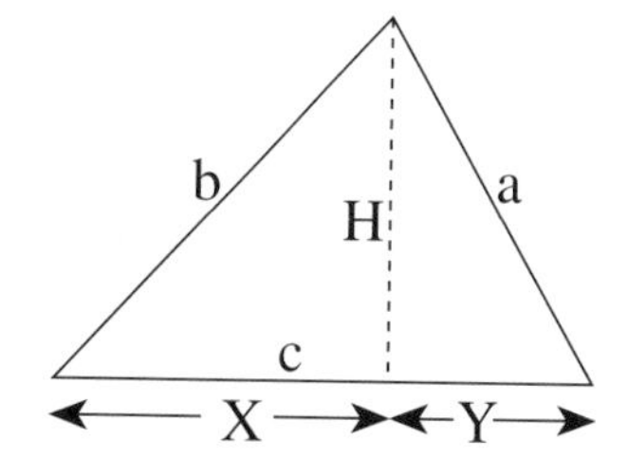

Basic mathematics for oblique triangles :

$$X = \frac{c}{2} + \frac{b^2 - a^2}{2c}$$

$$Y = \frac{b^2 + c^2 - a^2}{2c}$$

$$H = \sqrt{b^2 - x^2}$$

Figure 9-9. Taking inside micrometer readings between 3 tooling balls (triangular arrangement) or 4 tooling balls (pyramid arrangement).

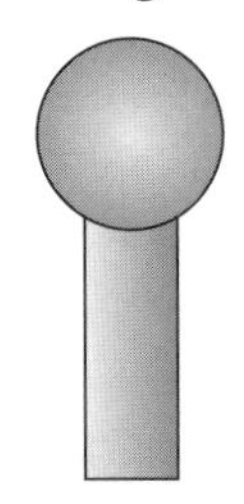

Tooling balls can be purchased from machine tool suppliers or can be 'homemade' as shown below. If standard tooling balls are used, holes must be drilled in the machine case and baseplate for installation. The 'homemade' design can be attached to machine case and baseplate with epoxy or dental cement and then removed when the survey is complete.

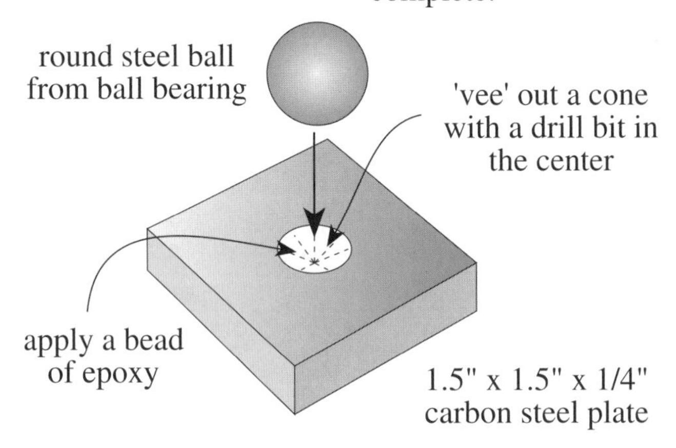

Figure 9-10. 'Homemade' tooling ball..

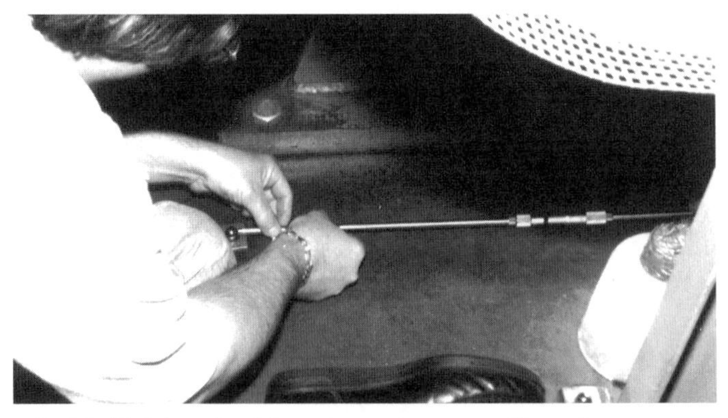

Figure 9-11. Taking inside micrometer readings between tooling balls.

These measurements can then be triangulated mathematically into vertical and lateral components (using the triangular arrangement) or into vertical, lateral, and axial component distances (using the four sided 'pyramid' arrangement). By comparing the coordinates of the tooling ball mounted on each end of all the machine cases from OL2R (or from R2OL) positional changes can be determined.

Key considerations for capturing good readings:

- remember that you will probably be dealing with oblique triangular arrangements and not right angle triangles (i.e. watch your math)
- it is important to have stable positions for the tooling balls
- the tooling ball on the machine case should be located as close as possible to the centerline of rotation since we are trying to determine where the shafts are going (if the bearing moves, the shaft is sure to move with it)
- recommend that concave tips be used at both ends of the inside micrometer to sit accurately on the round tooling balls when taking measurements
- keep the inside micrometer away from heat sources to prevent the mike from thermally expanding
- triangular tooling ball arrangements assume that there will be motion in the horizontal and vertical planes only
- have two or more people take measurements and compare notes to insure that the readings are identical (or at least close)
- capture a set of readings from off-line to running conditions and another set of readings from running to off-line conditions to determine if there is a consistent pattern of movement

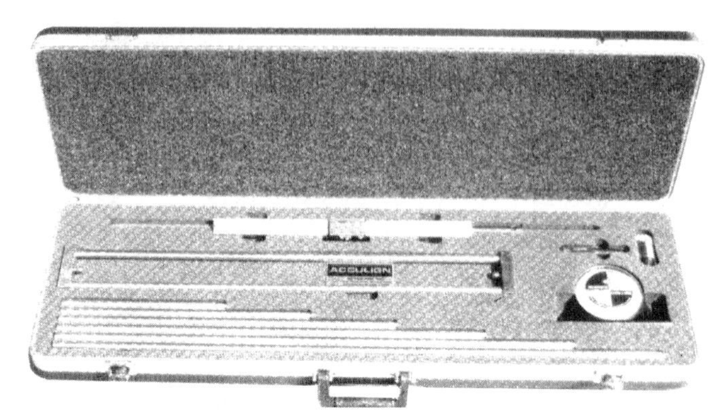

Figure 9-12. Acculign kit. Photo courtesy of Acculign Inc., Willis, TX.

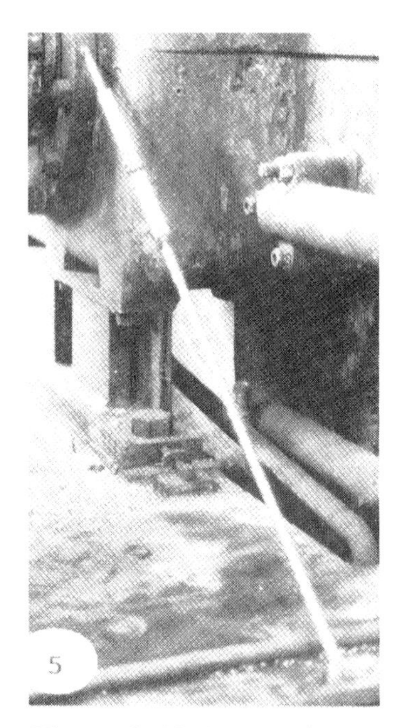

Figure 9-13. Measuring distances with the Acculign kit. Photo courtesy of Acculign Inc., Willis TX.

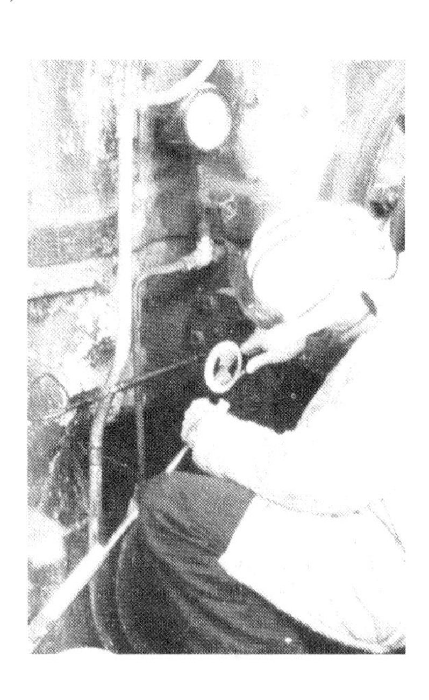

Figure 9-14. Measuring angles with the Acculign kit. Photo courtesy of Acculign Inc., Willis TX.

Advantages -

- relatively inexpensive
- somewhat easy to set up for vertical movement

Disadvantages -

- mathematics somewhat tedious, particularly on four-sided pyramid arrangements
- caution must be taken during running measurements since one end of the inside micrometer is near a rotating shaft
- if one or more tooling balls disengage from their positions (i.e. it worked out of its hole or the epoxy gave away), you'll probably have to start over

Figure 9-12 shows some specialized tooling used for the measurements and figures 9-13 and 9-14 show typical distance and angle measurements being taken on rotating machinery.

Proximity probes with water cooled stands (machine case to baseplate category)

This technique was originated and popularized by Charlie Jackson and has been successfully employed on many rotating machinery drive systems. The proximity probes are attached via a bracket to a water-cooled pipe stand which is firmly anchored to the machinery foundation near each bearing. To maintain a stable reference point, water should be circulated through the pipe stand or the pipe should be insulated and filled with a water - glycol or antifreeze solution to prevent as little dimensional change as possible to the pipe stand itself from radiant heat emitted from the machinery. The probes are mounted to a bracket attached to the pipe stand and positioned to monitor a metal block (usually steel) affixed to each end of every machine case in the drive train. OL2R movement can be monitored in the horizontal, vertical, and axial directions. The probes could also be positioned to monitor the movement of the shaft directly since it is really the position of the shaft that you are trying to determine from OL2R conditions.

Key considerations for capturing good readings:

- insure that the pipe stands are rigidly attached to a stable reference point on the frame or foundation and that they maintain a constant temperature throughout the OL2R measurement process

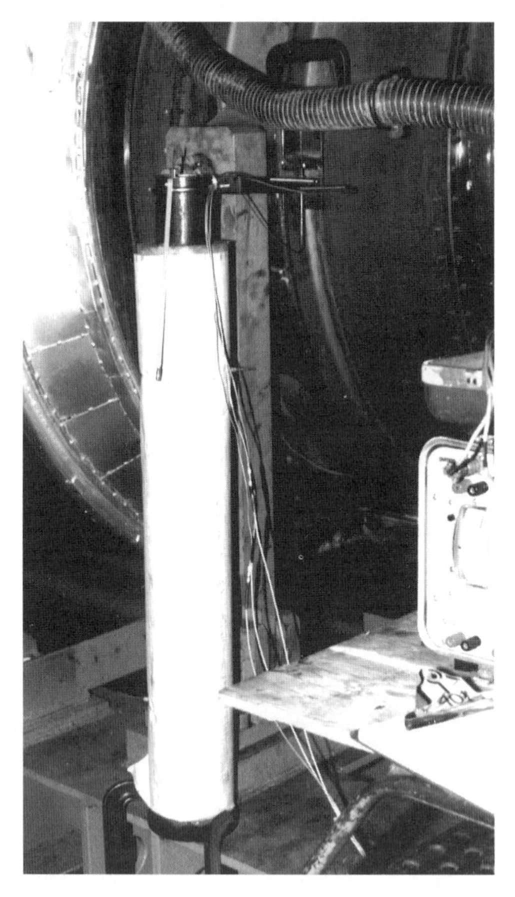

Figure 9-15. Water-cooled pipe stand and proximity probes monitoring exhaust end of gas/power turbine.

- insure that the probe tips are far enough apart to prevent any 'cross-field' effects from one probe to another that will affect accurate gap measurements
- probes should always be statically calibrated to the same type of material being observed since the gap vs. voltage characteristics are different from one material to another
- if the direction of machinery movement is not known when the probes are initially gapped, some adjustments may be necessary after the first attempt in case the target (or shaft) is moving too close or too far away from the probe tip to keep the probe within its linear range

Figure 9-16. Water-cooled stands with proximity probes. Photo courtesy of Charlie Jackson, Texas City, TX.

Figure 9-17. Close up view of proximity probes and target surface monitoring motion in all three directions.

Figure 9-18. Water-cooled stands with proximity probes. Photo courtesy of Charlie Jackson, Texas City, TX.

- standard probes (200 mv/mil sensitivity) are usually good for gap changes near 80 mils, and some manufacturers can supply special probes able to measure up to a half inch of gap change

Advantages -

- extreme accuracies possible with a good set-up
- capable of monitoring motion in all three directions (vertical, lateral, axial)
- continuous monitoring possible

Disadvantages -

- pipe stands must be mounted at both ends of each machine case
- cannot measure any change in the machinery foundation itself

Figure 9-19. Water-cooled stands with proximity probes. Photo courtesy of Charlie Jackson, Texas City, TX.

- somewhat expensive since pipe stands have to be fabricated, and probes, cables, proximitors, readout devices, and power supplies have to be purchased
- potential for inaccurate measurement when monitoring the shafts directly, particularly if a considerable amount of movement occurs (you're taking a reading on a curved surface)

Optical alignment equipment (remote reference point to machine case category)

Optical tooling levels and jig transits are one of the most versatile measurement systems available to determine rotating equipment movement. Figures 9-20 to 9-22 show the different types of optical instruments for alignment. This section will deal specifically with their ability to measure OL2R movement, but in

Figure 9-20. Jig transit (left) and tilting level (right) on stands.

Figure 9-21. Jig Transit. Photo courtesy Brunson Instrument Co., Kansas City, MO.

Figure 9-22. Universal transit square. Photo courtesy Brunson Instrument Co., Kansas City, MO.

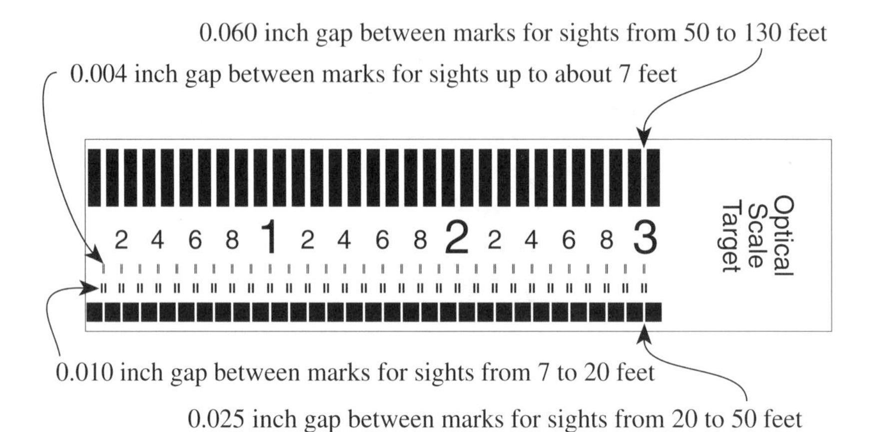

Figure 9-23. 3" scale target.

Figure 9-24. 10" scale targets attached to compressor case to monitor vertical movement.

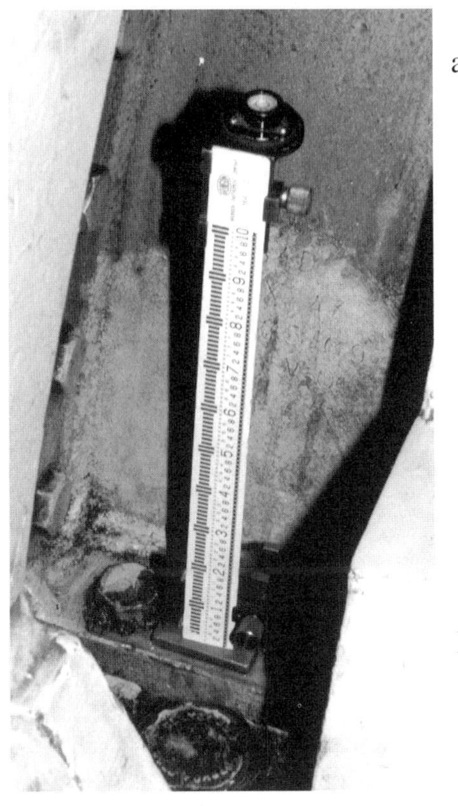

Figure 9-25. 10" scale target and level bubble with magnetic base attached to electrical generator bearing to monitor vertical movement.

Figure 9-26. Optical micrometer. Photo courtesy Brunson Instrument Co., Kansas City, MO.

no way will begin to explain their full potential for many other uses such as leveling foundations, squaring frames, roll parallelism, and a plethora of other tasks involved in level, squareness, flatness, vertical straightness, etc.

The scale targets used for optical tooling are shown in figure 9-23 to 9-25. There are generally four sets of sighting marks on the scales for centering of the crosshairs when viewing through the scope. An optical micrometer is attached to the instrument and can be positioned in either the horizontal or vertical direction. The adjustment wheel is used to align the crosshairs between the scale sighting lines on the targets. When the micrometer wheel is rotated, the crosshair appears to move up or

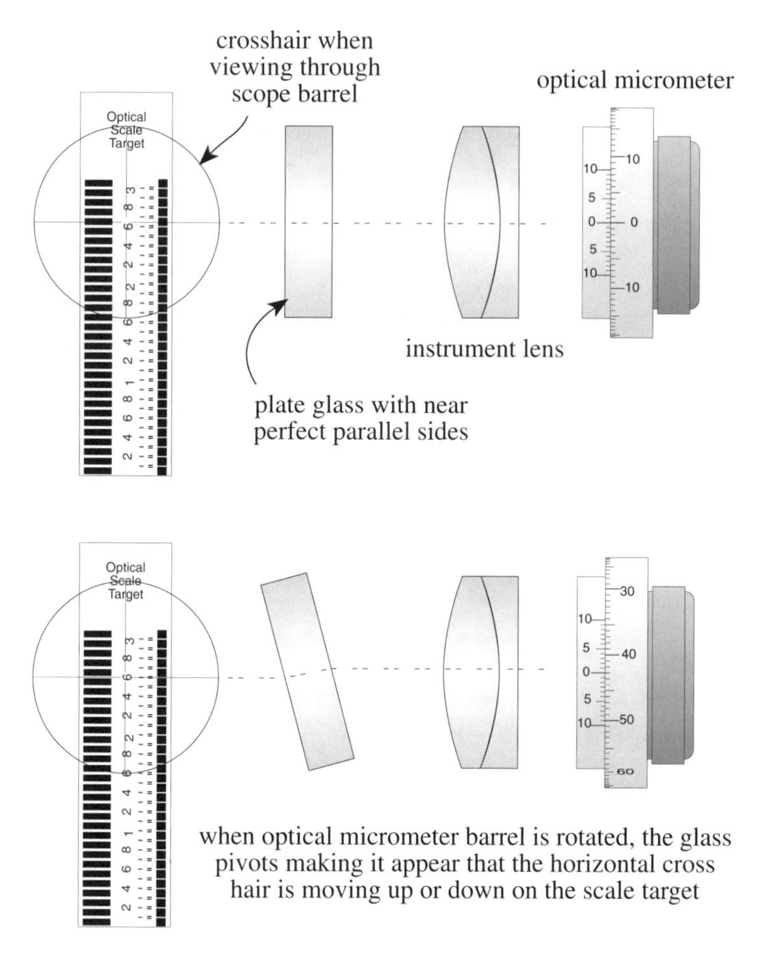

Notice in the upper drawing that when the optical micrometer is in the zero position, the horizontal crosshair is between 2.6 and 2.7 on the scale target but the crosshair is not exactly aligned with any of the marks. By rotating the micrometer drum, the horizontal crosshair is aligned at the 2.6 mark on the scale target. The inch and tenths of an inch reading is obtained off the scale target, the hundredths and thousandths of an inch reading is obtained off the micrometer drum position.

Figure 9-27. How an optical micrometer works.

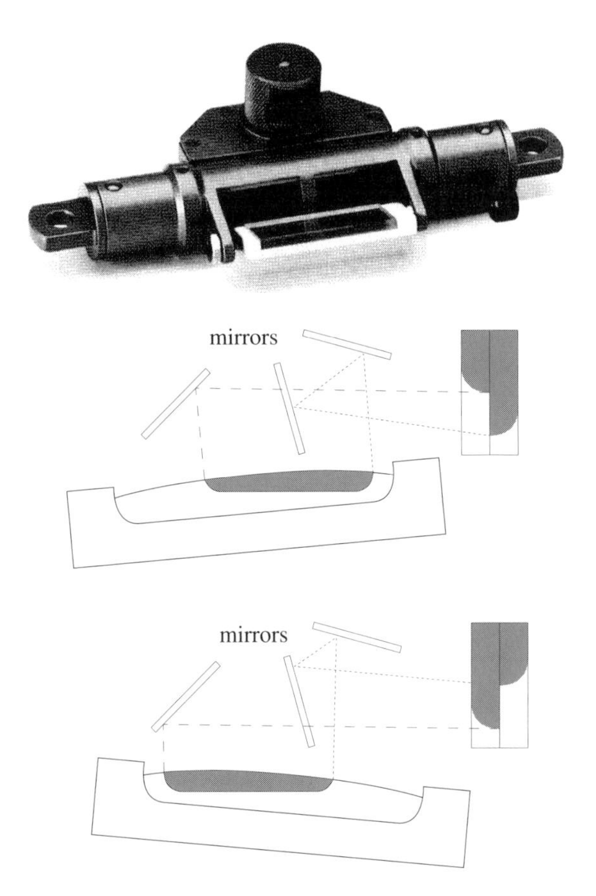

Figure 9-28. How a split coincidence level bubble works. Photo courtesy Brunson Instrument Co., Kansas City, MO.

down along the scale target (or side to side depending on the position of the micrometer). Once the crosshair is lined up between a pair of sighting marks, a reading is taken based on where the crosshair is sighted on the scale and the position of the optical micrometer. The inch and tenth of an inch reading is visually taken by observing where the cross-hair lines up with the scale target numbers, the hundredth and thousandths of an inch reading is taken on the micrometer drum as shown in figure 9-27.

The Peg Test

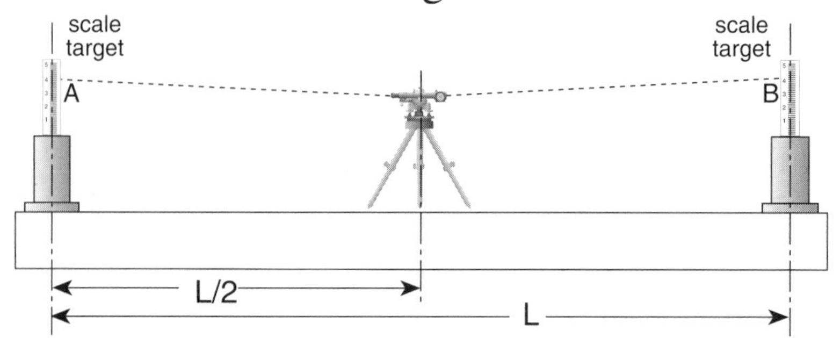

1) Set two scales apart by distance L (usually 40 ft.) and the telescope exactly half way between both scales. Use a plumb bob attached to the instrument stand in line with the azimuth axis to aid in positioning the instrument exactly at the half way point. Accurately level the instrument using the split coincident level.

2) Alternately take four readings on scale no. 1 (reading A) and scale no. 2 (reading B). Record and average these readings.

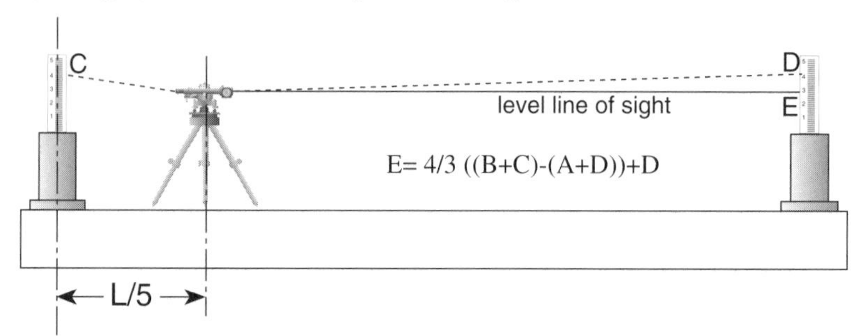

3) Move the scope to the 1/5 L position, level the scope, and alternately take four readings on scale no. 1 (reading C) and scale no. 2 (reading D). Record and average these readings.

4) If the instrument is calibrated, A minus B should equal C minus D. Should this not be the case, the coincident level calibration adjustment nuts can be adjusted to position the leveled line of sight to be set at reading E. Consult the manufacturer of the scope to assist in making this adjustment. Be extremely cautious when doing this to prevent damage to the instrument and coincident leveling system.

Note: Prior to using any optical instrument, it is recommended that the Peg Test be performed to insure measurement accuracy. At 40 ft., the accuracy of the scope is plus or minus 0.0024".

Figure 9-29. Coincidence level calibration test.

The extreme accuracy (one part in 200,000 or 0.001" at a distance of 200 inches) of the optical instrument is obtained by accurately leveling the scope using the split coincidence level mounted on the telescope barrel as shown in figure 9-28.

Prior to using any optical instrument, it is recommended that the Peg Test be performed to insure level accuracy as illustrated in figure 9-29. Figures 9-30 and 9-31 show the basic procedure of how to properly level the instrument. If there is any change in the split coincidence level bubble gap during the final check, go back and perform this level adjustment again. It might take a half an hour to an hour to get this right, but it is time well spent. It is also wise to walk away from the scope for about 30 minutes to determine if the location of the instrument is stable and to allow some time for your eyes to uncross. If the split coincident bubble has shifted during your absence, start looking for problems with the stand or what it is sitting on. Correct the problems and re-level the scope.

I cannot overemphasize the delicacy of this operation and this equipment. It is no place for people in a big hurry with little patience. If you take your time and are careful and attentive when obtaining your readings, the accuracy of this equipment will astonish you.

Optical parallax

As opposed to binoculars, 35 mm cameras, and some microscopes that have one focusing adjustment, the optical scope has two focusing knobs. There is one knob used for obtaining a clear, sharp image of an object (e.g. the scale target) and another adjustment knob that is used to focus the crosshairs (reticle pattern). Since your eye can also change focus, adjust both of these knobs so that your eye is relaxed when the object image and the superimposed crosshair image is focused on a target.

How to level optical tilting levels and jig transits

1) Set the instrument stand at the desired sighting location, attach the alignment scope to the tripod or instrument stand and level the stand using the 'rough' circular bubble level on the tripods (if there is one on the tripod). Insure that the stand is steady and away from heat sources, vibrating floors, and curious people who may want to use the scope to see sunspots.

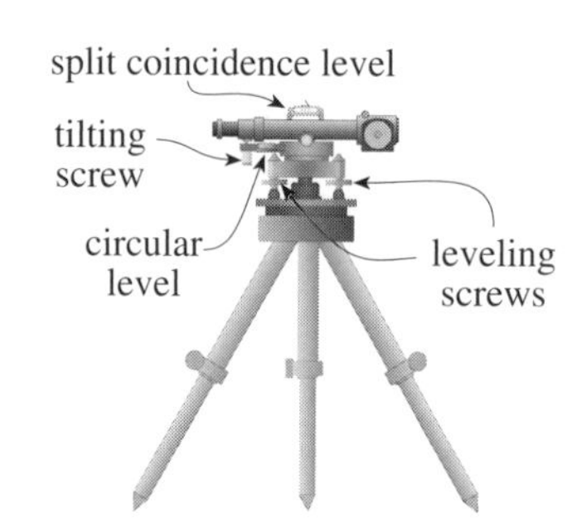

2) Rotate the scope barrel to line up with two of the four leveling screws and adjust these two leveling screws to roughly center the split coincidence level bubble in the same tilt plane as the two screws that are being adjusted as shown.

3) Rotate the scope barrel 90 degrees to line up with the other two leveling screws to completely center the bubble in the circular level as shown.

4) If the circular level is still not centered, repeat steps 2 and 3.

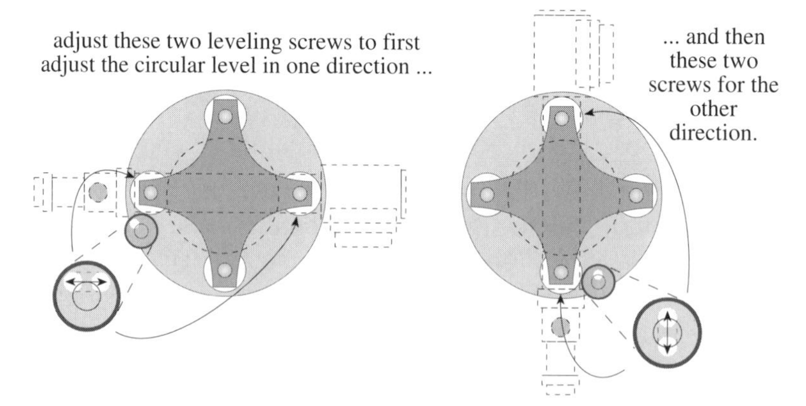

Figure 9-30. How to level optical instruments.

How to level optical tilting levels and jig transits

5) Once again rotate the scope to line up with two of the leveling screws as covered in step 2. Adjust the tilting screw to center the split coincidence level on the side of the scope barrel as shown.

6) Rotate the scope barrel 180 degrees and note the position of the two bubble halves. Adjust the two leveling screws in line with the scope barrel so that the gap between the two bubble halves is exactly one half the original gap.

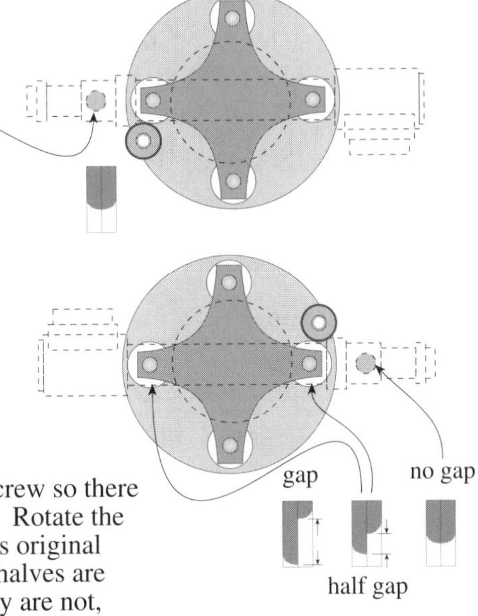

7) At this point, adjust the tilting screw so there is no gap in the two bubble halves. Rotate the scope barrel back 180 degrees to its original position and see if the two bubble halves are still coincident (i.e. no gap). If they are not, adjust the two leveling screws and the tilting level screw again as shown and rotate the scope barrel back 180 degrees until there is no gap when swinging back and forth through the half circle. The two leveling screws should be snug but not so tight as to warp the mounting frame.

8) The last step is to rotate the scope barrel 90 degrees to line up with the two remaining leveling screws yet to be fine adjusted. Follow the same procedure as outlined in step 6 and 7 above. When these adjustments have been completed, the split coincidence bubble should be coincident when rotating the scope barrel through the entire 360 degrees of rotation around its azimuth axis.

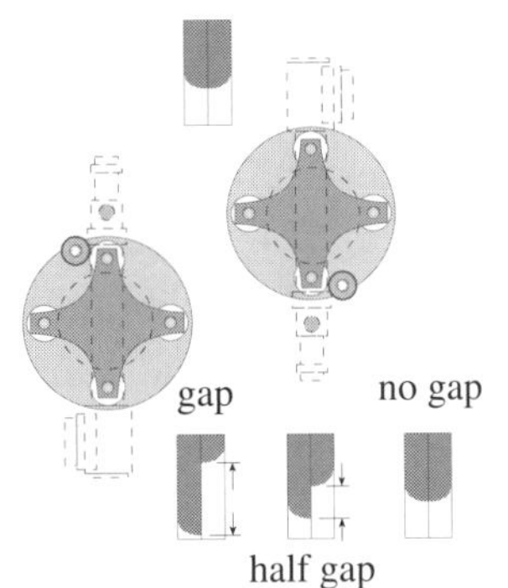

Figure 9-31. How to level optical instruments.

Adjusting the focusing knobs:

1) With your eye relaxed, aim at a plain white object at the same distance away as your scale target and adjust the eyepiece until the crosshair image is sharp.
2) Aim at a scale target and adjust the focus of the telescope.
3) Move your eye slightly sideways and then up and down to see if there is an apparent motion between the crosshairs and the target you are sighting. If so, de-focus the telescope and adjust the eyepiece to re-focus the object. Continue alternately adjusting the telescope focus and the eyepiece to eliminate this apparent motion.

Using optical tooling for measuring machinery movement

As described earlier, this OL2R method attempts to determine how points at the inboard and outboard ends of each machine are moving up or down and side to side. Selecting where the measurement points on each end of each piece of equipment is very critical. We are trying to determine where the centerlines of rotation are moving to and therefore must measure points as close as possible to the centerline of rotation of the shafts.

Either an optical tilting level or a jig transit can be used to measure the up and down movement of the scale targets. An optical tilting level will typically not work as well as a jig transit to measure side to side movement since the scope barrel is not capable of sweeping through a large enough vertical arc to capture the readings. Two people are required to measure side to side movement: the scale target holding person and the observer. Measuring side to side movement is a little more tedious since the scale targets have to be held by hand and there needs to be good communication between the scale target holding person and the observer to insure that the scale targets are at a precise right angle to the observer's line of sight.

Key considerations for capturing good readings:

- it is important to have stable platforms for the optical scale targets or detectors
- scale target should be located as close as possible to the bearings since we are trying to determine where the shafts are going (if the bearing moves, the shaft is sure to move with it)
- magnetic base holders should be used to hold the scale targets insuring a stable target position
- clamp-on circular level bubble sets should be used on the scale target to insure scale targets are in a pure vertical position
- if possible, try to keep the scale targets in place from off-line to running conditions
- keep the scale target holding platforms clean
- move slowly when working around the tilting level or jig transit and stand (if your foot or arm bumps it, it's probably not level any more)
- readings should be taken at night if equipment is located outdoors to prevent thermal instability of the tripod / stand when the sun alternately heats and cools the stand
- capture a set of readings from off-line to running conditions and another set of readings from RO2L conditions to determine if there is a consistent pattern of movement

Advantages -

- extreme accuracies possible (1 - 2 arcseconds) when precisely leveled
- somewhat easy to set up for vertical movement

Disadvantages -

- if the machinery is vibrating excessively, it appears that the scale targets cannot be focused when taking the running measurements

Optical alignment OL2R procedure for vertical (up/down) measurements

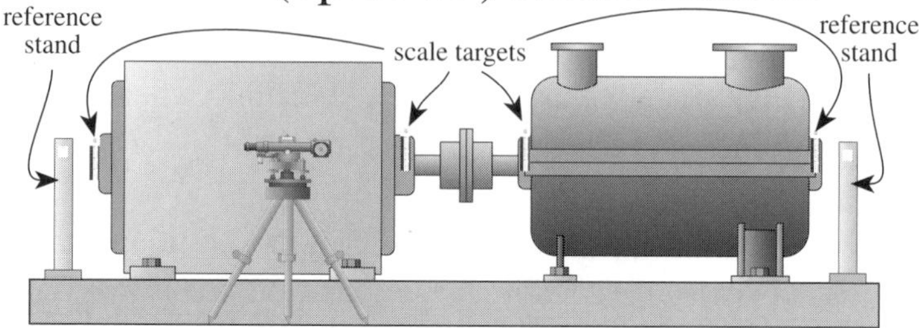

1) Check calibration of the instrument (see Peg Test).

2) Select suitable scale positions at the inboard and outboard ends of each piece of machinery in the drive train. The 'platforms' that the scale targets will be sitting on should be stable and slightly below the centerline of rotation and usually near the bearings. You can use 2" x 2" pieces of angle iron firmly affixed (preferably bolted or epoxied) to the machine case or bearing housing. Try to insure that the surface that the scale targets will sit on is relatively level and flat. It is also advisable to install reference stands at each end of the drive train and affix a 'stick-um' crosshair target to the reference stand. You can make these stands out of 3" or 4" pipe, fill them with a water glycol or antifreeze solution, insulate the pipe, bolt or clamp them to the frame or floor, and monitor the water temperature to insure thermal stability.

3) Set the optical instrument and stand at some remote reference point away from the drive train where a stable point in space can be established but close enough to maintain the maximum possible accuracy of the readings.

4) Accurately level the instrument and take a set of readings at each target scale mounted on the machinery when it is 'off-line' (i.e. not running) occasionally checking back to the reference targets at each end of the drive train to insure that you are maintaining the same vertical elevation (i.e. shooting through the same horizontal plane).

5) Run the machinery at normal conditions and allow the equipment to stabilize its position (this can take hours or even days).

6) Check the level accuracy and take a similar set of readings at each target scale occasionally checking back to the reference targets at each end of the drive train to insure that you are maintaining the same vertical elevation.

7) Compare the off-line set of readings to the running set of readings to determine the amount and direction of the movement of each scale.

Figure 9-32. How to take vertical OL2R measurements.

Optical alignment OL2R procedure for horizontal (side to side) measurements

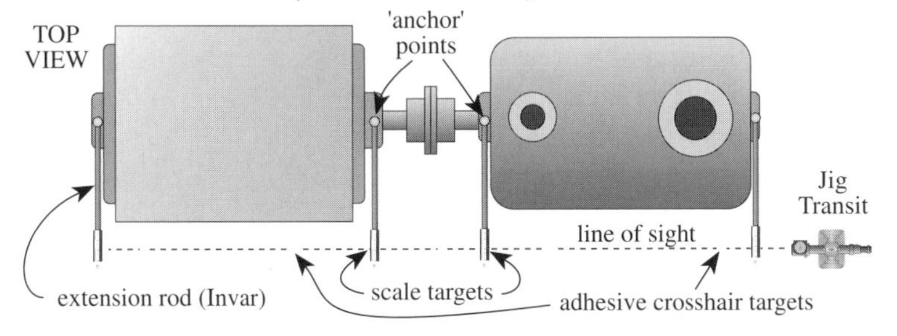

1) Check calibration of the jig transit (see Peg Test). Two people are required to do this procedure .. the scale target holding person and the observer.
2) Select suitable scale 'anchor' points at the inboard and outboard ends of each piece of machinery in the drive train. The points or 'anchors' that the scale targets (and probably extension rods) will be touching should be stable and directly above or below the centerline of rotation and usually near or at the bearings. You can use tooling balls firmly affixed to the machine case or bearing housing as the 'anchor' points to hold the scale target against for reference.
3) Set the jig transit and stand at some position along one side of the drive train where a stable point in space can be established insuring that measurements can be taken at each bearing location when the scale target / extension rod is placed at each 'anchor' point at every bearing location. Orient the optical micrometer on the scope barrel to allow variation in the position of the vertical crosshair when the micrometer barrel is rotated.
4) Accurately level the instrument, then loosen the vertical sweep axis screw allowing the scope to tilt up and down. Affix several (at least 2) adhesive crosshair targets to the foundation or floor along the full length of the drive train establishing a vertical reference line / plane. Take a set of readings at each bearing location by holding the scale target / extension rod when placed at each 'anchor' point when the machinery is 'off-line' (see 'waving scales') occasionally checking back to the adhesive crosshair targets attached to the foundation or floor to insure that you are maintaining the same horizontal position (i.e. keeping in the same vertical reference plane).
5) Run the machinery at normal conditions and allow the equipment to stabilize its position (this can take hours or even days).
6) Check the level accuracy and take a similar set of readings at each bearing location occasionally checking back to the adhesive crosshair targets attached to the foundation or floor to insure that you are maintaining the same horizontal position.
7) Compare the off-line set of readings to the running set of readings to determine the amount and direction of the movement of each scale target / bearing location.

Figure 9-33. How to take horizontal OL2R measurements.

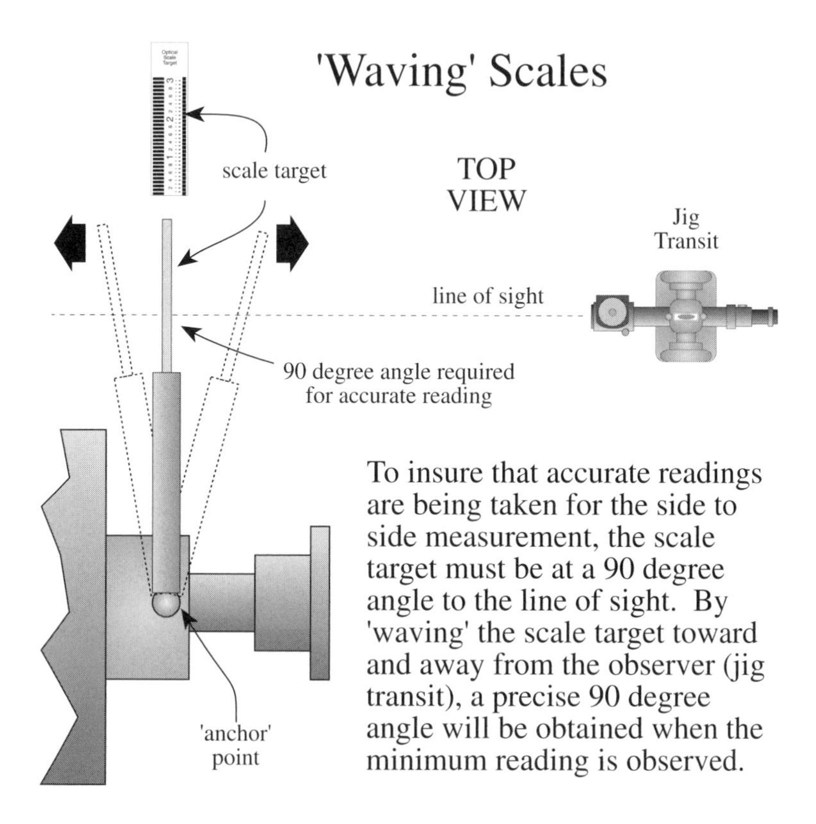

Figure 9-34. How to take accurate horizontal OL2R measurements.

Alignment bars with proximity probes (machine case to machine case category)

This device was invented in the early 1970's by Ray Dodd while working at Chevron. The machinery alignment bar OL2R system can be likened to electronic reverse dial indicators.

Two 'bars' are used, a 'probe' bar and a 'target' bar. The

The alignment bars are manufacturered from sections of tubing supported by triangular shaped braces. The proximity probes and targets can be positioned along the length of the tubing by means of a locking screw. This design attempts to minimize bar flutter from coupling windage when the bars are mounted inside a coupling guard.

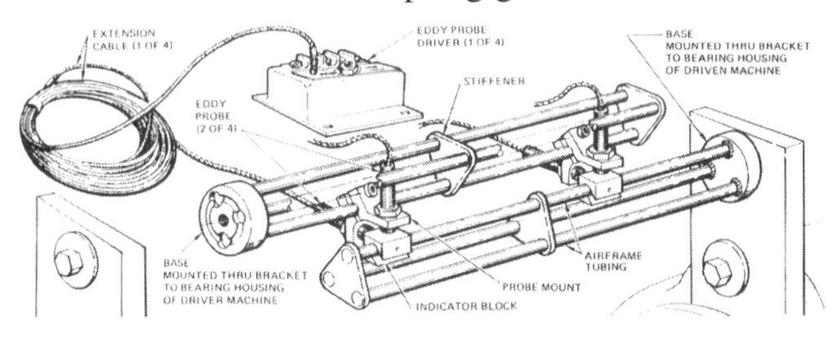

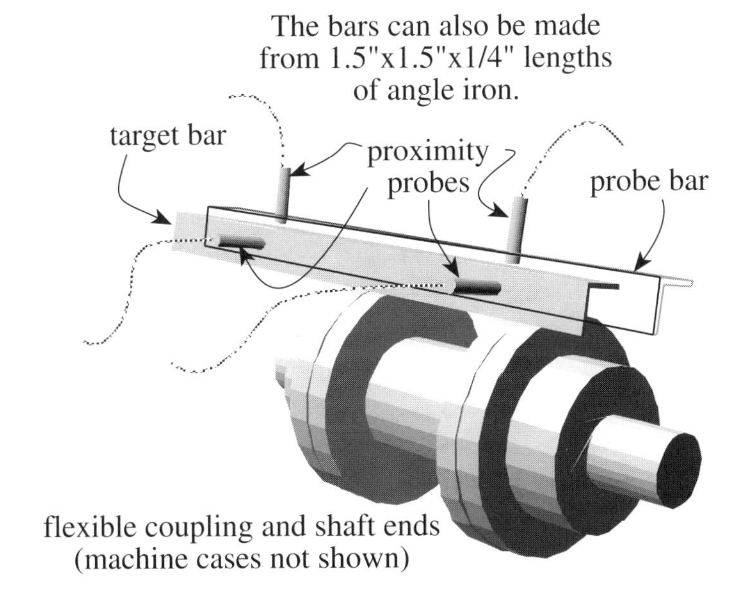

Figure 9-35. Alignment bar arrangements. Upper diagram courtesy of SKF Condition Monitoring, San Diego, CA.

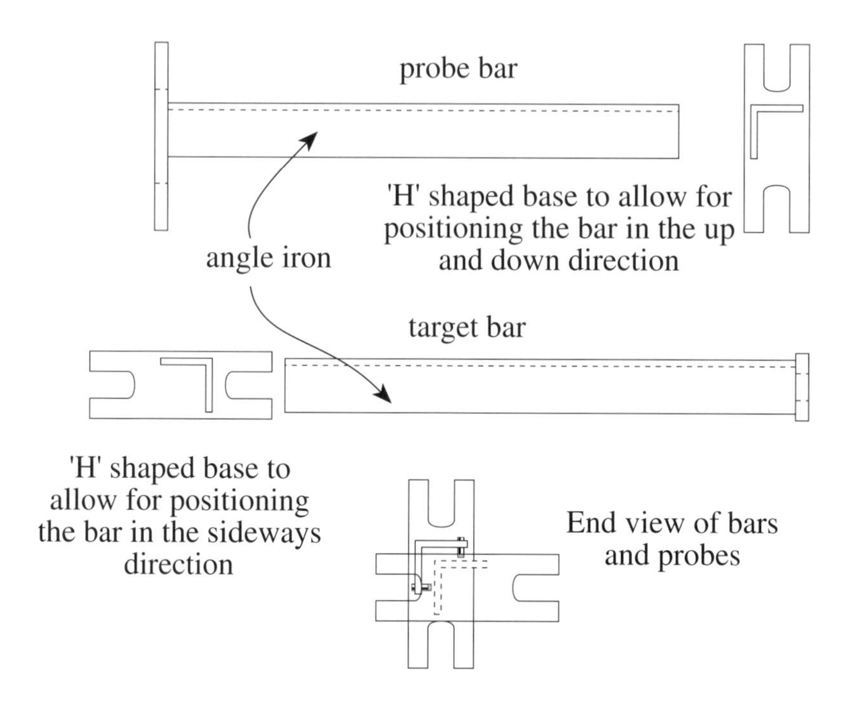

Figure 9-36. Fabricated alignment bar design.

'probe' bar is attached near the inboard (coupling end) bearing as close as possible to the centerline of rotation on one machine case. Four proximity probes are attached to this bar, two vertically oriented probes and two horizontally oriented probes. These probes sets are mounted at two locations along the length of the bar which observe two target blocks (usually steel) that are attached to another bar mounted to the other machine case near the inboard (coupling end) bearing of that machine. The probe bar and target bar do not touch each other.

Prior to starting the drive train, a set of gap readings are taken on each of the two vertically oriented probes and each of the two horizontally oriented probes. The drive system is then started up and operated under normal running conditions until the probe

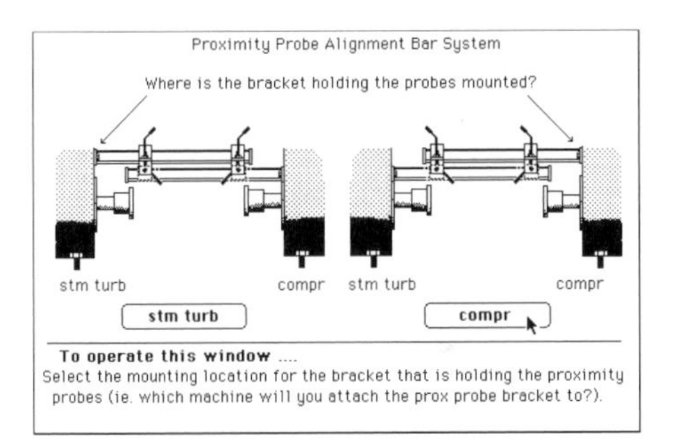

The probe and target bars can be mounted on either machine case but you have to know what's mounted where.

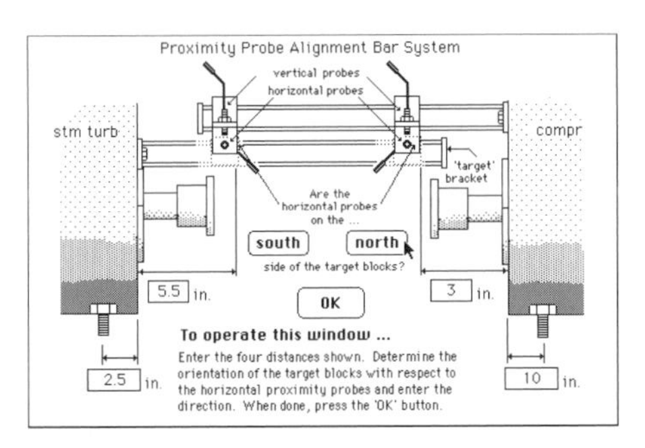

The distances from the inboard feet to the points where the proximity probes are taking readings on the target bar must be known, as well as the orientation of the horizontal probes.

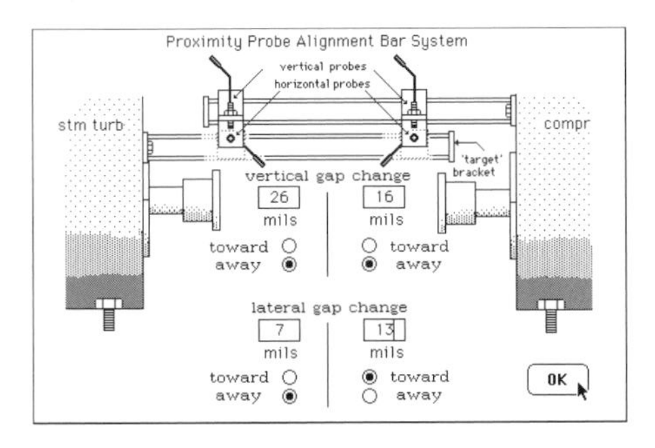

The gap changes from off-line to running (or vice versa) need to be recorded. Remember to capture not only the amount of the gap change but also the direction (increasing gaps means the probe moved away from the target).

Figure 9-37. How to set up and capture data using the alignment bar system. Software screen shots courtesy Turvac Inc., Cincinnati, OH.

Figure 9-38. Power supply, proximitors, and readout equipment for alignment bar system.

gaps have stabilized. Relative machinery casing movement can be determined by comparing the gaps on each probe before and after the equipment was running. Strip chart recorders (or similar devices) can be set up to monitor the rate of change of gap during warm up and on line operating conditions.

Key considerations for capturing good readings:

- the probe and target bars should be attached to each machine case as close as possible to the centerlines of rotation to accurately determine shaft motion, not casing expansion or bearing housing warpage
- it is important to have target surfaces at precise 90 degree angles to the proximity probes
- the bars can be positioned inside or outside the coupling guard / shroud with precautions taken to prevent excessive vibration of the bars from coupling 'windage' if mounted

inside the guard
- capture a set of readings from off-line to running conditions and another set of readings from running to off-line conditions to determine if there is a consistent pattern of movement

Advantages -
- fairly accurate measurements possible with proper set-up
- capable of measuring movement in vertical and horizontal directions
- can continuously monitor positions of machinery without disturbing sensors or bars
- if the machinery is vibrating excessively when taking the running measurements, the proximity probes average the oscillation effectively to insure accurate distances between the probe tips and the targets

Disadvantages -
- custom brackets and bars have to be designed, fabricated, and carefully installed on each machine case as close as possible to the centerline of rotation
- if the machinery is moving considerably from OL2R conditions, the proximity probes can bottom out on the target or move out of the linear range of the probes
- somewhat expensive since custom braces and bars have to be fabricated, and probes, cables, proximitors, readout devices, and power supplies have to be purchased

Laser - detector systems and PIBZLT mounts (machine case to machine case category)

Laser / detector systems can also be used to measure off - line to running machinery movement in the machine case to machine case measurement category.

In a very simple setup, where small amounts of relative movement between machinery cases is present, a laser (or laser-detector depending on what system you use) could be mounted at or near the centerline of rotation of one machine near the inboard bearing, and the detector (or prism depending on what system you use) could be mounted at or near the centerline of rotation of the other machine near its inboard bearing. With the machinery off-line, the laser / detector system can be oriented to have the laser

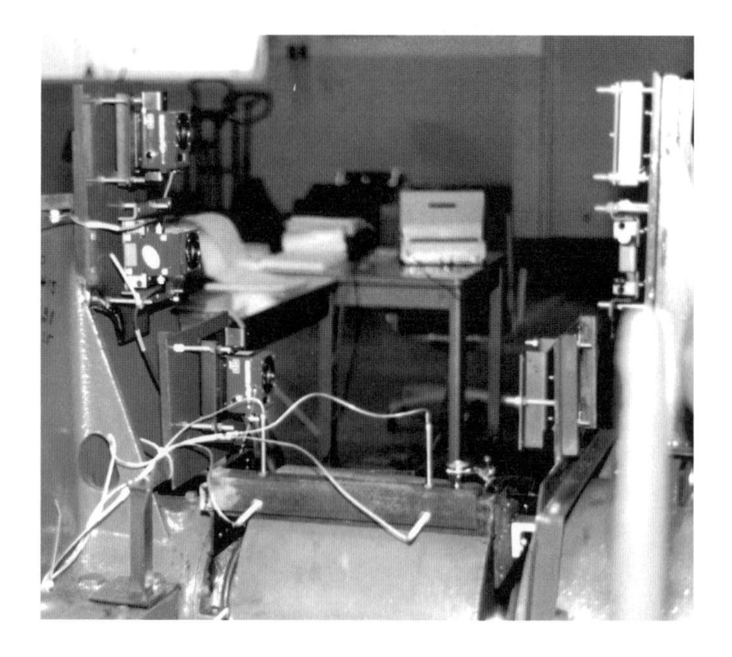

Figure 9-39. Laser detector / prism systems mounted on custom fabricated bases and positional adjustment devices.

strike in the center of the detector target(s). The laser / detector system is kept in place on the machine cases and the unit is started up and run until movement stops. If movement occurs between the machine cases, the laser will be striking at a different position in the photodetector.

With the myriad of laser shaft alignment systems in existence, it is somewhat amazing to find out that very few people have tried to use their systems for this purpose. The underlying reluctance seems to originate from the difficulty in mounting the lasers and detectors to the machine cases. Since there is such a wide variation in machinery design, custom brackets are usually required to hold the equipment in place. These 'brackets' must not only hold the laser and detector in a stable position, but the laser and detector need to have positional adjustments for centering the beam. In addition, if a considerable amount of movement occurs between the machinery cases, the laser beam moves completely out of the detector target surface area (remember most photodiodes currently used are only 10mm x 10mm in size) inhibiting complete and accurate OL2R measurements. Additionally, if laser/detector prism systems are used, two sets of laser/detector prisms are needed to capture all the required information (vertical offset, vertical angularity, horizontal offset, and horizontal angularity).

However, there are suitable, accurate, repeatable, and fairly inexpensive mounts that can be purchased that overcome all of these problems. Malcolm Murray (Murray & Garig Tool Works, Baytown, TX) has devised special mounts that allow you to attach different types of lasers, detectors, and prisms, and remove them repeatedly with good remounting precision. In addition, adjustments can be made to easily center the laser in the detector with provisions to measure the tilt and pitch of the mounting surface before and after the machinery has moved from off-line to running conditions allowing for re-zeroing. These mounts have effectively been named: Plug-In, Back-Zeroing, Laser-Target mounts or PIBZLT mounts.

Figure 9-40. Plug-in, back zeroing, laser / target mount. Courtesy Murray & Garig Tool Works, Baytown, TX.

Key considerations for capturing good readings:

- the laser / detector mounts should be attached to each machine case as close as possible to the centerlines of rotation to accurately determine shaft motion, not casing expansion or bearing housing warpage
- accurate positioning mechanisms for laser and detector are required on mounting fixtures
- capture a set of readings from off-line to running conditions and another set of readings from running to off-line conditions to determine if there is a consistent pattern of movement

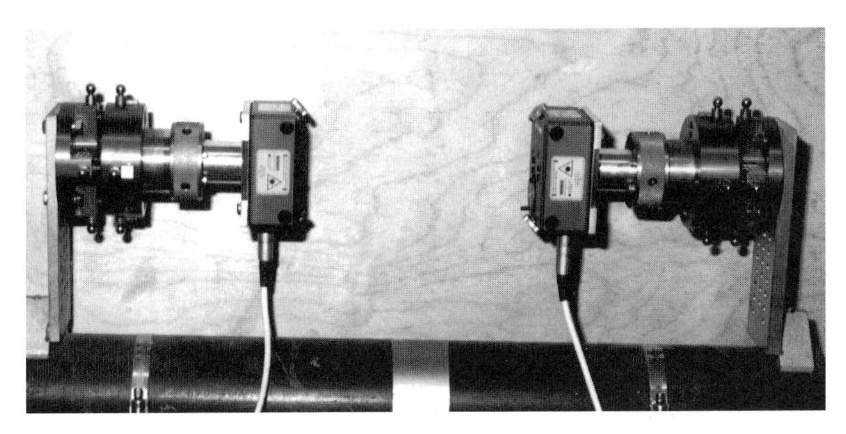

Figure 9-41. Plug-in, back zeroing, laser / target mounts holding FixtureLaser AB components. Photo courtesy Murray & Garig Tool Works, Baytown, TX.

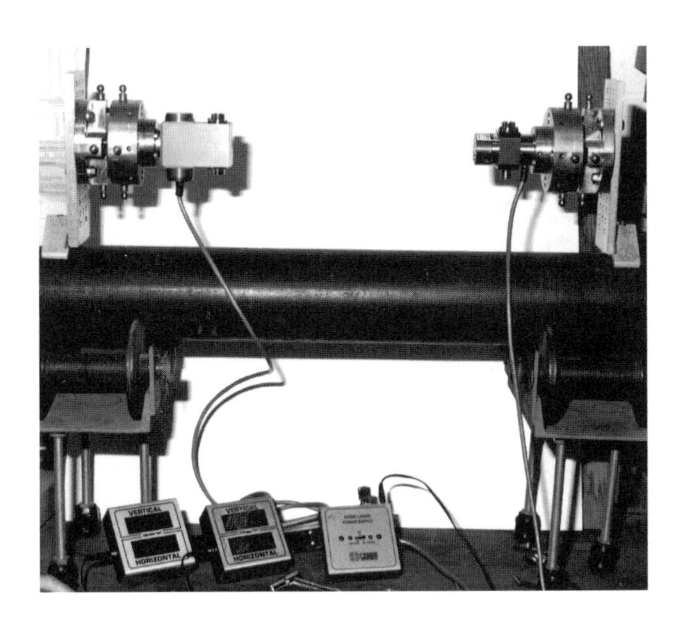

Figure 9-42. Plug-in, back zeroing, laser / target mounts holding Hamar laser components. Photo courtesy Murray & Garig Tool Works, Baytown, TX.

Figure 9-43. Plug-in, back zeroing, laser / target system on steam turbine and pump. Courtesy Murray & Garig Tool Works, Baytown, TX.

Figure 9-44. Measuring tooling ball gap with micrometer. Courtesy Murray & Garig Tool Works, Baytown, TX.

Advantages -

- fairly accurate measurements possible with proper setup
- capable of measuring movement in vertical and horizontal directions
- can continuously monitor positions of machinery
- if the machinery is vibrating excessively when taking the running measurements, the photodetector can average the position of the center of the 'bouncing' laser beam fairly accurately

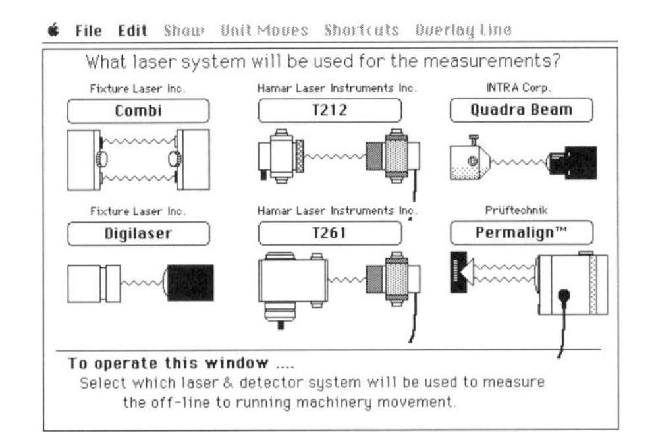

Determine which laser system will be used with the mounts. The data capture sequence will be slightly different depending on which system is used.

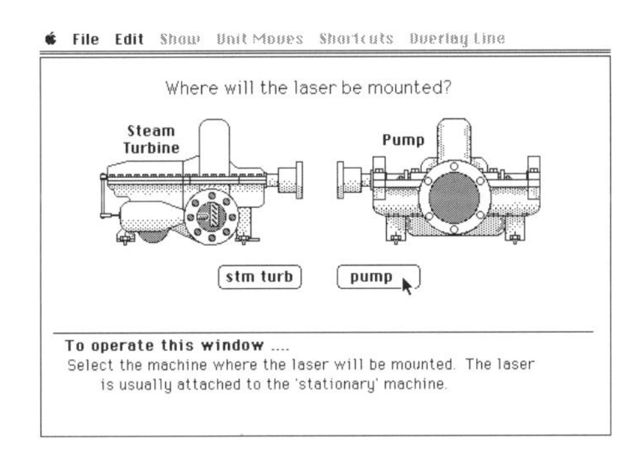

Select where the laser(s) and where the detector(s) will be mounted.

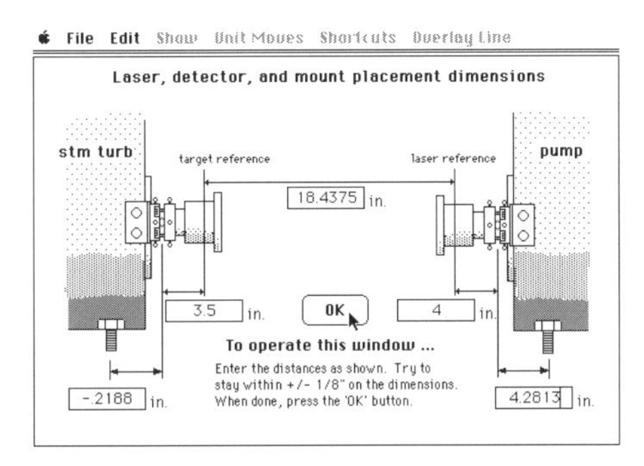

Measure the distances from the mounting locations of the laser(s) and detector(s) to the pivot point in the PIBZLT mount and then to the inboard feet of the machinery (or other reference points on the drive train).

Figure 9-45. Plug in, back zeroing, laser / target mount general procedure. Software screen shots courtesy Turvac Inc., Cincinnati, OH.

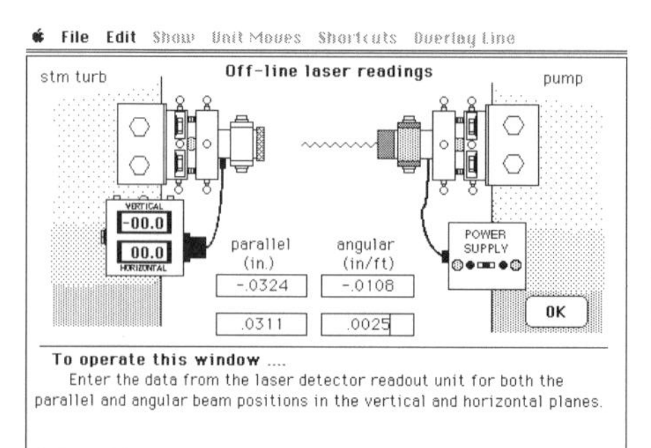

Zero the laser beam(s) into the detector(s) by positioning the tilt / pitch adjustment screws on the mounts with the machinery off-line.

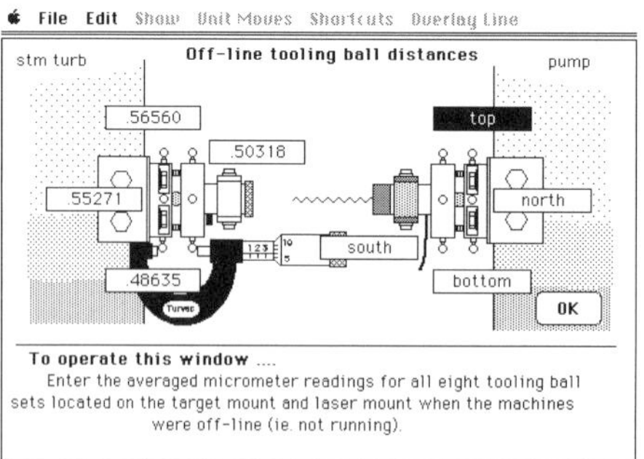

After the beam has been centered, measure the gaps at all eight tooling ball points (4 on each mount) and record the dimensions.

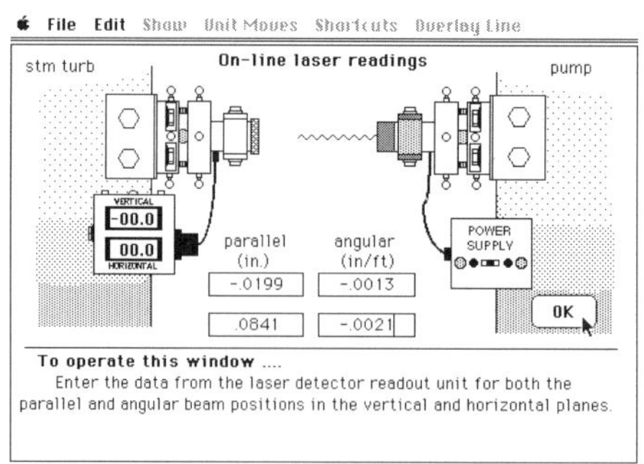

Start the machinery up and allow sufficient time for the readings to stabilize. Re-zero the beam by adjusting the tilt / pivot screws.

Figure 9-46. Plug in, back zeroing, laser / target mount general procedure. Software screen shots courtesy Turvac Inc., Cincinnati, OH.

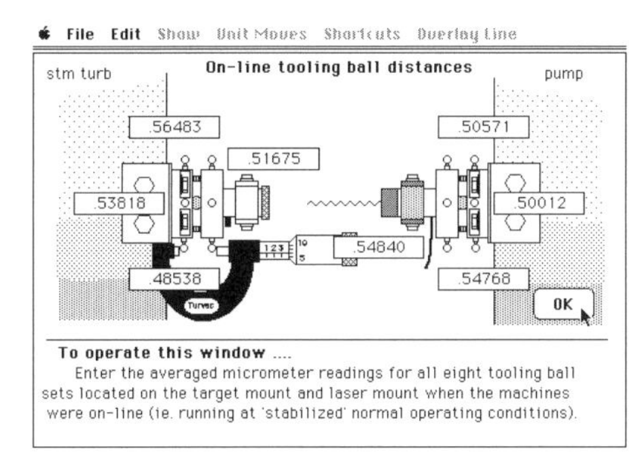

If the beam has moved out of the range of the detector, re-zero the laser(s) into the detector(s) by adjusting the tilt / pivot adjustment screws on the mount. Measure the distances between all eight tooling ball pairs and record the distances.

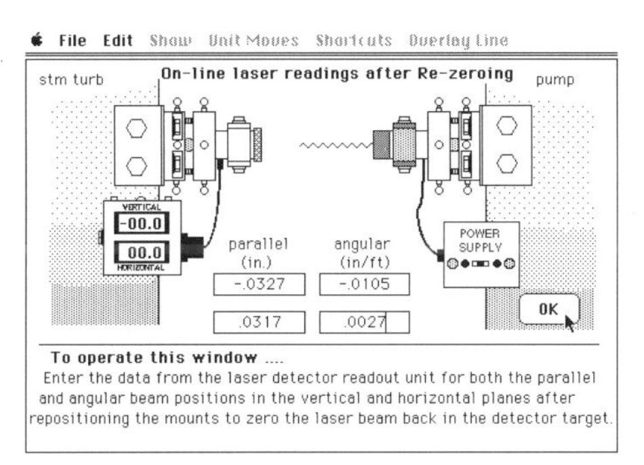

In the event that you are unable to exactly re-zero the beam, record the X-Y coordinates of the beam(s) in the detector(s).

Note : The procedure shown may vary slightly depending on the system used.

Figure 9-47. Plug in, back zeroing, laser / target mount general procedure. Software screen shots courtesy Turvac Inc., Cincinnati, OH.

Disadvantages -

- other light sources, steam, water vapor, etc. can affect the ability to capture stable readings
- if the machinery is moving considerably from OL2R conditions, the laser beam can move out of the range of the photodiode surface area unless the mounts have the capability of being repositioned knowing what the before and after positions were

- custom mounts have to be designed and fabricated or purchased and carefully installed on each machine case as close as possible to the centerline of rotation
- somewhat expensive since lasers, detectors, readout devices, and mounts have to be purchased

Ball-rod-tubing connector system (machine case to machine case category)

The ball - rod - tubing connector system (referred to as the BRTC system) was developed by Turvac Inc., Cincinnati, OH. It utilizes the shaft to coupling spool measurement principles adapted to measure off-line to running machinery movement.

Two base blocks are attached to each machine case near their inboard bearings. Two short rods with a round ball attached to the end of each rod are indexed through a hole in each base block and clamped into place. Telescoping tubing is affixed to each ball end. A brace is attached to each rod that spans out to the tubing connector that hold two proximity probes (a vertical and a horizontal probe). Proximity probe target surfaces are attached to the tubing connector at both ends.

Gap readings are taken at all four proximity probes with the machinery off-line. The BRTC system remains attached to each machine case and the equipment is started up and allowed to stabilize in its final operating position. Gap readings are taken again during operation.

Key considerations for capturing good readings:

- the base block should be attached to each machine case as close as possible to the centerlines of rotation to accurately determine shaft motion, not casing expansion or bearing housing warpage
- during operation, one rod end should be allowed to 'float' to allow for any axial movement of the machine cases and

With the machinery off - line, attach each rod end base to the machine cases as close as possible to their centerlines of rotation. Install the tubing connector, proximity probe targets, prox probe holding arms, and the vertical and horizontal proximity probes. Measure the gaps at all four probes.

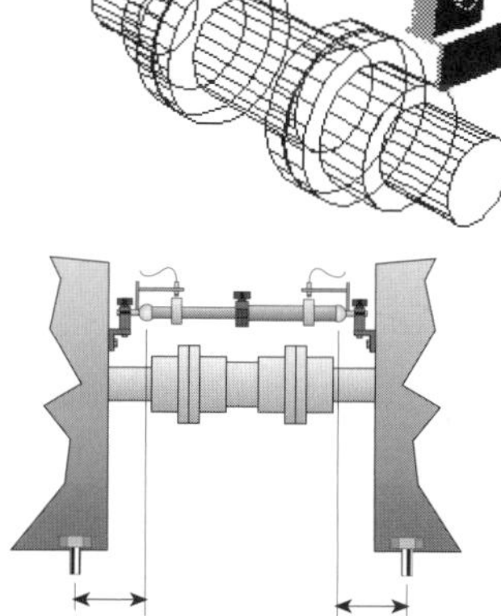

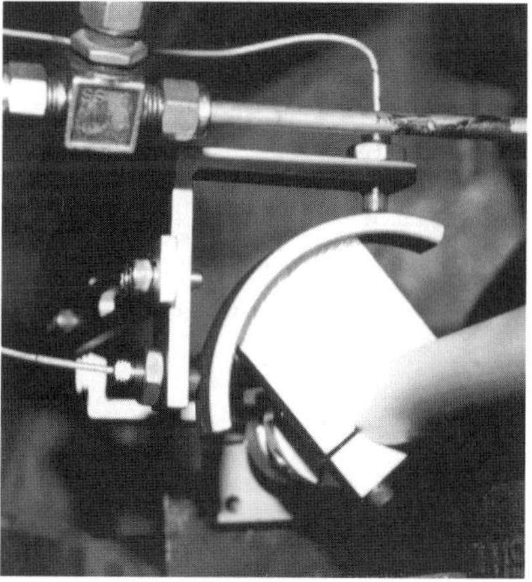

Measure the distances between the inboard and outboard bolts of both machines and the distances shown above. You should clamp one (but not both) of the rod ends firmly and allow the other end to float in the event that there is axial movement of the machine cases toward or away from each other. Allow sufficient time for the machinery to stabilize in its final operating position and firmly lock the 'floating' rod end in its clamp. Measure the probe gaps again. Refer to the shaft to coupling spool technique in Chapter 7 for an explanation of the basic measurement principle.

Figure 9-48. Ball - rod - tubing connector system general arrangement and procedure.

then locked in place to capture running prox probe gaps
- capture a set of readings from off-line to running conditions and another set of readings from running to off-line conditions to determine if there is a consistent pattern of movement

Advantages -
- fairly accurate measurements possible with proper setup
- capable of measuring movement in vertical and horizontal directions
- can continuously monitor positions of machinery without disturbing sensors or bars
- if the machinery is vibrating excessively when taking the running measurements, the proximity probes average the oscillation effectively to insure accurate distances between the probe tips and the targets
- the rod ends, probes, targets, and connector tubing can be removed and reinstalled without disturbing the position of the base blocks attached to the machinery cases
- if the machinery is moving considerably from OL2R conditions and the proximity probes bottom out on the target or move out of the linear range of the probes, the probe span can be changed (shortened or lengthened) to accommodate this excessive movement
- the tubing connector must be able to freely pilot / rotate at each ball end

Disadvantages -
- custom brackets and bars have to be designed, fabricated, and carefully installed on each machine case as close as possible to the centerline of rotation
- somewhat expensive since custom braces and bars have to be fabricated, and probes, cables, proximitors, readout devices, and power supplies have to be purchased

Vernier - Strobe system (direct shaft to shaft measurement category)

The vernier - strobe system, invented by Malcolm Murray (Murray & Garig Tool Works, Baytown, TX), utilizes the face-face measurement principles adapted to measure off-line to running machinery movement. Two small vernier scale sets are firmly attached across each flexing point in a coupling. A set of readings are taken at each vernier at the 12, 3, 6, and 9 o'clock shaft positions when off-line. The machinery is started up (with the verniers still attached to the coupling) and allowed to stabilize in its final operating position. A variable rate strobe light is then aimed at the coupling area to visually 'freeze' the shaft positions. By varying the strobe rate slightly, the vernier scales can be 'rotated' to the 12, 3, 6, and 9 o'clock shaft positions when readings are taken during running conditions.

Key considerations for capturing good readings:

- the vernier scales must be firmly attached to the coupling hubs and spool piece and appropriate measures taken to counterbalance the weight of the scales to maintain proper balance characteristics
- during operation, it is advisable to use a camera to capture the vernier scale readings for safety
- a metal mesh coupling guard should be used during operation
- in the event that the bottom reading (6 o'clock) can't be taken, particularly during running conditions, ensure that accurate 90 degree rotations are made for the 3 and 9 o'clock readings applying the validity rule to determine the bottom reading
- capture a set of readings from off-line to running conditions and another set of readings from RO2L conditions to determine if there is a consistent pattern of movement

Vernier - Strobe Technique

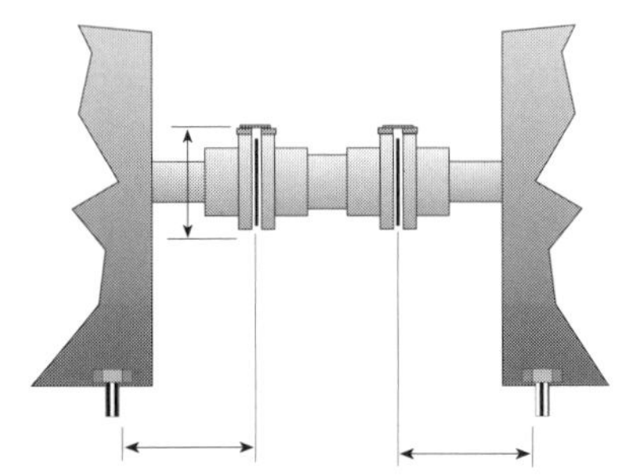

With the machinery off - line ... attach a vernier scale set across each flexing point on the coupling and measure the distances from the inboard and outboard feet on both machines and the distances as shown above. Manually rotate the shafts and take measurements at each vernier scale at the 12, 3, 6, and 9 o'clock positions and record the data.

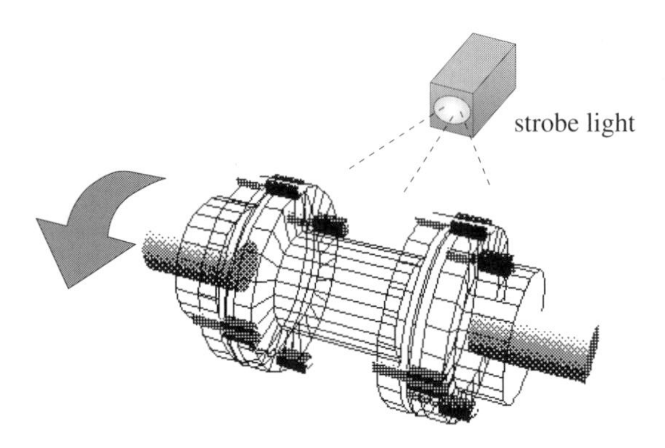

Start the machinery up and allow sufficient time for the machinery to stabilize in its final operating position. With a variable rate strobe light, 'freeze' the position of each vernier in the top, sides, and bottom position and observe the readings at each position of the vernier.

Compare the position of each shaft with respect to the coupling spool before and after. Refer to the face-face technique explained in Chapter 7.

Figure 9-49. Vernier - Strobe system general arrangement and procedure.

Figure 9-50. Vernier - Strobe system attached to flexible disk coupling. Photo courtesy of Murray & Garig Tool Works, Baytown, TX.

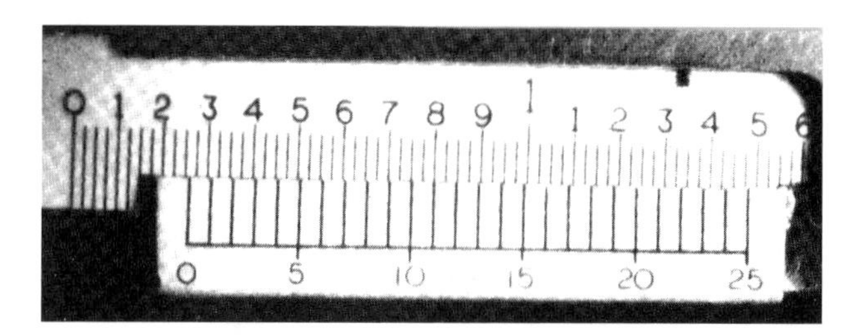

Figure 9-51. Close up of vernier scale 'strobe-frozen' during rotation at 1800 rpm (reading of 0.261"). Photo courtesy of Murray & Garig Tool Works, Baytown, TX.

Advantages -

- direct shaft measurements
- fairly accurate measurements possible with proper setup
- capable of measuring movement in vertical and horizontal directions
- can continuously monitor positions of machinery shafts
- relatively inexpensive

Disadvantages -

- potential for bodily harm if the vernier scales separate from the shaft during operation
- can only be used on couplings with a connector spool piece (e.g. flexible disk pack or gear type couplings)

Instrumented coupling systems (direct shaft to shaft measurement category)

The instrumented coupling system is manufactured by Indikon Corp. and this system utilizes the face-face measurement principles adapted to measure off-line to running machinery movement.

Four proximity probes are housed inside the coupling spool piece. Two axially positioned probes are attached at each end of the coupling spool observing target surfaces that are attached to the ends of each shaft. Two non-contacting, rotating / stationary coil sets transmit power to the probes and capture the signal from each

Figure 9-52. Instrumented flexible coupling system. Photo courtesy of Indikon Corp.

probe. The instrumented coupling transmits the proximity probe gap information during off-line and running conditions.

Key considerations for capturing good readings:

- recommend that manufacturer's installation instructions be followed very carefully to insure proper operation
- capture a set of readings from off-line to running conditions and another set of readings from running to off-line conditions to determine if there is a consistent pattern of movement

Advantages -

- direct shaft measurements
- fairly accurate measurements possible with proper setup
- capable of measuring movement in vertical and horizontal directions
- can continuously monitor positions of machinery shafts

Disadvantages -

- can only be used on couplings with a connector spool piece (e.g. flexible disk pack or gear type couplings)
- relatively expensive

Aligning rotating machinery to compensate for off-line to running machinery movement

Once accurate measurements have been collected and analyzed on how the equipment moves in the field, the machine elements can then be properly positioned during the off-line shaft to shaft centerline alignment process to compensate for this movement to insure collinear shaft centerlines during operating conditions. Depending on how the OL2R data was collected, there are different methods used to interpret the information to finally obtain the desired ‘off-line’ shaft positions.

Determining the desired off-line shaft positions when using the machine case to baseplate or machine case to remote reference point methods

If you employed one of the following techniques to measure OL2R movement, the data you collected shows how each end of the machinery moved from off-line to running conditions.

- calculating machine case thermal expansion using the strain equation
- inside micrometer / tooling ball / angle measurement devices
- proximity probes with water cooled stands
- Optical alignment equipment
- laser - detector measurement systems

Graph paper similar to what is used for the graphing / modeling techniques covered in Chapter 8 can be used to show the desired off-line shaft positions. The graph centerline will represent the final position of the shafts which is often referred to as the 'hot operating position' or running shaft position. If the machinery shafts move from off-line to running conditions, lines will be drawn on the graph paper to represent what position they should be in when off-line, so that when they move during operation, they will come in line with each other (i.e. end up on top of the graph centerline).

Along the graph centerline, mark where the OL2R measurements were taken at the inboard and outboard ends of each piece of machinery. Other critical points such as the dial indicator (or laser - detector) reading point locations and foot bolt points can be shown. Once the desired off-line shaft positions are drawn, 'shoot-for' dial indicator readings can be determined for the shaft positions when off-line.

It should become apparent by this time that if you are using dial indicators and brackets that have sag and that the shafts shouldn't be in line with each other when off-line, you should never want to 'spin zeros' for the dial indicator readings.

An off-line to running machinery movement survey has been conducted on a motor driven compressor using optical alignment techniques. Scale targets were positioned at the inboard and outboard ends of both machines to capture both vertical and horizontal movement of the centerline of the machine cases. The locations of the scale targets are illustrated in the diagram (note that the scale targets are not located at the same axial positions as the hold down bolts). The data indicates what movement occurred from off-line to running conditions.

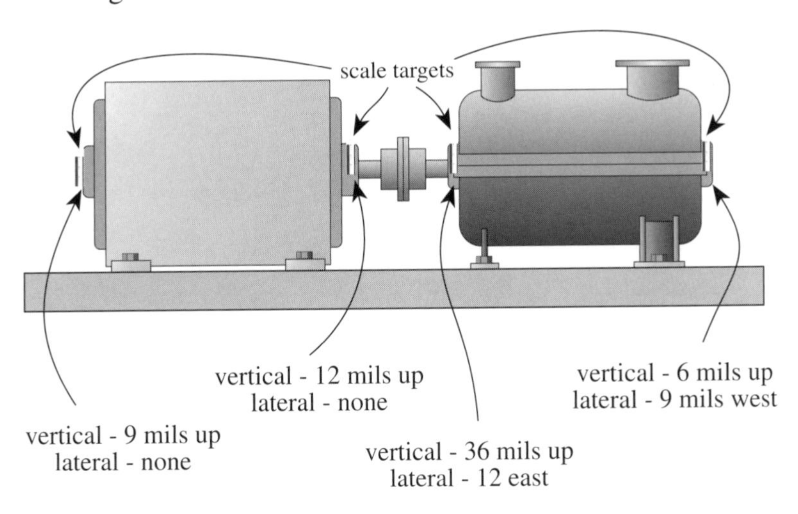

Figure 9-53. Sample problem using the machine case to baseplate or machine case to remote reference point OL2R methods.

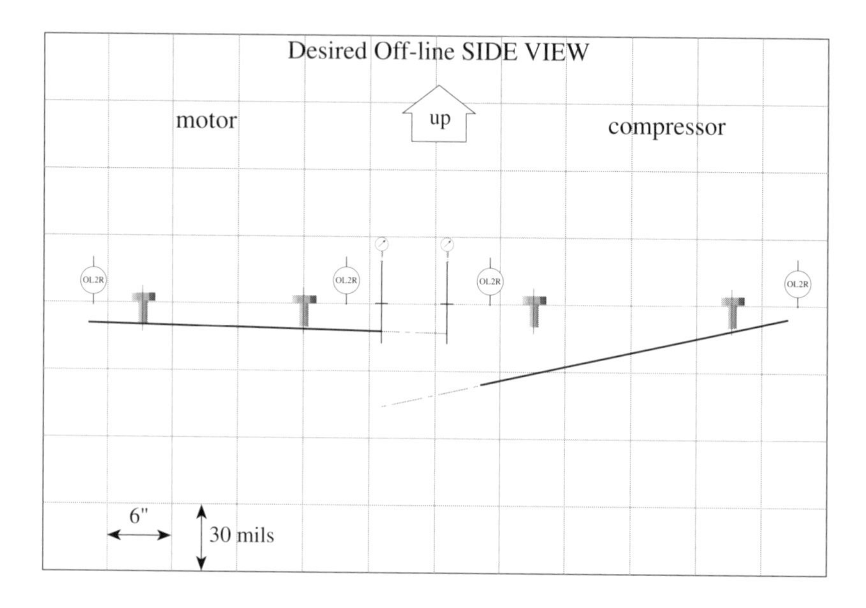

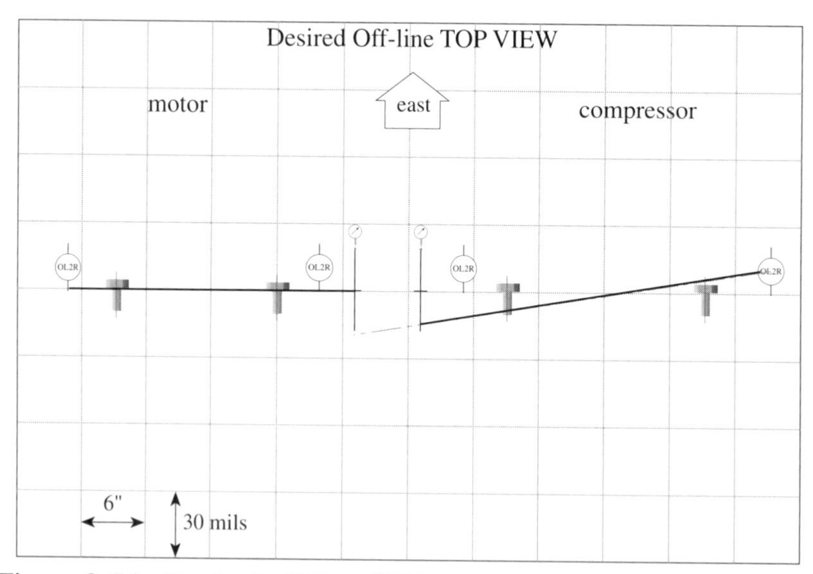

Figure 9-54. Desired off-line SIDE and TOP views of machinery shafts from example shown in figure 9-53. These are the positions the shafts should be in when off-line to assure collinearity when running.

Determining the desired off-line shaft positions when using any of the machine case to machine case OL2R techniques

If you have employed one of the following techniques to measure OL2R movement, the data you have collected shows how one machine case moved with respect to the other machine case from off-line to running conditions:

- alignment bars or custom fixtures with proximity probes
- laser / detector systems with custom fabricated brackets or special mounting systems
- Ball-Rod-Tubing connector system

Graph paper similar to what is used for the graphing / modeling techniques covered in Chapter 8 can be used to show the desired off-line shaft positions. The graph centerline will represent the final position of the shafts which is often referred to as the 'hot operating position' or running shaft position. In these OL2R methods, it is not known how each machine moved from OL2R conditions with respect to a fixed point in space (as opposed to the previously covered methods which do). What is known is how one machine saw the other machine move. Therefore one of the two machine cases / shafts is used as a 'reference' shaft and its position is placed directly on top of the graph centerline. The other machine case / shaft is then drawn on the graph paper to reflect how it moved with respect to the 'reference' shaft.

Along the graph centerline, mark where the OL2R measurements were taken at the inboard and outboard ends of each piece of machinery. Other critical points such as the dial indicator (or laser - detector) reading point locations and foot bolt points can be shown. Once the desired off-line shaft positions are drawn, 'shoot-for' dial indicator readings can be determined for the shaft positions when off-line.

If the alignment bar system was used to determine the

An off-line to running machinery movement survey has been conducted on a steam turbine driven boiler feedwater pump using the alignment bar technique. The 'target' bar was attached to the steam turbine and the 'probe' bar was attached to the pump. Four proximity probes (2 vertical & 2 horizontal) were affixed to the probe bar and gapped at mid-point in their linear range when the equipment had been off-line a sufficient amount of time to allow for the units to attain ambient temperature. The machinery was started up and the probe gaps were monitored until the gap readings did not change indicating equilibrium conditions during operation. The data indicates what movement occurred from off-line to running conditions.

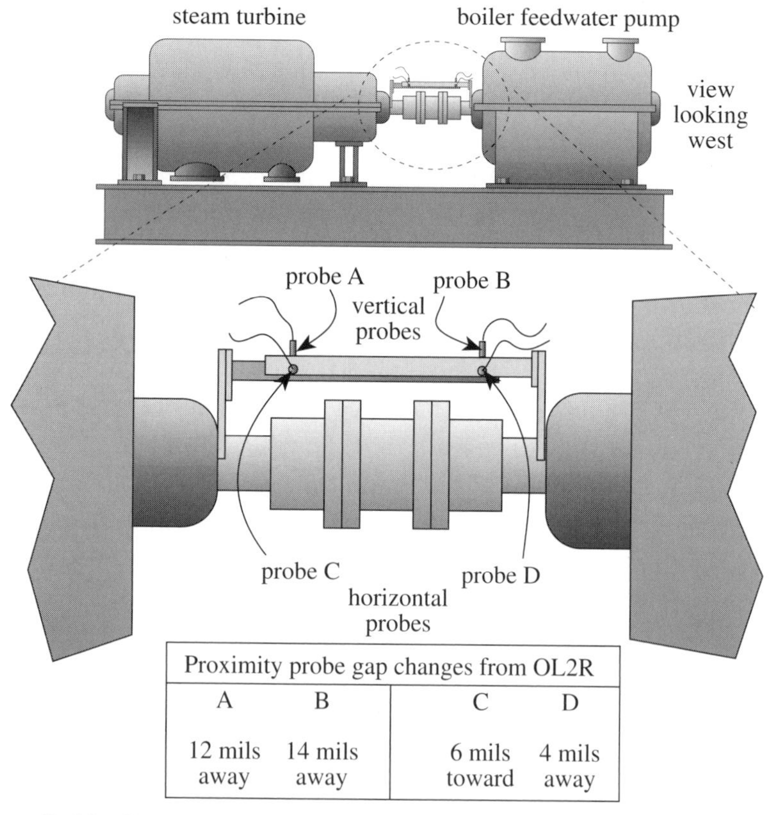

Proximity probe gap changes from OL2R			
A	B	C	D
12 mils away	14 mils away	6 mils toward	4 mils away

Figure 9-55. Sample problem using the alignment bar system (one of the machine case to machine case OL2R methods).

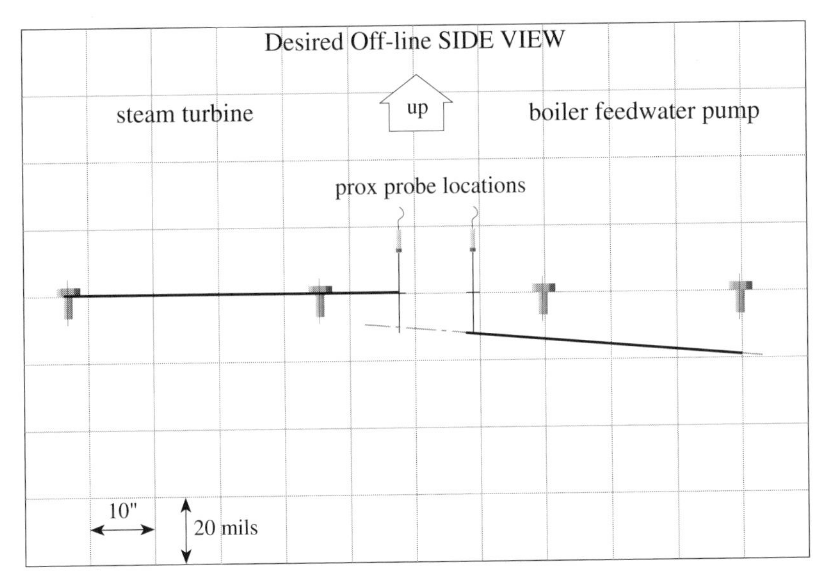

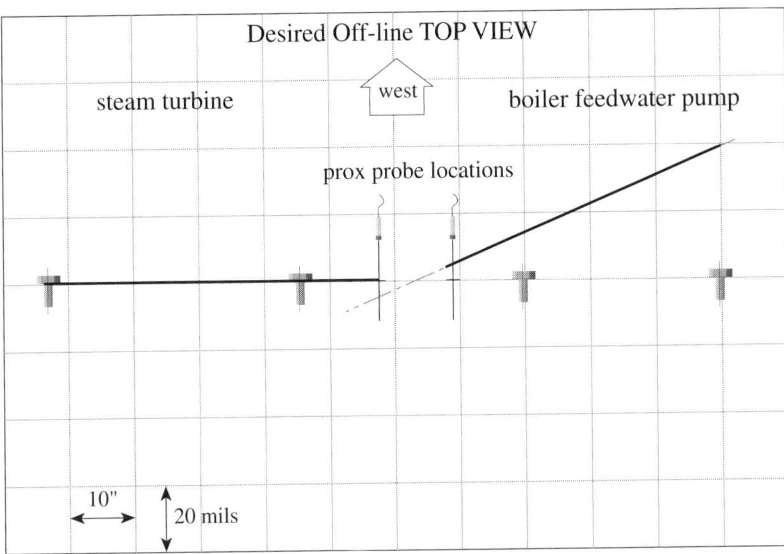

Figure 9-56. Desired off-line SIDE and TOP views of machinery shafts from example shown in figure 9-55. These are the positions the shafts should be in when off-line to assure colinearity when running.

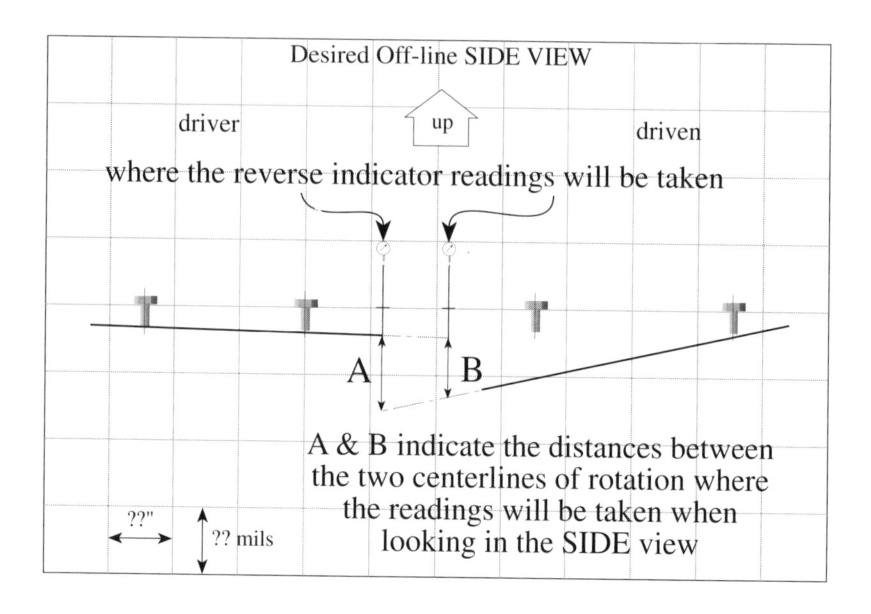

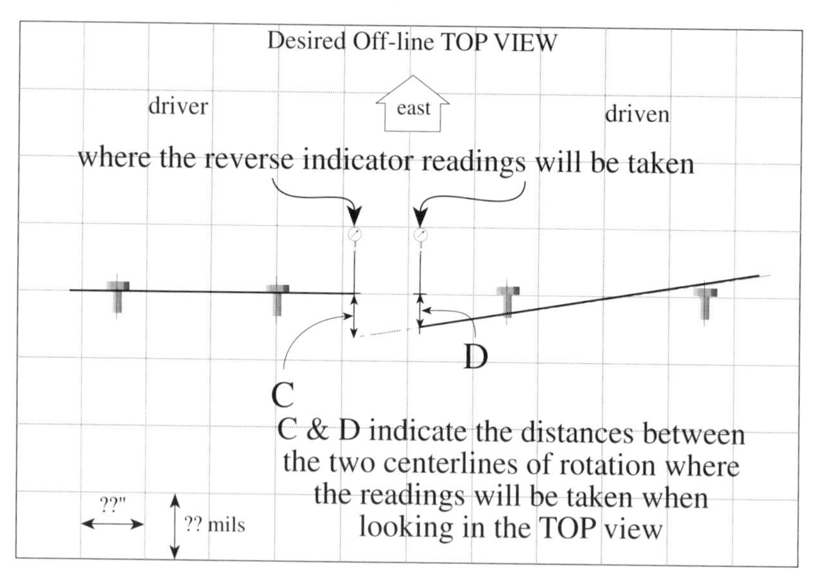

Figure 9-57. Plotting the OL2R movement data to extract the proper gaps that should occur at the dial indicator reading locations when aligning the machinery using the reverse indicator method.

machinery movement, the graph setup would look like figure 9-56. A little bit of thought is going to have to be put forth to recall how the probes were positioned when reading the targets and what decreasing or increasing gaps mean when setting up the chart. It's too easy to make a mistake here by misinterpreting the movement data, so it is wise to make sure both the amount of movement and the direction of movement is correct before calculating the 'shoot-for readings'.

How to determine the 'shoot-for' off-line dial indicator readings

So far in this chapter we have reviewed a number of methods to determine how machinery will move from off-line to run-

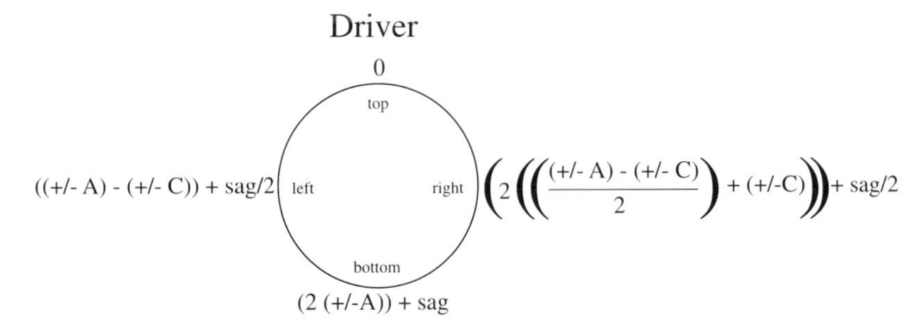

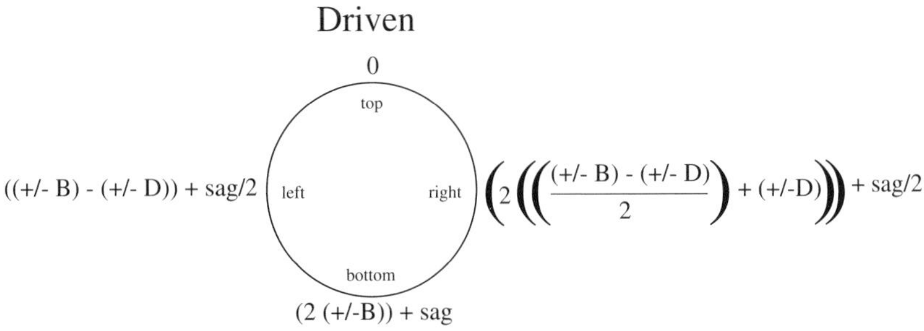

Figure 9-58. Equations for determining the 'shoot-for' dial indicator readings when aligning the machinery using the reverse indicator method.

ning conditions. In addition, we have been able to take this data and plot the information onto a graph showing where the shafts should be when the equipment is not running. As you can see, if the machinery cases do not move the same amount at the inboard and outboard ends, the shaft centerlines should not be collinear (i.e. in line with each other) when off-line. Since the shafts should not be in line with each other when off-line, what should the dial indicator readings be to insure the that shafts are in the desired off-line positions?

Here we will review how to calculate the desired 'shoot-for' reverse dial indicator readings. Similar techniques can be employed to determine the dial indicator readings if you are using any of the other shaft alignment methods (face-rim, shaft to coupling spool, double radial, and face-face).

Apply these general procedures to determine what the 'shoot-for' readings will be when offset aligning machinery to compensate for OL2R movement.

1) Graph the desired off-line offset shaft positions of both the driver and driven units. Figure 9-57 shows two units plotted in both the vertical and horizontal planes. In the vertical plane, the driver and driven centerlines are positioned below the chart centerline to indicate that they move upward when going from the off-line to running conditions. The amount of movement of these centerlines is based on the data collected from any of the OL2R measurement techniques explained in this chapter. In the horizontal plane (TOP view), the driver didn't move sideways at all and the outboard end of the driven unit moves to the west and the inboard end moves to the east.
2) Based on the chosen scale factor from top to bottom on the chart, record the A and B gaps as shown in figure 9-57 for the vertical position.
3) Determine whether the bottom readings taken on each

shaft are positive or negative.

RULE:
If the actual centerline of a unit is toward the bottom of the graph with respect to a projected centerline, the reading will be POSITIVE.
If the actual centerline of a unit is toward the top of the graph with respect to a projected centerline, the reading will be NEGATIVE.

In other words, try to visualize what is going to happen to the dial indicator stem as it traverses circumferentially from top to bottom on the shaft or coupling hub of each machine. Is it going to move outward (negative) or inward (positive)? Figure 9-57 shows that the driver centerline appears to be higher from the vantage point of the driven unit, therefore the dial indicator stem will move outward as it rotates to the bottom of the driver coupling hub producing a negative reading (gap A). From the vantage point of the driver unit however, the dial indicator stem will move inward as it rotates to the bottom of the driven coupling hub producing a positive reading (gap B).

4) Based on the chosen scale factor from top to bottom on the chart, record the C and D gaps as shown in figure 9-57 for the TOP view.
5) Determine the appropriate sign (positive or negative) for the side readings by applying the same rule in step 3.
6) Apply the appropriate gaps at A, B, C, and D into the equations shown in figure 9-57.

The OL2R survey showed the motor and compressor moved the amounts and directions shown below. The off-line positions of the motor and compressor shafts were measured using the reverse indicator technique as shown below. After the readings were captured, the bracket was measured for sag (-10 mils).

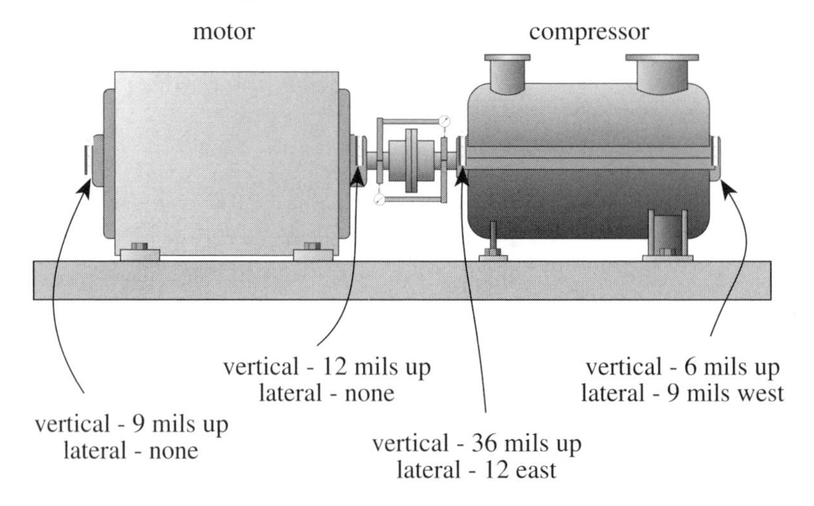

Reverse Indicator readings

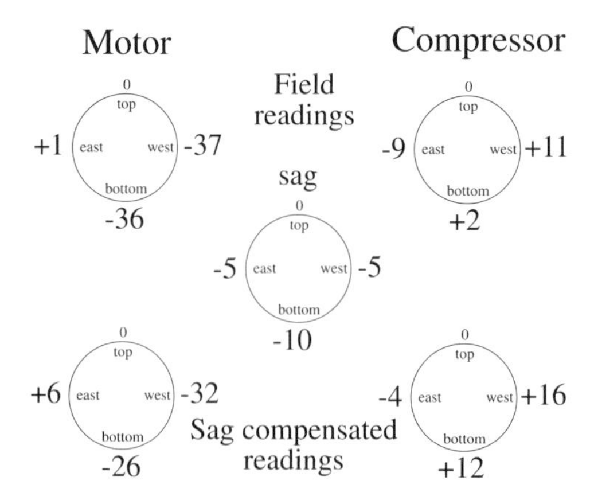

Figure 9-59. Example problem aligning a motor and compressor that moves from OL2R conditions.

Aligning shafts for running conditions (aka 'hot operating shaft positions')

The graphing / modeling techniques shown in Chapter 8 illustrated how to align two shafts with each other to insure they are collinear when off-line. What if you do not want the shafts to be collinear when they are not running, but want them to be in a specific desired off-line position similar to what is shown in figure 9-61?

The trick to offset aligning rotating machinery shafts is to shift the position of the shafts to where they will be when they are running and align the running shaft positions (also known as the 'hot operating shaft positions'). Once the shafts have been shifted to their running positions, the overlay line can then be used to determine the appropriate vertical and lateral repositioning movements required to put the shafts in the desired off-line positions.

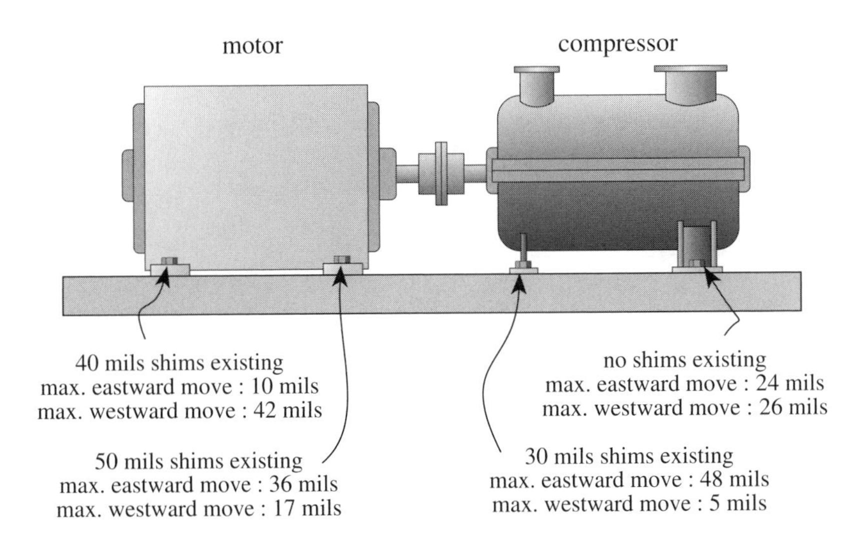

Figure 9-60. Observing and recording the allowable movement envelope on the motor and compressor.

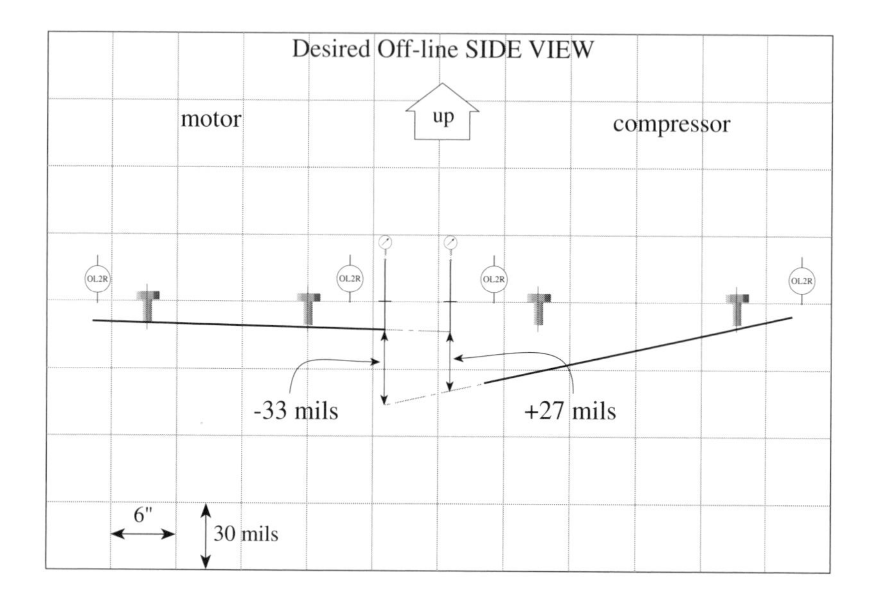

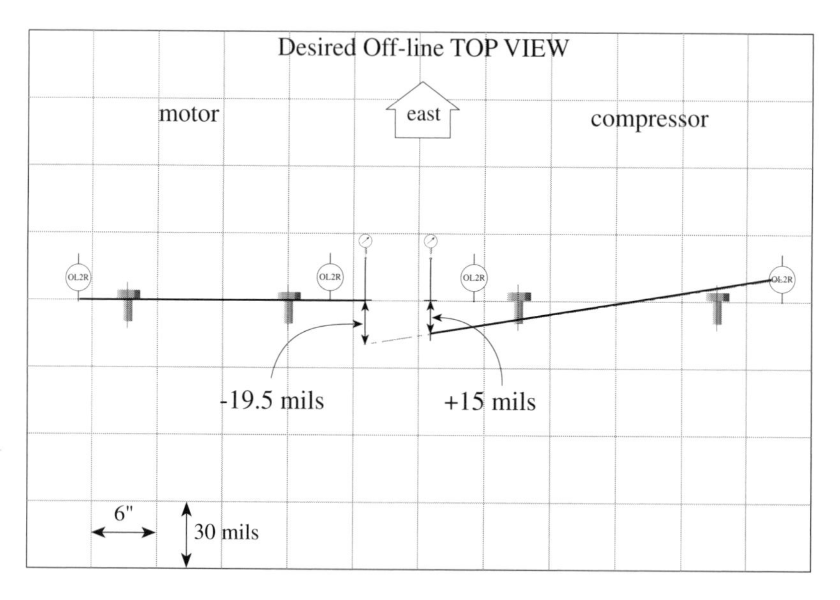

Figure 9-61. Graphs showing the desired off-line shaft positions of the motor and compressor.

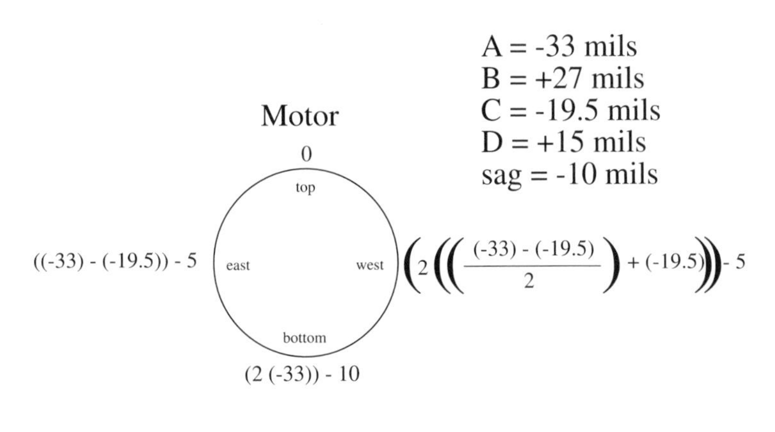

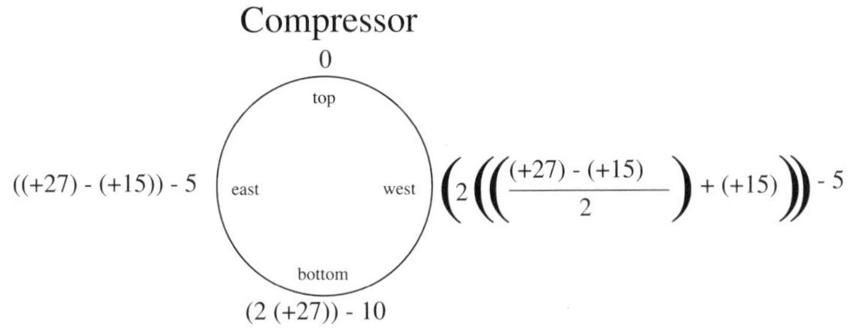

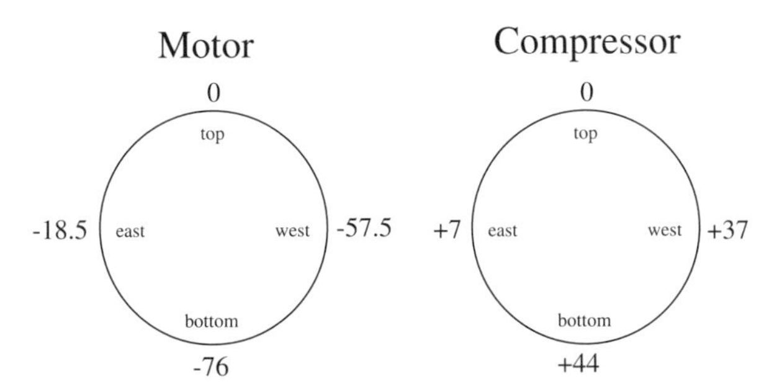

Figure 9-62. Calculating the 'shoot-for' reverse indicator readings for the motor and the compressor (see figures 9-57 & 9-58).

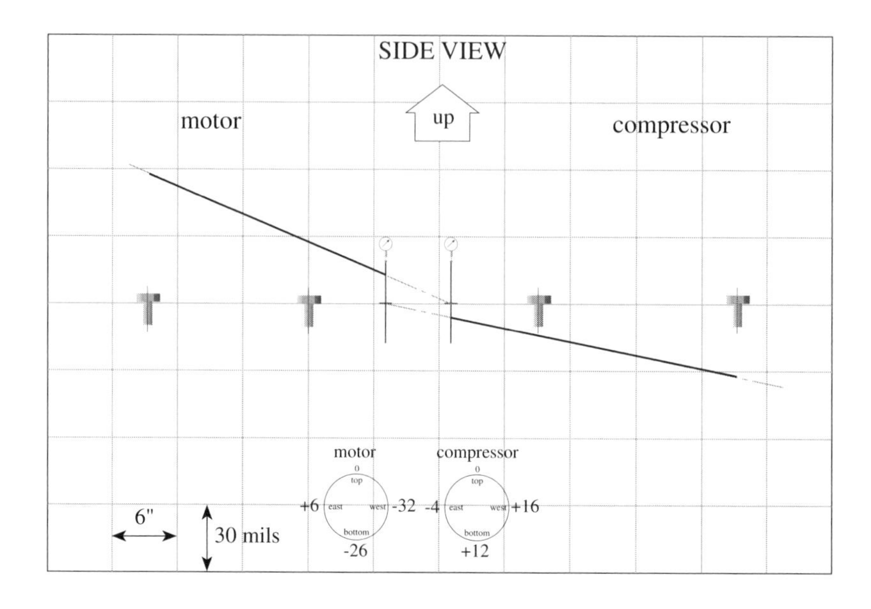

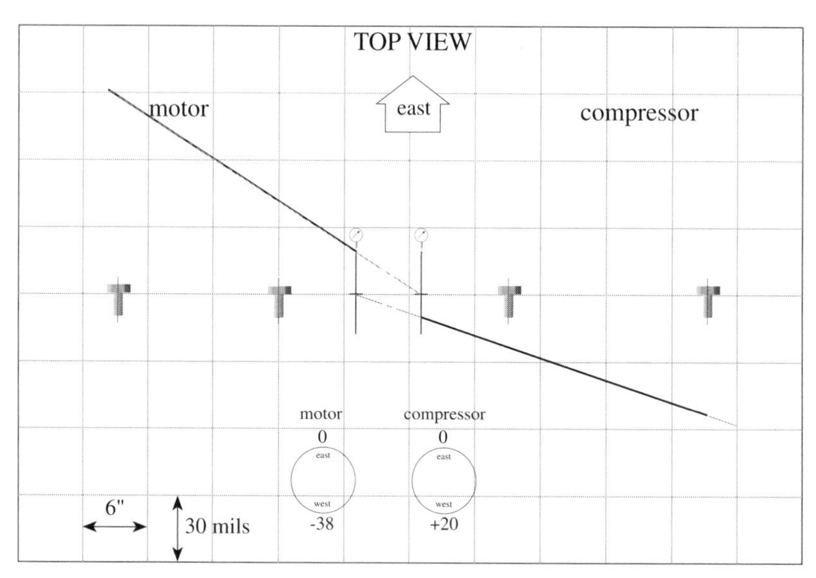

Figure 9-63. Modeling the actual motor and the compressor shaft positions using the point to point reverse indicator modeling method.

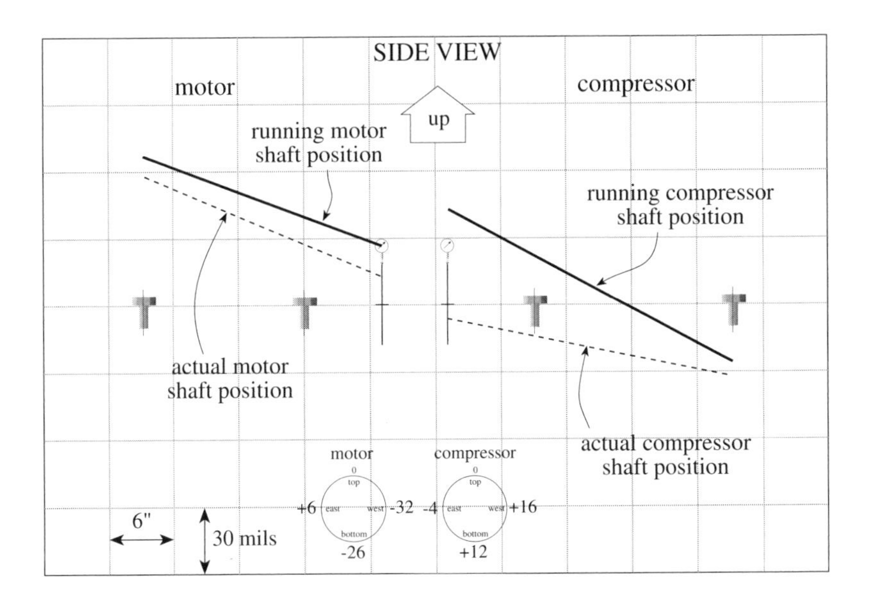

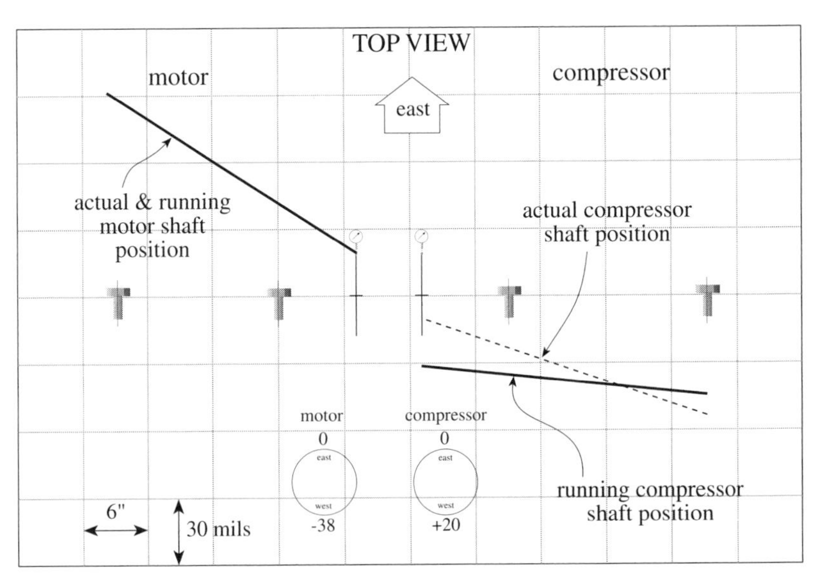

Figure 9-64. Shifting the positions of the shafts to show the running positions on the motor and the compressor.

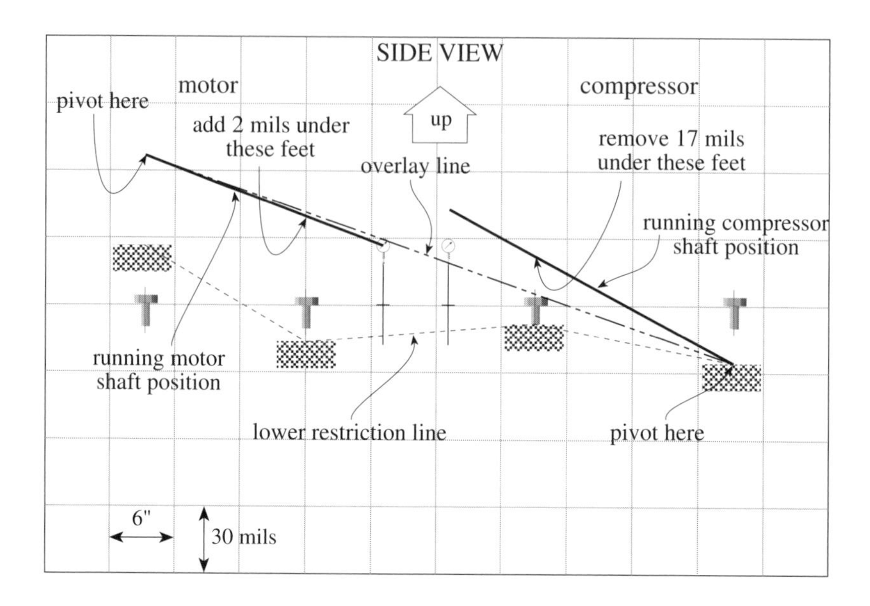

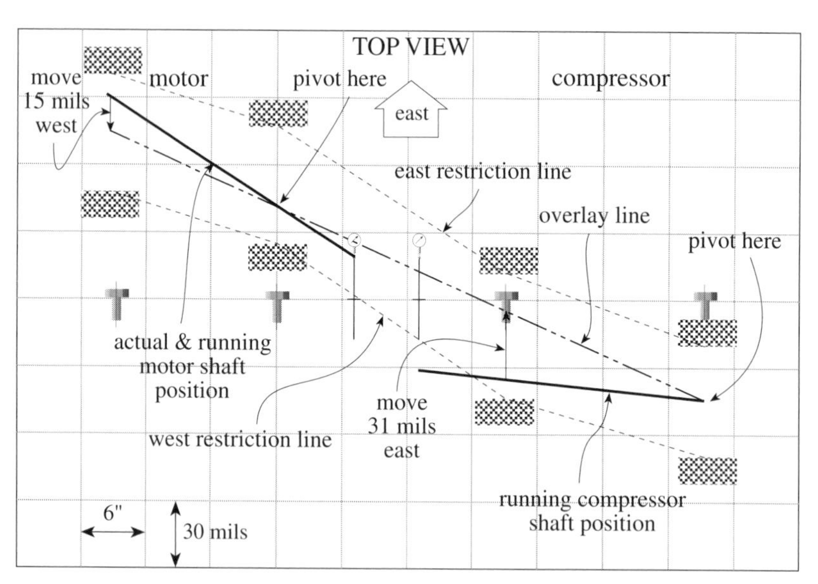

Figure 9-65. Plotting in the vertical and lateral movement restrictions on the motor and the compressor and projecting an overlay line for some possible movement solutions.

Figures 9-59 through 9-64 illustrate the type of information that needs to be gathered on a two element drive train. In summary:

- gather specific information on the drive train such as how the machinery will move from off-line to running conditions, piping strain problems, how many shims exist under the machinery feet, how far each unit can be moved sideways at the feet, what positions the shafts should be in when off-line, and what the 'shoot-for' readings are
- what positions are the shafts actually in when off-line?
- how will the movement restrictions affect the final chosen alignment line?

All of this data gathering, modeling, and calculations seems quite tedious, but in many cases this extra effort may spell the difference between machinery that seems to be plagued with problems or equipment that operates successfully for long periods of time.

Bibliography

Yarbrough, C. T., "Optical Checks Put Plant in Line for Low-Maintenance Future", Iron Age, January 20, 1960.

Hanold, John, "Application of Optical Equipment for Installing and Checking Large Machinery", A.S.M.E., paper no. 61-PET-26, Aug. 9, 1961.

Kissam, Philip, **Optical Tooling for Precise Manufacturing and Alignment**, McGraw Hill Co., New York, NY, 1962.

McGrae, **Optical Tooling in Industry**, Hayden Book Co., New York, NY, 1964.

Koenig, Eugene, "Align Machinery by Optical Measurement", Plant Enginering, May, 1964, pgs. 140-143.

Baumann, Nelson P., Tipping, William E. Jr., "Vibration Reduction Techniques for High-Speed Rotating Equipment", A.S.M.E., paper no. 65-WA/PWR-3, Aug. 5, 1965.

Yarbrough, C.T., "Shaft Alignment Analysis Prevents Shaft and Bearing Failures", Westinghouse Engineer, May, 1966.

Nelson, Carl A., "Orderly Steps Simplify Coupling Alignment", Plant Engineering, June, 1967, pgs. 176-178.

Jackson, Charles, "Alignment with Proximity Probes", A.S.M.E., paper no. 68-PET-25, Sept, 26, 1968.

Jackson, Charles, "Successful Shaft - Hot Alignment", Hydrocarbon Processing, Jan., 1969.

Blubaugh, R.L., Watts, H.J., "Aligning Rotating Equipment", Chemical Engineering Progress, April, 1969, Pgs. 44-46.

O'Kelley, J.F., "Optical Shaft Elevation Measuring", Power Engineering, Oct., 1969, pgs. 36-37.

Optical Alignment Manual, No. 71-1000, Keuffel And Esser Co., Morristown, N.J., 1969.

"Optical Alignment - A Maintenance Service to Reduce Your Machinery Downtime", bulletin no. 371-5000GP, Dresser Machinery Group, Dresser Industries, Houston, Texas, 1969.

Barnes, E.F., "Optical Alignment Case Histories", Hydrocarbon Processing, Jan. 1971, pgs. 80-82.

Yarbrough, C. T., “Extracting the Lemon from a Large Drag Line”, Open Pit Mining Association 27th Annual Meeting, June 1971.

Jackson, Charles, "How to Align Barrel-type Centrifugal Compressors", Hydrocarbon Processing, Sept., 1971, pgs. 189-194.

Mitchell, John S., "Optical Alignment - An Onstream Method to Determine the Operating Misalignment of Turbomachinery Couplings", Dow Industrial Service, June 1972.

Dodd, V.R., "Shaft Alignment Monitoring Cuts Costs", Oil and Gas Journal, Sept. 25, 1972, pgs. 91-96.

Lukacs, Nick, "Proximity Probe Applications for Troubleshooting Rotating Equipment Problems", I.S.A. paper no. 72-627, 1972.

Essinger, Jack N., "Hot Alignment Too Complicated?", Hydrocarbon Processing, Jan. 1974, pgs. 99-101.

Mitchell, John W., "What is Optical Alignment?", Proceedings - Third Turbomachinery Symposium, Gas Turbine Labs, Texas A&M University, College Station, Texas, 1974, pgs. 17-23.

Campbell, A.J., "Optical Alignment Saves Equipment Downtime", Oil and Gas Journal, Nov. 24, 1975.

Dodd, V.R., **Total Alignment**, The Petroleum Publishing Co., Tulsa, Okla., 1975.

VanLaningham, Fred L., "Distortion of Speed Changer Housings and Resulting Gear Failures", Proceedings - Fifth Turbomachinery Symposium, Gas Turbine Labs, Texas A&M University, College Station, Texas, Oct. 1976, pgs. 7-13.

"The K&E Optical Leveling Kit and How to Use It", bulletin no. T66-91222-66CT-3, Keuffel and Esser Co., Morristown, N.J., 1976.

Norda, Torkel, "Use Infrared Scanning to Find Equipment Hot Spots", Hydrocarbon Processing, Jan. 1977, pgs. 109-110.

Jackson, Charles, **The Practical Vibration Primer**, Gulf Publishing Co., Houston, Texas, 1979.

Applied Infrared Photography, Publication no. M-28, Eastman Kodak Co., Rochester, N.Y., May, 1981.

Baumeister, Theodore, Avallone, Eugene A., Baumeister, Theodore III, **Mark's Standard Handbook for Mechanical Engineers**, 8th edition, McGraw-Hill Book Co., 1221 Ave. of the Americas, New, York, N.Y.

Mager, Michael, Miller, Roy, “Permalign Movement Exercise #15 Joy Air Compressor”, service report Owens Corning Fiberglas, Kansas City, MO., October 1989.

Teskey, William, "Dynamic Alignment Project", University of Calgary, November 1994.

10
Aligning V-Belt Drives

V-belt driven equipment poses a different type of alignment problem than equipment directly coupled together. The advantages of V-belt driven equipment are that the alignment is not as critical as directly coupled rotating machinery and that V-belts provide an inexpensive means to increase or decrease the speed of the driven unit without resorting to more expensive means such as gear boxes or variable speed drivers. Special V-belt sheaves can be installed for right angle drive systems. Positive drive belts (aka cog or gear belts) provide no slippage as found in improperly adjusted standard V-belts and can transmit the same amount of horsepower as standard belts without excessive belt tension minimizing bearing loads.

Although crude in comparison to direct drive alignment measurement methods, V-belt alignment is accomplished with a straight edge or string. The objective is to insure that the shaft centerlines are parallel to each other.

This chapter will illustrate a graphing / modeling method that utilizes the 'T' bar overlay to show the relative positions of the shafts and arrive at an acceptable repositioning scheme.

- insure that the bearings on both machines are not damaged
- inspect the condition of the belts insuring that they are not cracked, glazed, stretched, etc.
- check for any 'soft foot' conditions between the machinery feet and the baseplate at all of the hold down bolts
- check the wear of the 'V' in the sheave with an appropriate wear indicator

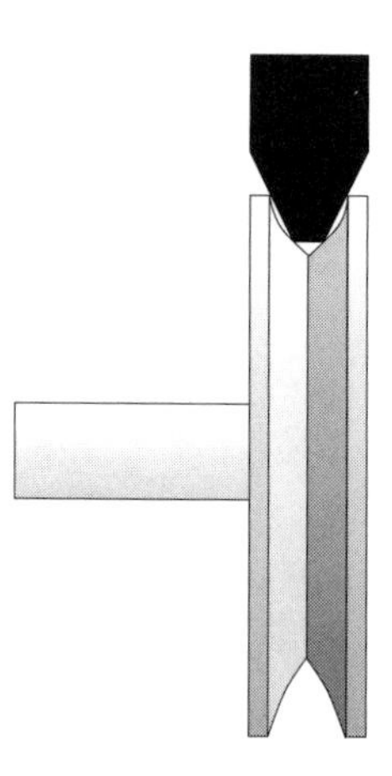

Figure 10-1. Preliminary checks.

• insure that the face runout does not exceed 1 mil for every six inches of sheave diameter

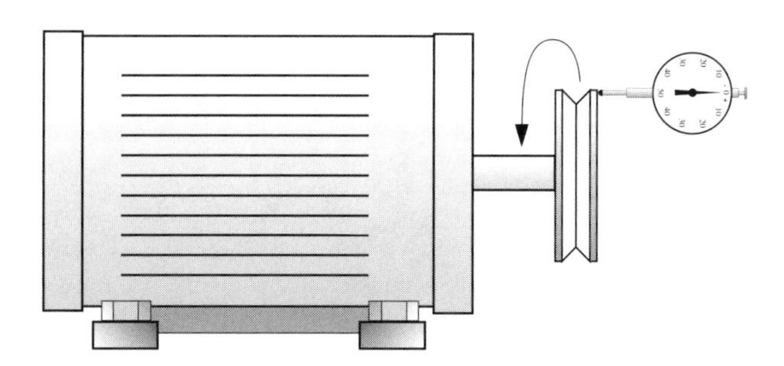

• insure that the radial (rim) runout does not exceed 5 mils

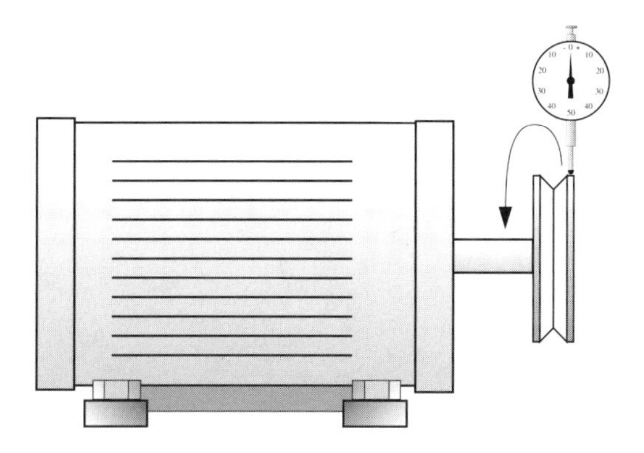

Figure 10-2. Preliminary checks.

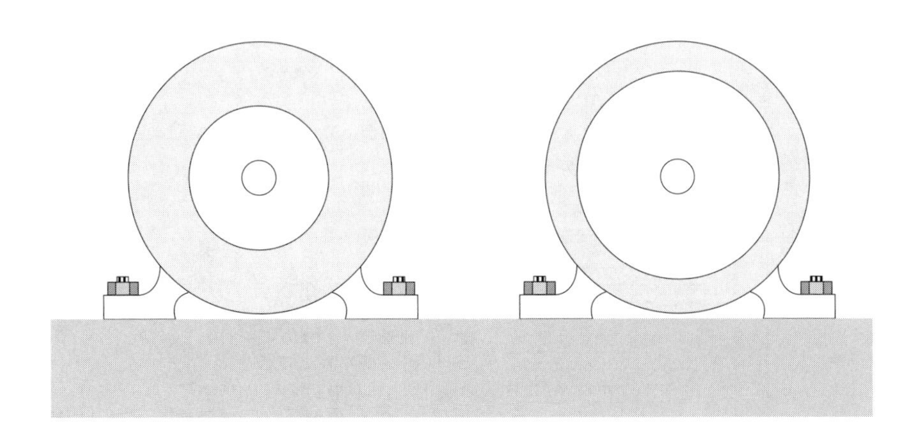

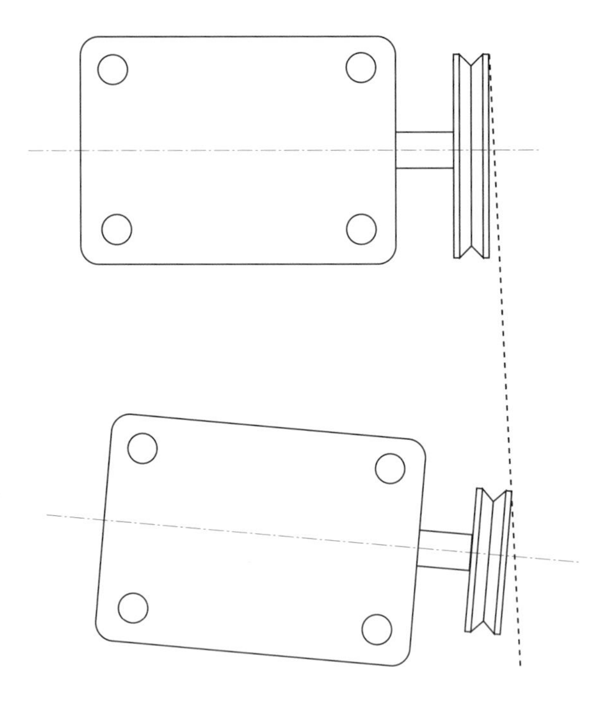

Figure 10-3. Typical sheave misalignment.

Modeling V-belt alignment problems

- Graph paper is recommended (20 division per inch) to set up the graph model.
- Use two 'T' bar overlays (see Face-rim graphing method in Chapter 8) to represent each shaft / sheave.
- Select the appropriate scaling factors to fit the machinery onto the graph paper.

Basic Procedure

- Start by scaling one of the two shafts onto the graph (in this example, the fan shaft was drawn directly onto the graph first).
- Draw a reference line to represent the distance between the two shaft centerlines.
- Use a ruler or straightedge to draw a line starting at the point on the first shaft drawn on the graph (in this case, the fan shaft) to go through the point on the outer edge of the sheave where the straightedge was touching and a point that represents the gap that was measured at the other side of the sheave (in this case, 20 mils of gap was measured across the fan sheave). Insure that this line stretches past the position of the other shaft (in this case, the motor shaft.
- Use the 'T' bar overlay to represent the motor shaft and position the motor shaft to reflect the gap that was measured across the sheave on the motor (in this case, a 30 mils gap was measured on the motor sheave on the fan side). Transfer the position of the 'T' bar onto the graph and draw the other shaft's position (again in this case, the motor shaft).
- Reposition the 'T' bar over each shaft until the desired positions of both shafts are satisfactory.

• measure the distance between the outboard and inboard bolts of both machines
• measure the distance from the inboard feet to where the straightedge or string/wire will be placed to capture the gap readings on the sheaves
• measure the diameter of the sheaves (i.e. the distance across the sheaves where the gap readings will be taken)
• measure the distance between the shaft centerlines

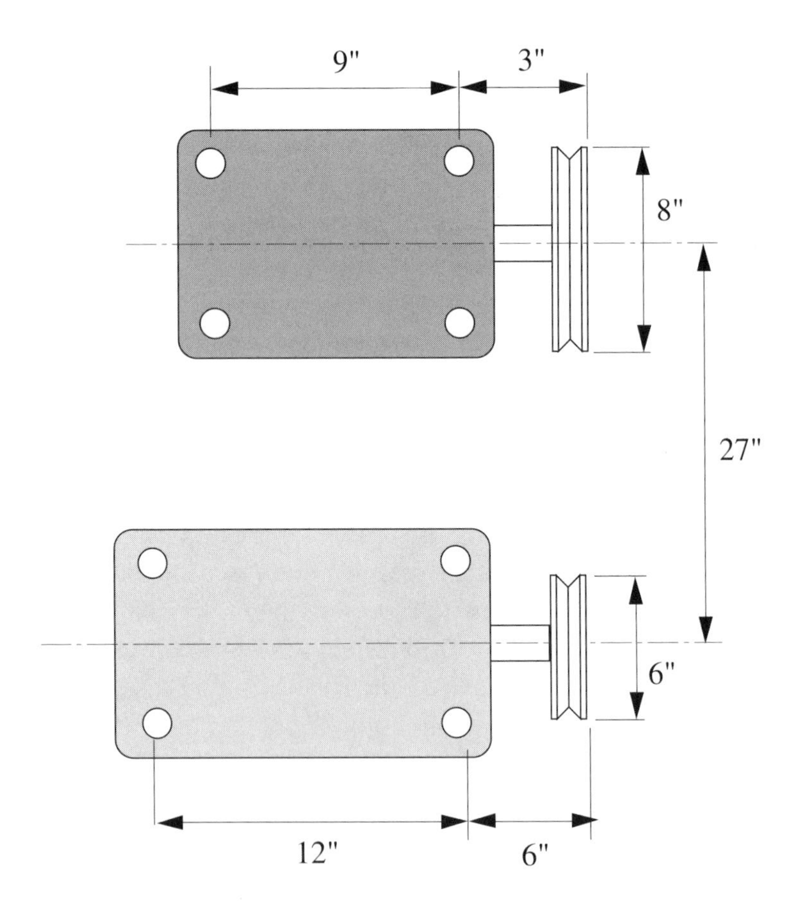

Figure 10-4. Typical measurements needed for graphing.

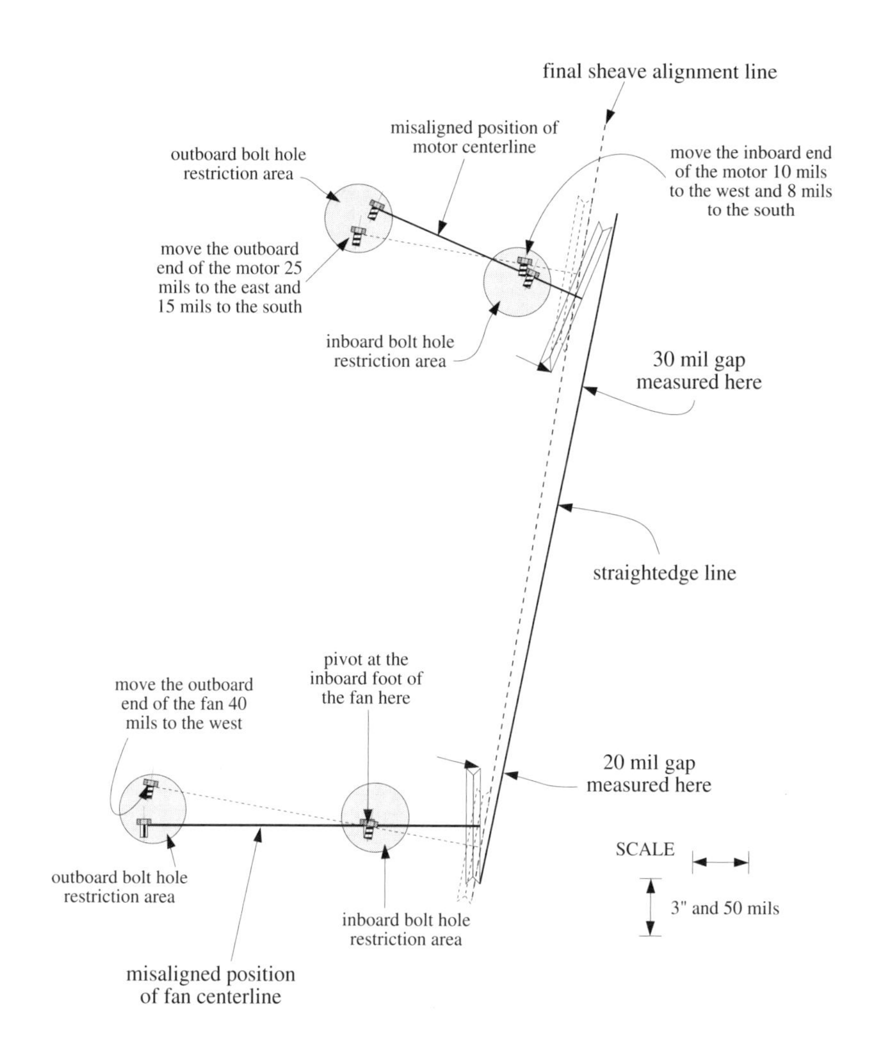

Figure 10-5. V-belt alignment sample graph.

Bibliography

Power Transmission Belt Drives - Installation, Maintenance & Trouble-shooting Guide, Goodyear Tire & Rubber Co., publication# 575000-3/86.

Installation & Maintenance Booklet, Goodyear Tire & Rubber Co., publication# T-630.

Installation and Maintenance of V-Belt Drives, T. B. Wood's Sons Co., Chambersburg, PA. Pub. # VBD-IM 128720K.

11
Moving Machinery in the Field

I frequently wonder how far we have really advanced in the area of machinery alignment.

In the previous chapters we have seen the various types of devices used to measure shaft alignment, covered graphical / modeling techniques used to determine which way and how much machinery casings should be moved, and reviewed a host of methods used to measure off-line to running machinery movement. A tremendous amount of effort on the part of hundreds of people over a span of decades has been put into the measurement methodology of shaft alignment. Millions of dollars have been invested in developing these three core aspects of alignment: mechanical measurement devices, calculating the required moves on the machinery, and determining how machinery will move from off-line to running conditions. Development in these three areas continues at a feverish pace. But is this where the majority of our efforts should be spent in the future?

Take a moment and refer back to figure 1-5 in Chapter 1 showing the amount of time typically spent on the overall alignment job. Notice that there are 7 basic tasks performed in the alignment process.

- safety tag & lockout
- preliminary checks
- measure shaft positions
- calculations
- decide moves
- reposition machinery
- run & check

If you eliminate the last step (run & check) and examine the amount of time taken to perform each of the other 6 tasks as illustrated in Figure 11-1, you easily can see that the vast majority of time in virtually every alignment job is spent on just two of the tasks – the preliminary checks and repositioning the machinery.

Something is critically wrong here! The vast majority of research and development work in shaft alignment is spent trying to figure out how to do the following three tasks better and faster:

- measure shaft positions
- calculations
- decide moves

Why are we spending so much time, effort, and money on something that occupies less than 14% of the overall time performing an alignment job? What difference is it going to make in the overall scheme of things if we reduce the amount of time it takes to do these three tasks (measure shaft positions, calculations, decide moves) from fifteen minutes to five minutes if it takes us somewhere between 2 to 4 hours to reposition the machinery?

One of the most bizarre sights I have witnessed in industry is to watch someone have an expensive laser shaft alignment measurement system attached to machinery shafts and stand next to the machines with a sledge hammer beating the equipment sideways to get it into alignment.

Figure 11-1. Standard lateral movement tool (hammer) produces undesirable results on motor foot. Photo courtesy Murray & Garig Tool Works, Baytown, TX (they didn't do this by the way!).

Question : What is the number one tool used in industry to lift a piece of machinery to install shims under the feet?

Answer : A screwdriver or pry bar.

Sounds like a joke doesn't it? In a very real sense, it is a joke that isn't very funny. People continue to use wood boards,

Shaft alignment in a 3-dimensional world

axial position

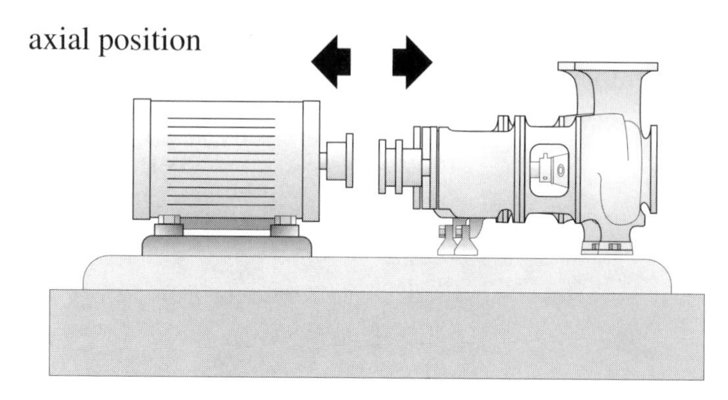

horizontal position

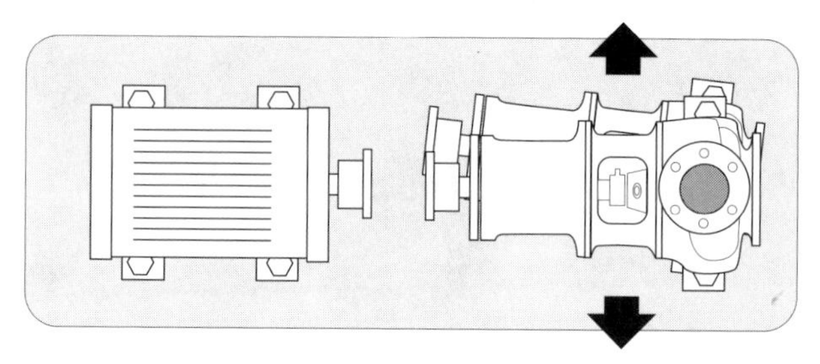

vertical position

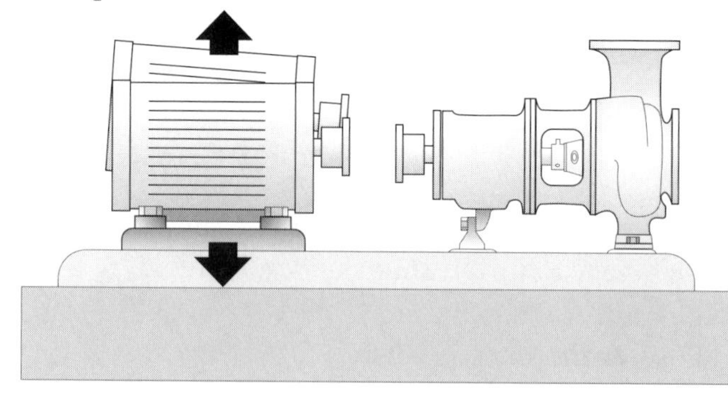

Figure 11-2. Shaft alignment is a three-dimensional problem so the machinery must be adjusted in all three planes.

crow bars, fork lift trucks, chain falls, mallets, pipes, conduits, cranes, and wedges to lift machinery. Remember from Chapter 2 that the lever was first explained by Aristotle in 400 B.C.? Well, my fellow readers, that's about as far as we have progressed. Shameful, isn't it?

We have been using flat spacers (e.g. shim stock) to vertically reposition machinery since the Dark Ages (and probably before). There have been no significant advancements made in how to effectively reposition machinery for 50 centuries! Yet today we are using optical encoders, lasers / photodiodes, proximity probes, charge couple devices and computers to measure and calculate machinery moves. Of the 140 worldwide patents researched for this book only 3 patents have been filed that focus on the aspect of repositioning machinery. Something is seriously wrong with this narrow-minded portraiture of our 'achievements'.

All of the measuring tools and all of the calculators and computers in the world will not move, reposition, and align one single piece of machinery in existence! Dial indicators, lasers, encoders and the full gamut of measuring tools don't align machinery ... people do!

Judging the quality of an alignment measurement tool by how minute a measurement can be made is fruitless. What good will it do you to have an alignment measurement system that can measure to one-ten thousandths of and inch (0.0001") if you cannot purchase shim stock that thin?

The true 'art' of shaft alignment pivots on your ability to control what you want to happen to the machine casings to bring them into an aligned state. You might be able to accurately measure shaft positions within one-thousandths of an inch, but can you control the position of the machinery to that level of precision? If you cannot, all of your valiant efforts during the measurement process are for naught since you can't achieve the 'true' final goal of properly positioning the machinery to achieve acceptable alignment tolerances.

Step 1 -
Roughly adjust (+/- 100 mils) the axial position of the shafts.

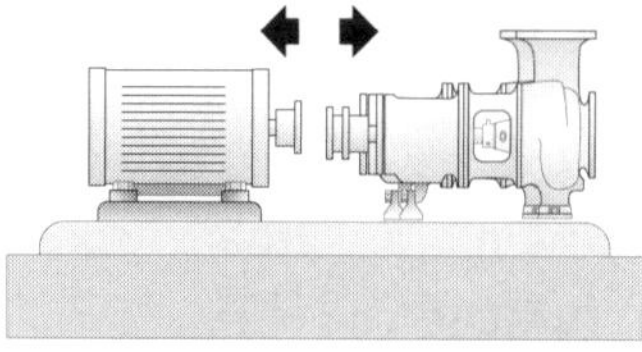

Step 2 -
Make a farily large adjustment to either or both of the machines in the up and down direction by adding or removing shims.

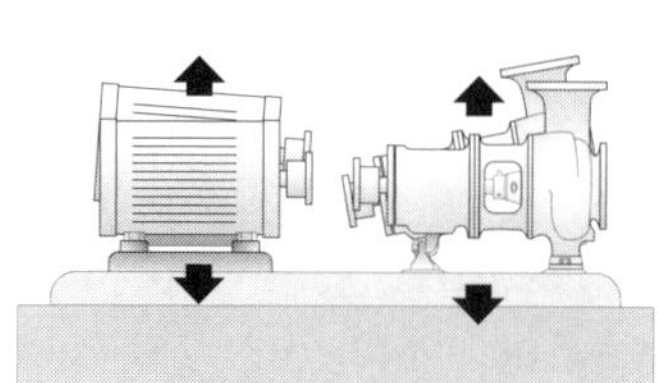

Step 3 -
Make a farily large adjustment to either or both of the machines in the side to side direction by translating the inboard and outboard ends.

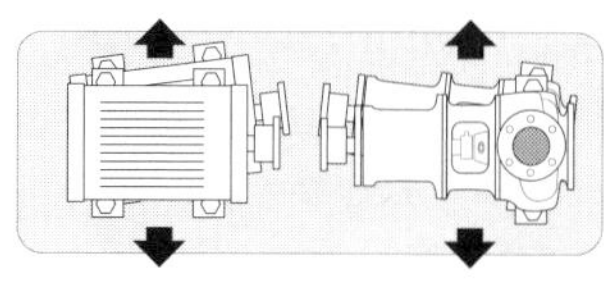

Step 4 -
Make a 'trim' adjustment (if necessary) to either or both of the machines in the up and down direction by adding or removing shims.

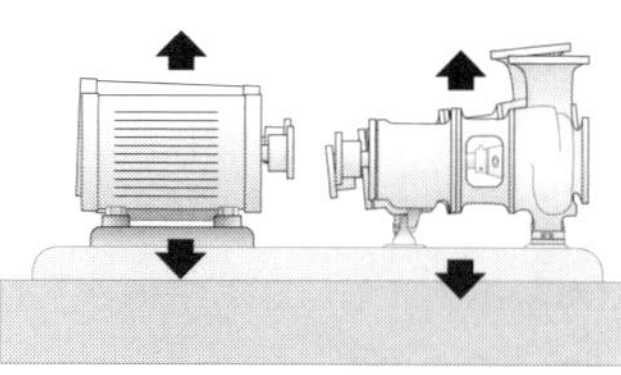

Step 5 -
Make a 'trim' adjustment (if necessary) to either or both of the machines in the side to side direction by translating the inboard and outboard ends.

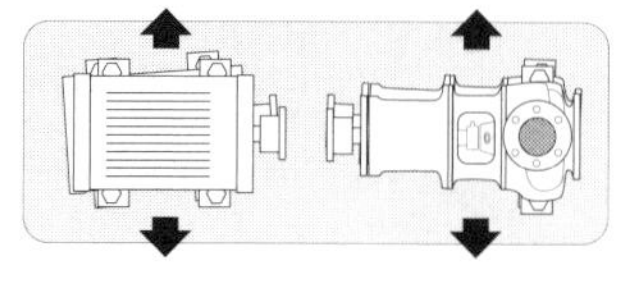

Step 6 -
Make any final adjustments to the axial position of the shafts.

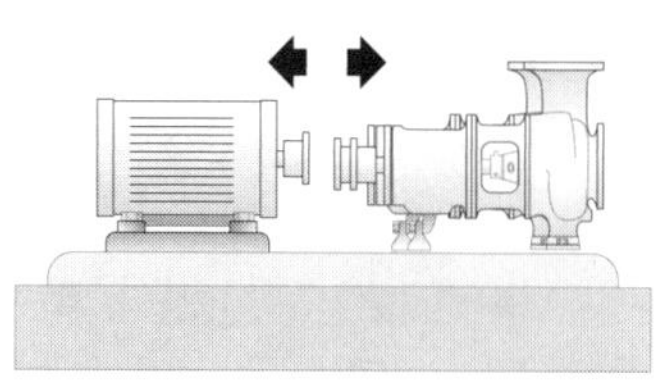

Note : If you are adding or removing shims three or more times and adjusting the machinery sideways, something is probably wrong (see Chapters 6, 7, 8, and 9).

Figure 11-3. Basic step by step sequence for aligning rotating machinery.

Accurately repositioning machinery is the ultimate test of your expertise in shaft alignment and it will be your most challenging one. Hard work, sweat, dirty and bloody hands, perseverance, discipline, patience, and practice are the core elements of success here.

Machinery Positioning Basics

Once the shaft positional measurements have been taken, the most efficient way to reposition machinery casings is as follows:

1 - Roughly adjust the axial position of the machinery insuring that the shaft end to shaft end distance is +/- 50 mils of the final desired dimension.

2 - Make a large move in the up and down direction by adding or removing shim stock between the machinery feet and the baseplate.

3 - Make a fairly large move in the lateral (side to side) direction on one or all the machines.

4 - Make a small (trim) move in the up and down direction by adding or removing shim stock between the machinery feet and the baseplate.

5 - Make a small (trim) move in the lateral (side to side) direction on one or all the machines.

6 - Adjust the axial position of the machinery insuring that the shaft end to shaft end distance is +/- 5 mils of the final desired dimension.

Axial Spacing

For many applications where rotors are supported in antifriction bearings, the shaft end to shaft end distance measurement is fairly straightforward. For rotors with sliding type radial and thrust bearings, the actual position of the rotors must be taken into account. For example, electric motors with armatures supported in sliding type bearings will seek 'magnetic center' when the field is applied. Therefore, motors should be run solo to determine where magnetic center is prior to setting the axial spacing. Centrifugal compressors, turbines, and other driven machinery will typically operate against their 'active' thrust bearing. During the alignment process, their rotors should be seated against the active thrust bearing prior to setting the axial spacing.

The axial distance should be measured as close as possible to the centerline of rotation of each shaft unless otherwise specified by the equipment or coupling manufacturer. If no tolerance is given, the rule of thumb is to hold this distance to +/- 0.010" of the recommended dimension. The importance of properly maintaining this gap cannot be overstressed since coupling 'lock-up' conditions occur as often from improper gap distances as from excessive vertical or lateral misalignment. The purpose of the coupling is to transmit rotational force, not thrust forces from one machine to another.

Shim Stock

The most commonly used device to change the height or pitch of a machine case is shim stock. Shim stock is flat sheets of metal rolled to thicknesses ranging from 0.001" to 0.100" (1 to 100 mils). Shim stock over 100 mils is generally referred to as 'spacer' or 'plates'. Shim stock is commonly available in brass, carbon steel, or stainless steel, but can be made out of virtually any metal if desired. The accuracy of the thickness of shim stock is typically

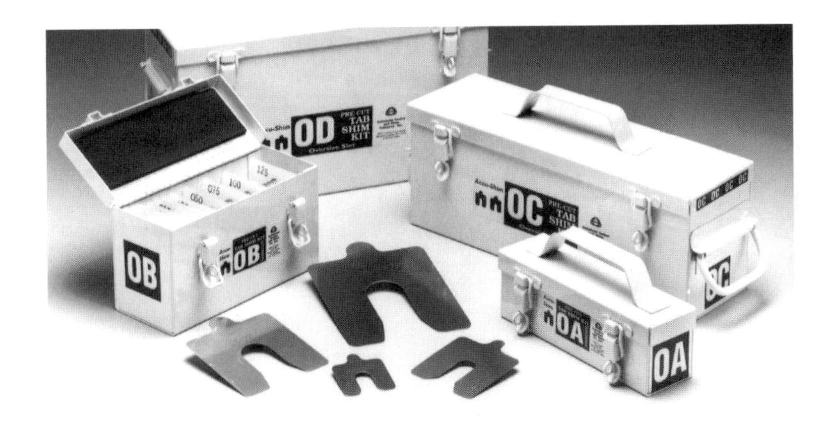

Figure 11-4. Pre-cut shim stock kits. Photo courtesy of Industrial Gasket & Shim, , PA.

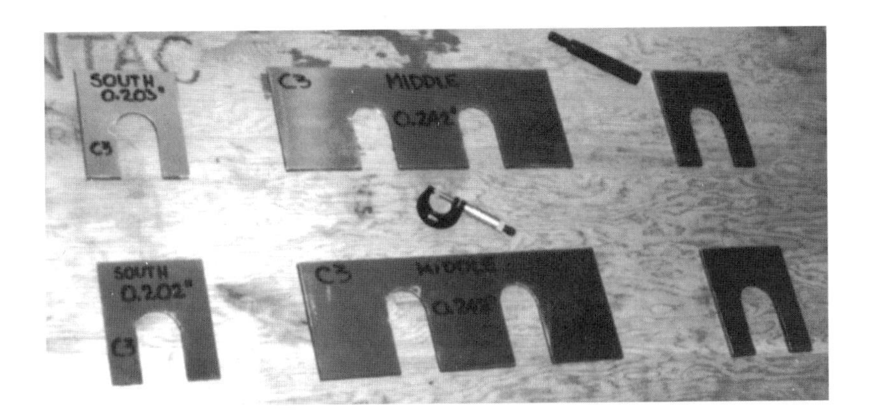

Figure 11-5. Custom milled, single thickness 'U' and 'W' shaped 'spacers' or 'plates' to be installed under 7000 hp synchronous motor.

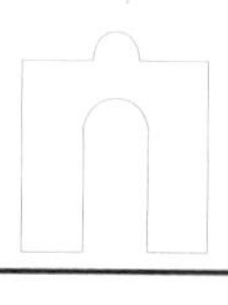

Standard Pre-cut
Shim Stock Sizes

Shim Size	Dimensions
A	2" x 2"
B	3" x 3"
C	4" x 4"
D	5" x 6" (of 5" x 5")

Table 11-1. Standard pre-cut shim stock sizes.

1 mil	0.001"
2 mil	0.002"
3 mil	0.003"
4 mil	0.004"
5 mil	0.005"
6 mil	0.006"
7 mil	0.007"
8 mil	0.008"
9 mil	0.009"
10 mil	0.010"
15 mil	0.015"
20 mil	0.020"
25 mil	0.025"
50 mil	0.050"
75 mil	0.075"
100 mil	0.100"
125 mil	0.125"

Table 11-2. Standard shim stock thicknesses.

+/- 5%. For example, a 50 mil shim could range anywhere from 48 to 52 mils. Shim stock is typically sold in widths of 6 inches in lengths of 10 feet.

Several companies offer pre-cut 'U' shaped shim stock and is sold in various sizes and thicknesses as shown in Table 11-1 and 11-2.

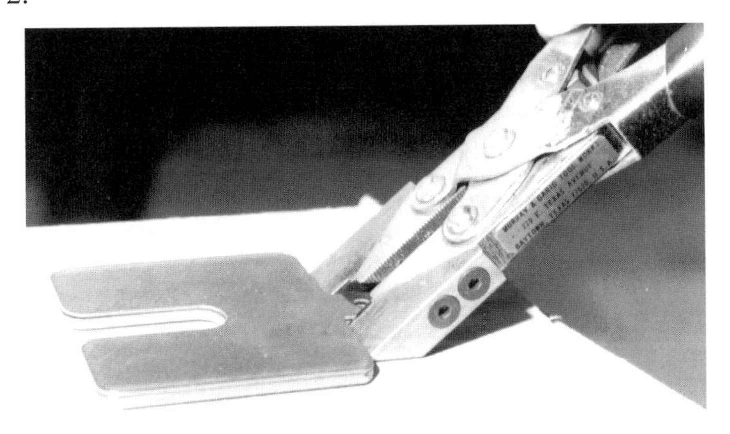

Figure 11-6. Shim pliers for installation and removal of shims. Photo courtesy Murray & Garig Tool Works, Baytown, TX.

Lateral Movement

Jackscrews or whatever devices that may be used to slide equipment sideways should be placed as close as possible to the foot points without interfering with tightening or loosening of foundation bolts. A typical jackscrew arrangement is shown in figure 11-7. Dial indicators mounted on the baseplate that are used to monitor sideways movement should usually be placed on the opposite side of the machine case from where the movement device (e.g. jackscrew) is located to keep the indicator from inadvertently being bumped as shown in figures 11-8 and 11-9.

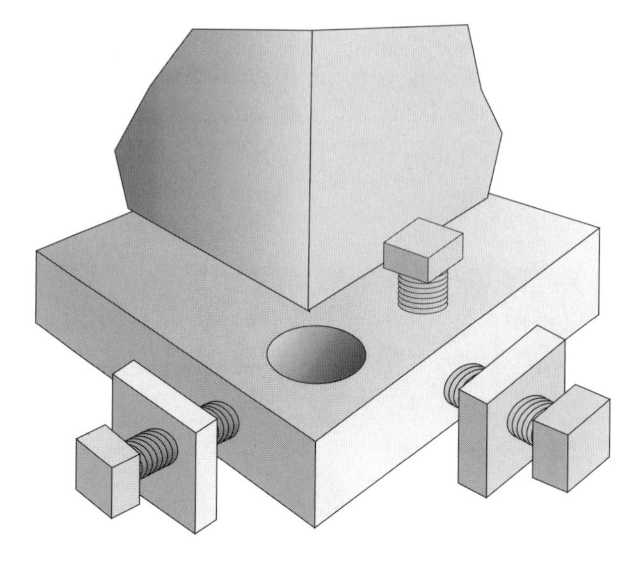

Figure 11-7. Ideal jackscrew arrangement enables machine to be lifted, translated laterally, and translated axially.

Figure 11-8. Monitoring any lateral moves by placing a dial indicator at the corner of a machine case directly in line with the foot bolt.

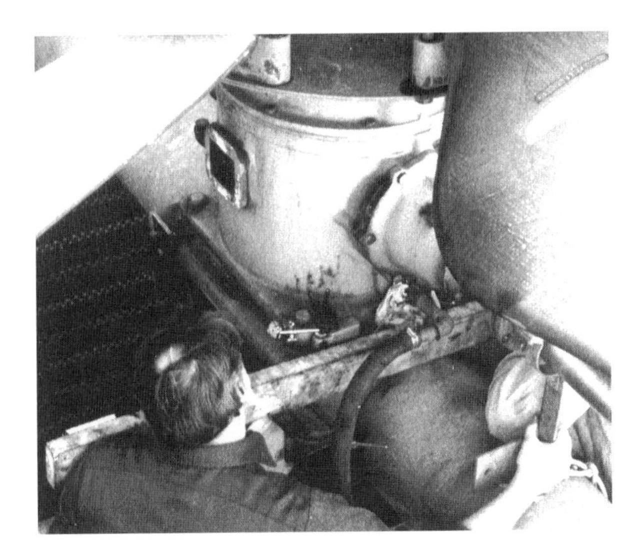

Figure 11-9. Monitoring a controlled lateral move by placing a dial indicator at the corner of a cooling tower gear.

Use a corner foot bolt as a 'pivot' point and move one end of a unit at a time when moving sideways as shown in figure 11-11. After the outboard end has been moved the entire amount, tighten one of the outboard bolts and loosen the inboard 'pivot' point bolt.

Monitor the movement of the inboard end either by placing a dial indicator at the side of the machine casing at the inboard foot or by using a dial indicator and bracket arrangement attached to one shaft, zero the indicator on one side of the coupling hub, then rotate the dial indicator and bracket arrangement 180 degrees and note the reading. Start moving the inboard end in the appropriate direction until the dial indicator on the coupling hub reads one half the original value. Zero the indicator again and rotate the dial indicator and shaft assembly back 180 degrees to the original zeroing point on the other side of the coupling hub and check the reading on the indicator. If necessary, continue moving the inboard end to get the dial indicator to read zero when swing-

Controlling your lateral moves

Machinery has a great tendency to misbehave when moving sideways. Quite often machine cases will translate or slide straight sideways rather than pivot at one end. For instance, if you are not watching what happens to the inboard end while you are moving the outboard end, a shift at the inboard end may have occurred that you didn't notice.

Here are some suggestions to help minimize potential problems:
- have indicators mounted to monitor any movement at the inboard and outboard ends of the machinery
- zero the indicators prior to loosening the foot bolts particularly with machinery that has piping or ductwork attached to it (if you see more than 2 mils of movement sideways on the machine case after loosening the bolts, you may have an uncorrected 'soft foot' problem or excessive static piping stress)
- use jackscrews wherever possible not only to move the machinery but also to hold one end in place when trying to position the other end
- if you don't have jackscrews ... use one of the foot bolts as a 'pivot' point , move the outboard end first, then use an alignment bracket and dial indicator to assist in positioning the inboard end

Figure 11-10. Suggestions for controlling lateral movement of rotating machinery.

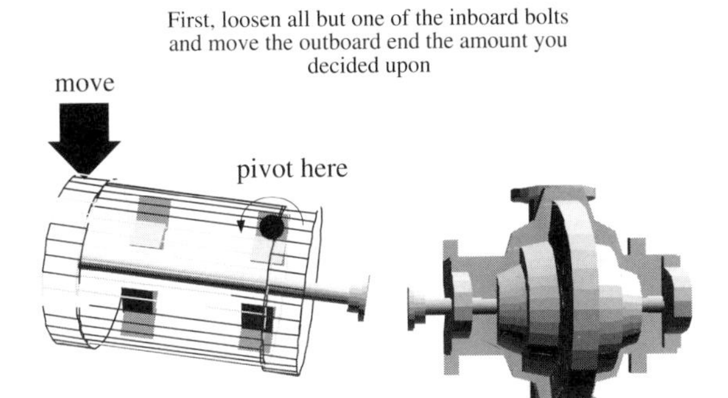

Next, tighten one of the outboard bolts, loosen the inboard bolt used as a 'pivot' point, mount the bracket and indicator onto one of the shafts, rotate the bracket/indicator over to one side, zero the indicator, and rotate to the other side and make a note of the reading

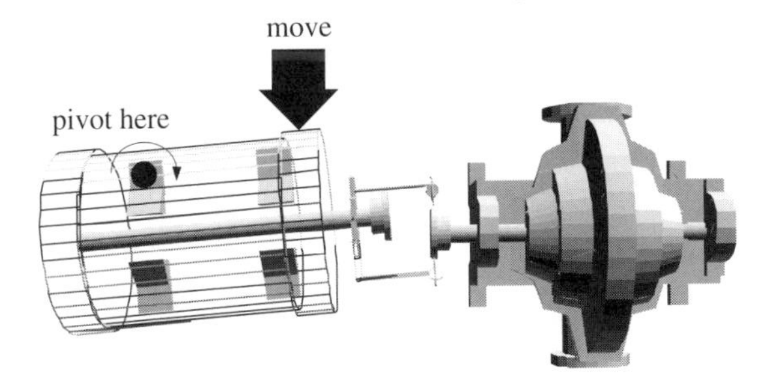

Finally, move the inboard end until the indicator is reading half of the original value (assuming that you want the shafts to be colinear when the units are off-line)

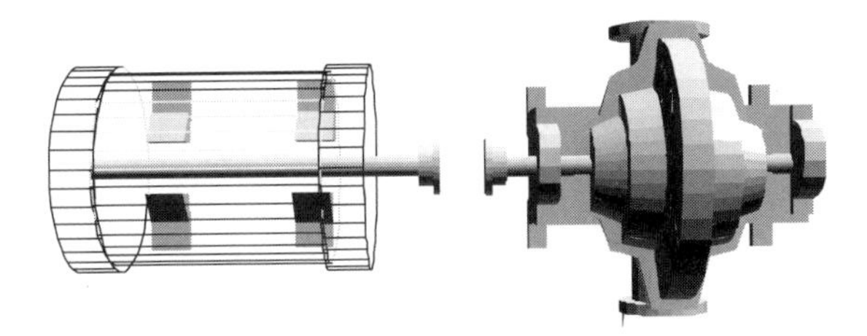

Figure 11-11. Suggested method of moving machinery sideways if jackscrews are not present.

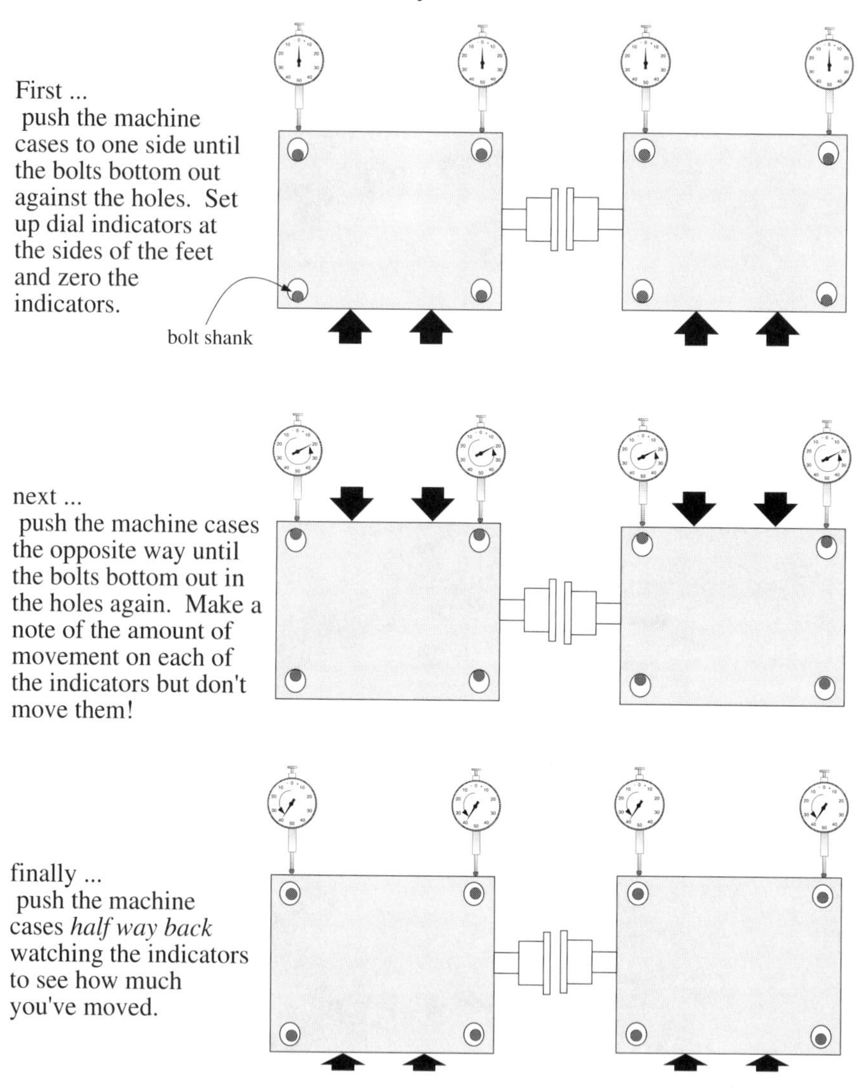

Figure 11-12. Determining the total allowable movement envelope of rotating machinery casings.

ing from side to side on the coupling hub (assuming there is no dynamic lateral movement in the machinery). It is also possible to move both ends simultaneously using indicators and jackscrews at each foot. For equipment with inboard to outboard foot distances of three feet or less, this seems to work alright with two people on the alignment job. Larger equipment usually requires four or more people to be effective.

On new installations it may be desirable to generate a total sideways movement 'map'. This will come in handy when calculating the necessary lateral moves to determine whether it is possible to move it as far as the calculation requires. An example of a typical movement map is illustrated in figure 11-12.

Once the map has been established, place each unit in the center of its allowable sideways travel and begin to take readings at this point. These 'maps' will prove invaluable when aligning multiple machine drive trains.

Vertical Movement

Lifting equipment is markedly more difficult than sliding it sideways, so it is desirable to make the minimum number of moves necessary to achieve the correct vertical position.

If good lateral alignment has been achieved, try to keep as many foot bolts tight or have the jackscrews tightened against the machine element to prevent the unit from moving back out of alignment when shims are being added or removed from the feet. Lifting equipment with a couple of foot bolts tightened can be a very delicate and challenging operation and must be performed with extreme caution. The idea is to lift the unit just far enough to slide shims in or out.

Types of Movement Tools

Hammers are probably the most widely used tool for moving machinery sideways. Even if this is the least desirable method, there are preferred techniques when using hammers to move a unit sideways.

1- use soft-faced hammers (plastic or rubber) instead of steel hammers.
2- If soft-faced hammers are not available, place a piece of wood or plastic between the hammer and the impact point on the piece of machinery to prevent damaging the case.
3- Take easy swings at first and then begin increasing the force. With practice, you can develop a 'feel' for how the unit moves at each impact. The more force that is used however, the greater chance there is to jolt the dial indicators that are monitoring the units movement rendering the readings useless.

Using hammers and steel wedges to lift equipment once again, is the least desirable method. If there is no alternative, here are some tips when using this technique.

1- Place the wedge as close to the foot that needs to be lifted without interfering with the process of adding or removing shims. The casing may distort enough to get the necessary shims in or out of that foot area without having to lift the entire unit.
2- Apply a thin film of grease or oil to both sides of the wedge.
3- It is fairly easy to install a wedge, but it is quite another thing to get it out from under a heavy piece of machinery. Provide some means of removal of the wedge.

Pry and Crow Bars

Like sledge hammers, pry and crow bars can be found in every mechanic's tool box and they invariably end up out at the alignment site 'just in case' they're needed. Consequently, for smaller, light equipment, pry bars end up being the most widely used device to lift equipment, yet it is the least desirable. A pry bar provides very little control in accurately lifting equipment and can slip from its position easily which can be very painful for your partner who is trying to remove old shims from under the feet with his fingers.

A pry bar can also be used to move the equipment sideways, assuming there is a leverage device near the feet. The leverage device, however, usually ends up being piping, electrical conduit, or a long piece of 2x4 supported against something else on the machinery frame or foundation.

Comealongs and Chainfalls

These devices can be used to both lift and move equipment laterally. The primary problem with this equipment is usually the lack of proper rigging or anchor points for the chainfalls or comealongs when moving sideways. There is also the problem of exceeding the capacity of the chainfall when rigged to lift the equipment. The better quality chainfalls and comealongs, however, provide improved control and safety over hammers and pry bars.

Hydraulic Jacks

There are many types of hydraulic jacks and kits that can readily be purchased at reasonable prices. When rigged properly, hydraulic jacks provide good control and safety when lifting or sliding equipment and are one of the preferred methods for moving rotating machinery.

Permanent Jackscrews

Although jackscrews are the most preferred method for moving machinery, they are not found frequently in industry mainly because of the cost and effort required to install them. It is indeed a shame that all rotating machinery manufacturers and users can't seem to find the time to provide these for their machinery.

Portable Jackscrews and Machinery Positioners

A considerable amount of imagination has gone into designing these clever devices that could be used in one form or another by most of industry for machinery alignment applications. If you have more than one of the same type of pump, motor, compressor, etc., it is recommended that devices such as shown in figures 11-13 through 11-17 be used for your specific application.

Figure 11-13. Custom fabricated removable positioner.

Figure 11-14. Positioner pin indexed in dowel pin hole for pulling machine sideways.

Figure 11-15. Custom fabricated removable positioner for large electric motor.

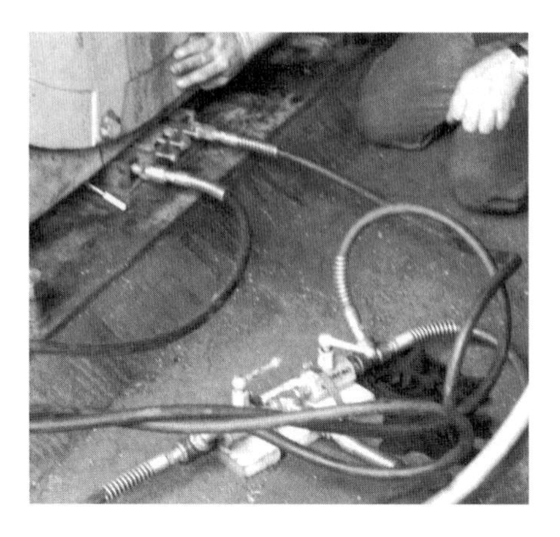

Figure 11-16. Hydraulic machinery positioner that lifts machinery upward and adjustment bolts for lateral or axial positioning.

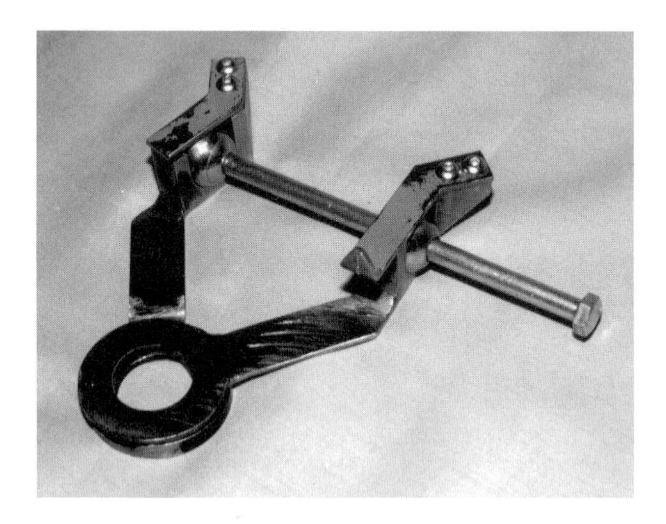

Figure 11-17. Prototype machinery positioner with automatic soft foot correction capabilities. Photo courtesy Turvac Inc., Cincinnati, OH, (patent pending).

Figure 11-18. Undercut bolts.

Figure 11-19. Always check to make sure all foot bolts are tightened down after each machinery move and when finished with alignment job.

The Do's of moving machinery

- Insure that the undersides of the machinery feet and the points of contact on the base are clean, rust-free, flat and parallel (use soft foot shims if necessary to obtain this condition). It may help if some lightweight oil is used between the foot points to facilitate lateral movement.
- Use pliers to add or remove shim stock from under machinery, not your fingers.
- Consider removing shims instead of continually adding them under the foot points. If possible, fabricate single thickness plates instead of stacking many thin shims together.
- Think about how piping or conduit may affect how you can position the machinery or how it can improve or eliminate static piping or conduit strains.
- Use jackscrews to keep a piece of machinery in its current sideways position if additional shims are needed under the feet. If jackscrews aren't installed, install shims under one end at a time by keeping one end bolted down and gingerly lift up the other end just enough to install the shims needed.
- Have dial indicators positioned along the sides of each unit when tightening the foot bolts down to determine if the unit is moving back out of alignment. This is usually due to a soft foot condition that has not been corrected.
- Insure that the foot bolts are tight after making a move on the machinery and be sure they are securely tightened when you're done with the job.

Figure 11-20. Recommendations for moving machinery.

The Don'ts of moving machinery

- Use pipe wrenches or a pry bar jammed between the coupling bolts to rotate machinery shafts. Use strap or chain wrenches or special shaft turning tools (see figure 11-22).
- Use carbon steel shim stock or use shims that have paint anywhere on the surface of the shim.
- Use a large number of thin shims to make a thick spacer under a foot.
- Loosen any foot bolts unless dial indicators (or some other measuring device) have been set up to monitor whether the machinery happens to move when the bolts are loosened.
- Undercut foot bolts. When some unsuspecting person tries to remove the bolt later on, it might shear off in the baseplate.
- Dowel your equipment.

Figure 11-21. Recommendations to prevent problems when moving machinery.

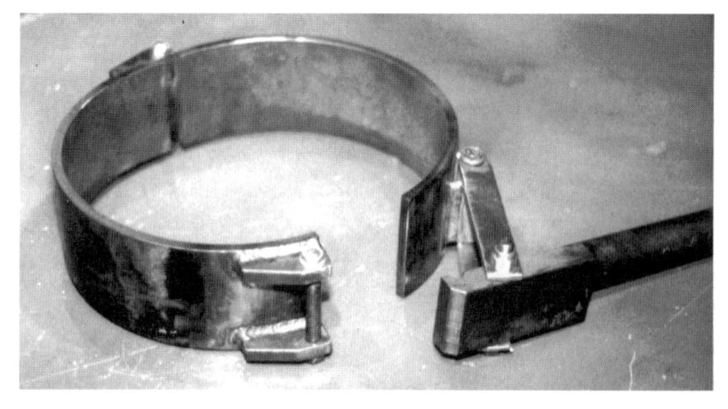

Figure 11-22. Custom made shaft rotation tool.

What to Do When The Alignment Isn't Working

Invariably, there will come a time when you're aligning a piece of machinery and nothing seems to be working or the alignment seems to be getting worse rather than better. Or perhaps you have improved gross amounts of misalignment, but just can't seem to bring the units within 1 mil per inch or better. You keep taking 5 - 10 mils of shim stock out from under the feet and then putting it back under the same bolting plane you just removed it from, or perhaps the machinery seems to keep wanting to move out of alignment sideways when you tighten the bolts down. After muttering obscenities to the equipment and tools you're working with, you look up and notice someone walking by looking at you as if you're deranged. What should you do now that you're stumped?

The Alignment Problems Troubleshooting Guide

1. Stop, and mosey off the job site for a few minutes. Go back to your break area and relax. First of all, you're going to have to realize there is something that you have overlooked. It might be staring you right in the face and you can't see it through the fog, or it could be one or several minor things acting together to cause a larger problem. As you go through these next steps, don't lie or pretend you did them, really go back and check each and every one again (or maybe for the first time). Remember, you might find more than one thing wrong, so don't stop if you happen to discover something halfway through this list.
2. Go back to Chapter 1 and read the section entitled Summary of an overall alignment job. You can skip step one and start concentrating on what's being said from step 2 on. Don't just read the words like you're reading a novel. Stop after each sentence and understand what each sentence

means. If you read a sentence and haven't done what was suggested, stop, mark your place, and do it now. If you don't understand what's being asked, take a few moments and find where that piece of information is in this book and read through it.

3. Pay particular attention to step 4 (the preliminary checks). Over half of the time that someone is having problems aligning the machinery, it's due to something in this step. Again, without trying to be excessively repetitive, don't pretend you did these. Even if you did, go back and check them again. If you find that there is an excessive amount of runout, or 'soft foot', or sloppy bearing fits, or warped baseplates, or deteriorating foundations, or excessive piping strain ... fix the problem. I fully understand what's involved in the simple sounding statement ... 'fix the problem'. Some of the problems you find are going to be time consuming, expensive, pain-in-the-neck problems to solve. I also understand that in many cases you are going to have to get other people involved and when you explain the problem(s) that you've found, there might be (will probably be) some (a lot of) hesitation on their (and your) part to correct what's been found. It's your choice. If you find something, and decide not to do anything about it, you can try to go on and do the best job you can, but don't expect great results and don't blame your alignment measurement system, your wrenches, the equipment manufacturer, your coupling manufacturer, the graphing or modeling technique, this book, or your barber for what's going to happen. You and the people who decided not to do anything should walk up to a mirror, stare straight ahead, and start blaming the person you see in front of you.
4. Whew! Did you get this far yet? OK, you should be feeling a little better by now (assuming you feel good about not finding any problems in the previous item). Now, take

a good look at your shaft alignment measurement system. If you're using dial indicators, remove them and with your finger, gently push the stem in trying to feel for any 'rough spots' or 'sticky spots' in the stem travel. If your indicator doesn't have smooth stem travel ... replace it. If you're using a laser system or optical encoder system (or any system with electronic sensors) check that all of the cables are good and that they are making good connections to and from the sensors. Check the batteries. Take a good look at your shaft clamping brackets. Are they firmly affixed to the shafts? Is the base of the bracket sitting in a keyway? Is the span bar tube clamped firmly to the bracket on dial indicator type systems? Is the laser and detector housing clamped firmly to the bracket posts? You may want to remove the system from the machinery ... get a 20"-30" long piece of Schedule 80 pipe... set the pipe on some 'V' blocks or pipe roller stands (or in a vise) ... clamp the brackets, indicators/sensors to the fixtures ... zero the indicators on top and roll the system to the bottom and watch what happens. On dial indicator systems ... does the bracket sag stay the same if you do this several time? On laser systems ... do the readouts stay at zero all the way around? Now, put the alignment measurement system back on the machinery shafts. Check to make sure the foot bolts are all tight and then take several sets of readings. Are the reading sets pretty much the same each times? Are you dead sure you're reading the dial indicator correctly? If there's any doubt, go ask someone you believe knows what they are doing and have them take a set of readings (without your help) and see what they get. Do the two sets of readings compare? Loosen the shaft brackets, rotate the entire shaft alignment measurement systems 120 degrees around the shaft and take another set of readings. Do the two sets of readings compare? If the flexible coupling is bolted in place, take a set of readings, loosen the coupling bolts up and take another set of readings. Are they the same? If you're moderately to severely misaligned, they probably will not be the same. Now why's that? Take a good look at figure 1-2. When rotating machinery shafts are connected together, even with a flexible coupling, the shafts begin to elastically bend due to the misalignment condition. Now take a good look at where the shaft alignment measurement system is going to be set up on the shafts. Notice that the brackets are going to be attached to a section of the shafts that are undergoing bending stresses. If the coupling was not connected, these bending stresses would not be present and the shafts would not be in the positions shown in figure 1-2, rather, the shafts would be in a position concentric to the centerline of the bearings that support each shaft. This phenomena is one of the major reasons why you 'chase your tail' when using alignment measurement systems that require you to keep the coupling in place when taking measurements. Look, the only way you're going to be able to determine what moves need to be made to correct the misalignment conditions is to make sure you are getting an accurate set of measurements that indicate where the centerlines of rotation are going to be when the shafts are in an 'unstressed' state of affairs. If you work with garbage measurements, you're going to get garbage results.

5. Ok, by now you should be sure you're getting really good measurement information on the positions of the machinery shafts when you're off-line. If you are using one of the graphing / modeling methods explained in Chapter 8, go back and check to see if you are graphing the measurements correctly. If there's any doubt, go ask someone you believe knows what they are doing and have them model the information (again, without your help please) and see

what they get. Do the two sets of graphs compare? If you are using a computer of some kind, take the raw readings from the measurement system, and graph the data. Does it compare to what the software program is telling you? If you are using a laser system, remove the laser system, set up some brackets and take some dial indicator readings, graph/model the readings and compare it to what the laser is telling you. Do the results match?

6. Well, well. Are you still feeling good about everything so far? Because now comes the ultimate test. Are the machines moving exactly where you want them to go? OK, you said the inboard end of the pump had to be lifted up 27 mils. Did you measure the shim pack thickness with a micrometer before you installed it? Did you install them under the right machine? Did you put them under the right bolt set? Did you put them under both of the inboard bolts? Are you pushing shims out that are currently under the feet when you install the new ones? Are you disrupting your soft foot corrections when you install the new shims? Did you tighten the bolts back down after the shims were installed? When you try to move the machinery sideways, is the machine case shifting sideways without you seeing it when you loosen the foot bolts? Did you set up some dial indicators on the sides of the machine case at the inboard and outboard ends to monitor for this movement? OK, you said the outboard end of the steam turbine had to be moved eastward 86 mils. Did you see if that end could be moved that much before you started beating on it with a hammer? Did you move the outboard end of the pump rather than the outboard end of the steam turbine? Did you move the outboard end to the east or did you really move it to the west? When you tighten the bolts back down, is the machine staying where you located it sideways? How do you know that? Did you set up some dial indicators on the

sides of the machine case at the inboard and outboard ends to monitor for this movement? Are you sliding shims out that are currently under the feet when you move the machine sideways?

7. Take another set of readings after each move that you make. If you are using a dial indicator system, are you 'spinning zeros' all the way around? You better not be! Unless, of course, you don't have any bracket sag and you actually want the two shaft centerlines to be exactly in line with each other. If there is some OL2R movement, did you back calculate what the 'shoot-for' readings were supposed to be? Did you graph the new positions of the shafts and are they in the 'desired off-line positions'? Did you determine if you are within acceptable alignment tolerances? Are the shaft to shaft distances correct? Are you trying to stick 1 mil of shims under the feet to get a perfect alignment? If so, button everything up, come back after you run the machinery for 24 hours and see if the shafts are in the exact same position that you left them.
8. Did you run the machinery and measure bearing temperatures, vibration levels, and other pertinent operating data? Did the vibration go up and is someone accusing you of doing a poor alignment job? If so, have them read Chapters 1 and 14.

It's this simple. If you're aligning good machinery, and the foundation / frame is stable, and there is no 'soft foot' or piping strain, and the shaft to shaft measurements are correct, and the graph / model is correct, and you moved the equipment exactly as the graph indicated, and you know where the machinery should be when off-line ... everything should work. This isn't black magic - it's measurement, mathematics, and control.

12
Aligning Multiple Element Drive Trains and Right Angle Drives

The majority of rotating machinery drive systems in the world consist of two separate machines mounted to a common base. For example, there are many electric motors driving pumps and fans in virtually every industrial plant around the globe.

There are, however, a considerable number of rotating machinery systems where there are three or more machines coupled together to form what is commonly referred to as a drive 'train'. Some examples of typical drive 'trains' are as follows:

- motor - gear - pump systems
- motor - gear - compressor systems
- high pressure steam turbines- intermediate pressure steam turbines - low pressure steam turbines - generators - exciter (HP-IP-LP-Generator systems)

- motor - generator - generator - generator systems (sometimes called 'Drag' lines or MG sets)
- steam turbine - gas turbine - gear - low stage compressor - gear - high stage compressor (commonly found in refining and chemical industries)
- drive motor - clutch - gear - pinion - brake - right angled gear - clutch - slow speed turning motor (commonly found in cement plants on kiln or ball mill drives)
- rack & pinion gear set - pinion drive shaft - right angled gear - motor - pinion drive shaft - rack & pinion gear set (found in automotive industries at vehicle assembly plants on 'body drop' overhead cranes and various other industries having cranes that translate in two directions)

The variation and mixture of rotating machinery drive trains in industry is as diverse as one could possibly imagine.

Figure 12-1. Motor - gear - compressor trains. Photo courtesy General Electric, Evendale, OH.

These machine cases can be horizontally mounted and coupled end to end like a railroad train or they can be arranged at a right angle and some have even been set up in a 'U' or zig-zag configuration.

In virtually every case, these drive trains are very expensive and frequently the heart and soul of the operation of the plant. Some multiple element drive trains fall in the small horsepower range (10-500 hp), but there is a large percentage of these drive trains that are 500 to 50,000+ hp costing millions of dollars. Consequentially, they are the most critical pieces of machinery in the operation and the ones that seem to get everyone very nervous when something goes wrong. Sometimes small problems with these drive trains turn out to be just minor distractions but when big problems occur, these systems turn out to be absolute nightmares for the people involved in correcting the malady. Most people would just as soon have pleasant dreams than nightmares so a lot of effort is expended toward getting the situation fixed correctly so that major problems occur infrequently.

Figure 12-2. Motor - generator set.

It is highly recommended that you become very adept at aligning two element drive systems before you try to tackle a multi-element drive train. Every shred of knowledge you have gained in your alignment experiences will be tested to your limits when one of these systems has to be installed or completely rebuilt.

The multiple element drive train alignment laws

- Don't move any machine case until you know where all of the centerlines of rotation are.
- Every machine case is movable. Some machine cases are just more difficult to move than others.
- Once you have determined where each centerline of rotation on every machine is, define the allowable movement 'envelope' before you begin repositioning the machinery.

Figure 12-3. Compressor - motor - compressor trains. Photo courtesy General Electric, Evendale, OH.

- Make sure you know (or can at least make a good guess) where the centerlines of rotation will move to when the machines are running under normal operating conditions.
- Calculate how much time you think it will take to align the drive train and multiply this number by 1.5.
- Never consider a drive train fully functional until you surpass 100 hours of operation.

Multiple Element Drive Train Graphing / Modeling Techniques

The graphing and modeling shaft alignment techniques covered in Chapter 8 illustrated the effectiveness of using these models to find the most reasonable solution to reposition rotating machinery casings to achieve collinear shaft operation. The advantages of graphing or modeling your alignment problems are:

- it is an accurate visual representation of the centerlines of rotation of the machinery you are attempting to align
- the graphs can be generated by people who don't have extensive backgrounds in geometry, trigonometry, or software programming
- it is inexpensive since all you need is paper and a pencil (sometimes with a large eraser)
- you can use the model to show the possible movement envelope in the up and down and side to side directions to prevent wasting time on grinding baseplates or machine casings or cutting bolt shanks down or filing foot bolt holes open
- by using an overlay line there can be an infinite number of possible alignment solutions available (again, some of the solutions make sense and others do not)
- it is an excellent mechanism to use when explaining to your co-workers or managers what your plan of action is and

why you chose the solution you did even if they may not understand exactly how the graph was generated
- the model can be used to determine if you are (or are not) within acceptable alignment tolerances
- the model is a permanent record of the alignment that can be kept for future reference that can show 'before and after' alignment positions and/or any moves in between

It should be somewhat obvious by this time that if you can show the relative positions of two machinery shafts on a graph / model, why can't you show three machinery shafts or even ten?

Well you can, and here's how.

Multiple Element Drive Train Modeling - One Set of Shafts at a Time

For example, if you are aligning a three element drive train such as a motor, gear, and compressor, one method is to generate a model that shows the SIDE view of the motor and gear, a TOP view of the motor and gear, a SIDE view of the gear and compressor, and a TOP view of the gear and compressor. From these four graphs, two more graphs can be generated that show the relative positions of all three units on the same graph as illustrated in figures 12-5 through 12-8. An interesting point to note in this example is that gear boxes obviously consist of more than one shaft, but because these shafts are housed in a single case, the input and output shafts can be shown as a single centerline of rotation. If you install 50 mils of shim stock under the outboard end of the gear case (sometimes this may be a row of bolts), the bullgear shaft, any intermediate shafts, and the pinion shaft will all lift up 50 mils.

Figure 12-5 shows a multiple element drive train consisting of a motor - gear - compressor assembly. A set of reverse indicator readings were captured between the motor and the gear and also

Figure 12-4. Motor - gear - compressor train. Photo courtesy General Electric, Evendale, OH.

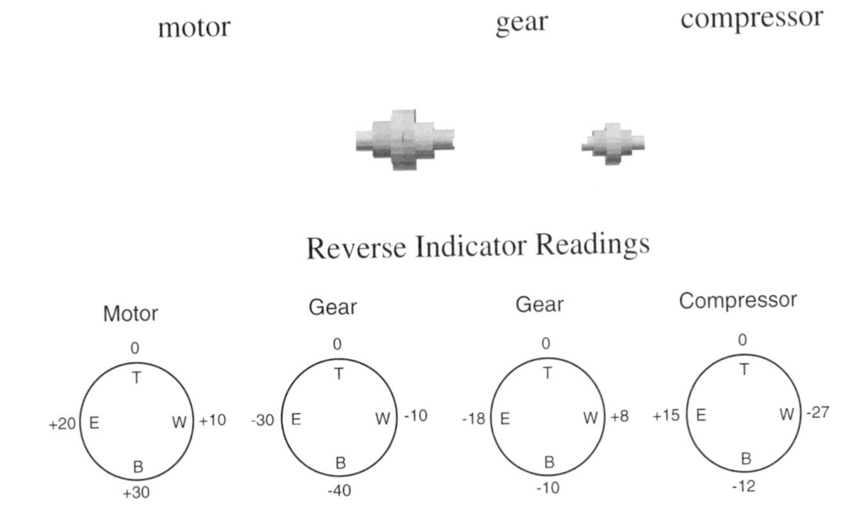

Figure 12-5. Motor - gear - compressor example problem.

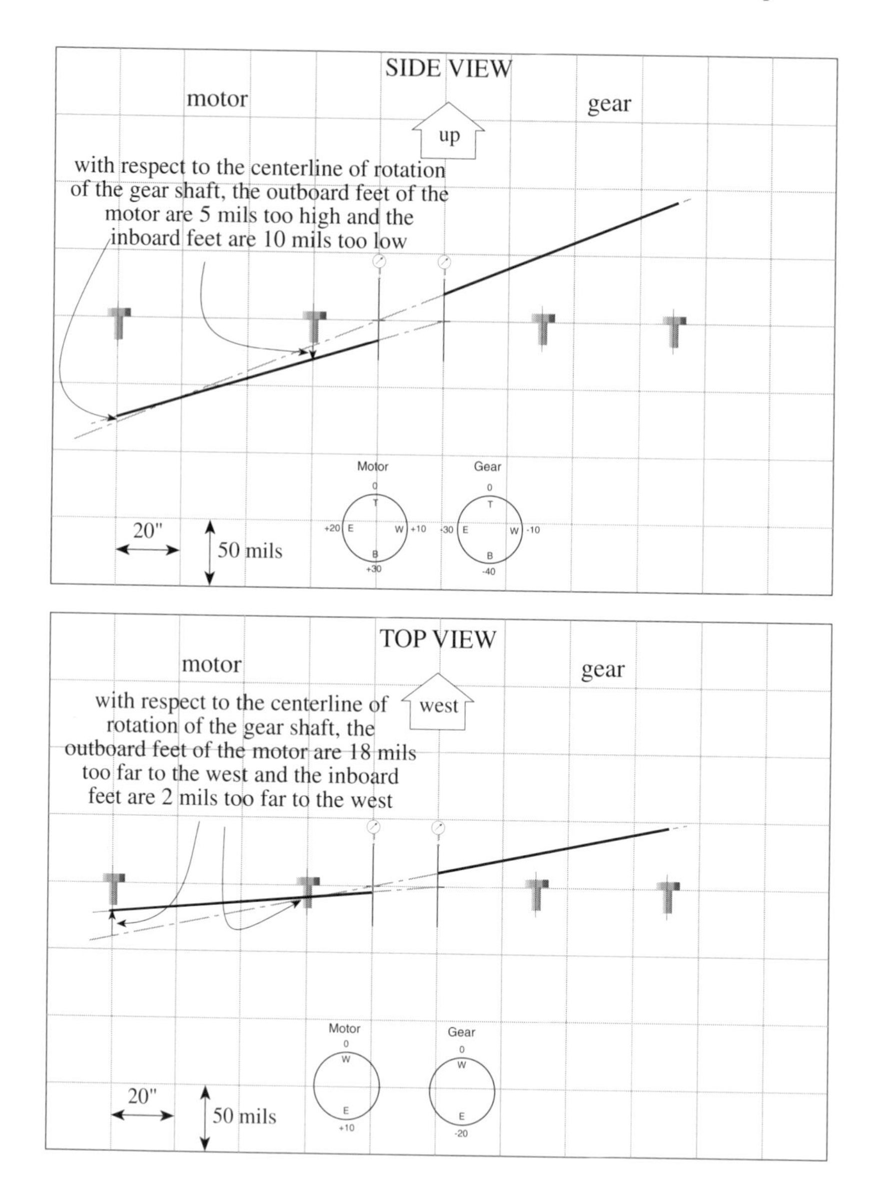

Figure 12-6. On separate sheets of graph paper, graph the SIDE and TOP views of the motor and gear using the point to point reverse indicator technique.

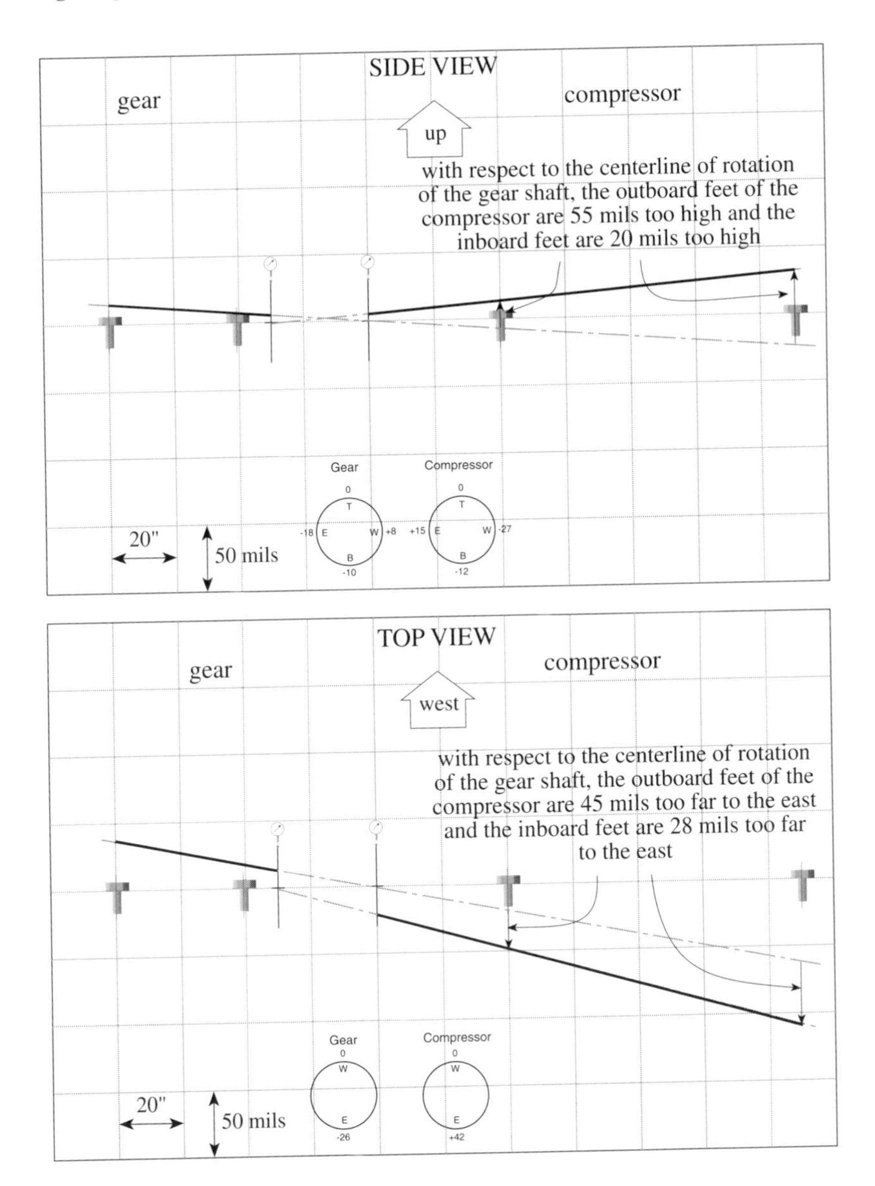

Figure 12-7. On separate sheets of graph paper, graph the SIDE and TOP views of the gear and compressor using the point to point reverse indicator technique.

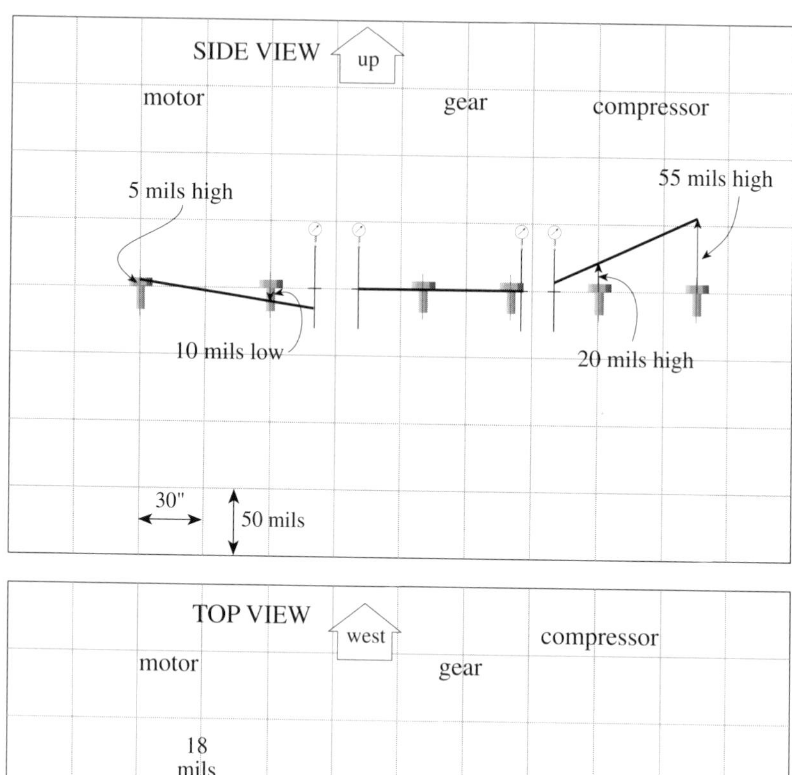

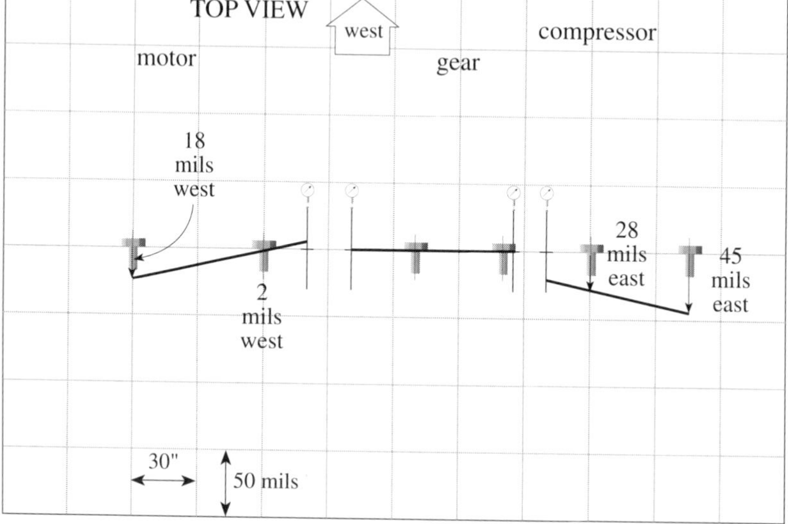

Figure 12-8. Again, on separate sheets of graph paper, transfer the relative positions of the motor and compressor shafts with respect to the gear shaft onto the multiple element model (see figures 12-5 and 12-6).

between the gear and the compressor. Two sheets of graph paper were then used to construct the relative positions of the motor and the gear shafts in the SIDE and TOP views using the point to point reverse indicator modeling technique as shown in figure 12-6. Two more sheets of graph paper were then used to construct the relative positions of the gear and the compressor shafts in the SIDE and TOP views using the point to point reverse indicator modeling technique as shown in figure 12-7.

Finally, two more sheets of graph paper were laid out to show all three elements on the graph and the gear shaft was placed directly on top of the graph centerline. By extending the invisible centerline of rotation of the gear shaft outward toward the motor shaft in figure 12-6, the distances between the gear's centerline and the centerline of the motor shaft at the inboard and outboard feet were measured and notes were made on the graph showing that the outboard feet of the motor were 5 mils too high and 18 mils too far to the west and that the inboard feet were 10 mils too low and 2 mils too far to the west. These ‘gaps’ were then transferred to the model showing all three machines as shown in figure 12-8. Additionally, by extending the invisible centerline of rotation of the gear shaft outward toward the compressor shaft in figure 12-7, the distances between the gear's centerline and the centerline of the compressor shaft at the inboard and outboard feet were measured and notes were made on the graph showing that the outboard feet of the compressor were 55 mils too high and 152 mils too far to the east and that the inboard feet were 20 mils too high and 75 mils too far to the east. These ‘gaps’ were then transferred to the model showing all three machines as shown in figure 12-8. Figure 12-8 now shows the relative positions of the motor, gear, and compressor shafts all on one model.

Multiple Element Drive Train Graphing - Modeling All the Shafts at One Time

Graphing one shaft at a time can be somewhat tedious and a slight waste of paper. Figures 12-9 and 12-10 show how the same motor, gear, and pump arrangement can all be modeled on one graph. In this case, use the line to point reverse indicator graphing technique instead of the point to point method. Place the gear shaft directly on top of the graph centerline and position the motor and pump accordingly based on the dial indicator readings on both ends of the gear.

It doesn't matter which set of readings are graphed first, the motor to gear or the gear to compressor. Notice that the positions of the shafts in figure 12-10 are identical to the positions of the shafts in figure 12-8 even though two different methods were used to generate the two graphs.

Again, just as in modeling two shaft systems, all that has been accomplished with the graph is that the positions of the shafts have been determined. At this point it is imperative that the restrictions in the up / down and east / west directions be determined and plotted onto the graph. Once this has been done, the final desired overlay line can be drawn onto the graph and the movement solutions of all of the bolting / translation planes can be determined and executed as illustrated in figure 12-11.

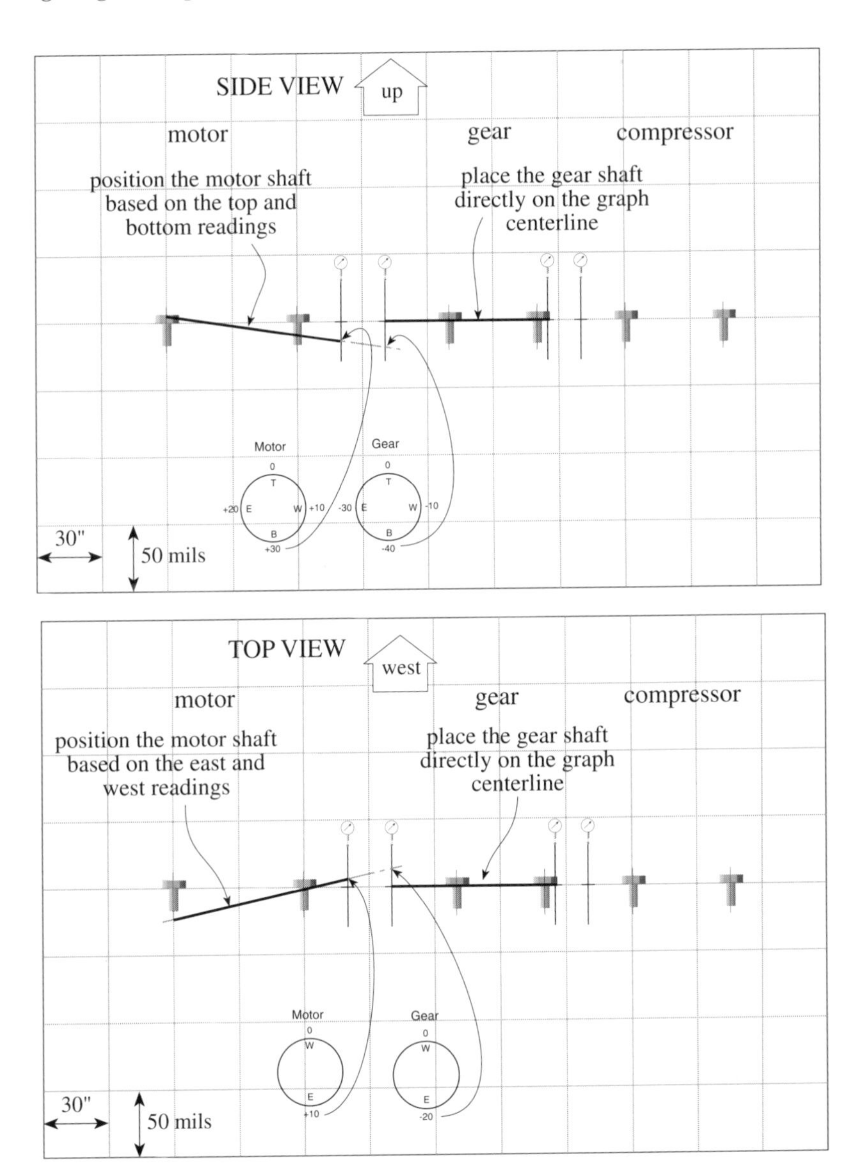

Figure 12-9. Modeling the position of the motor shaft with respect to the gear shaft using the line to point reverse indicator technique.

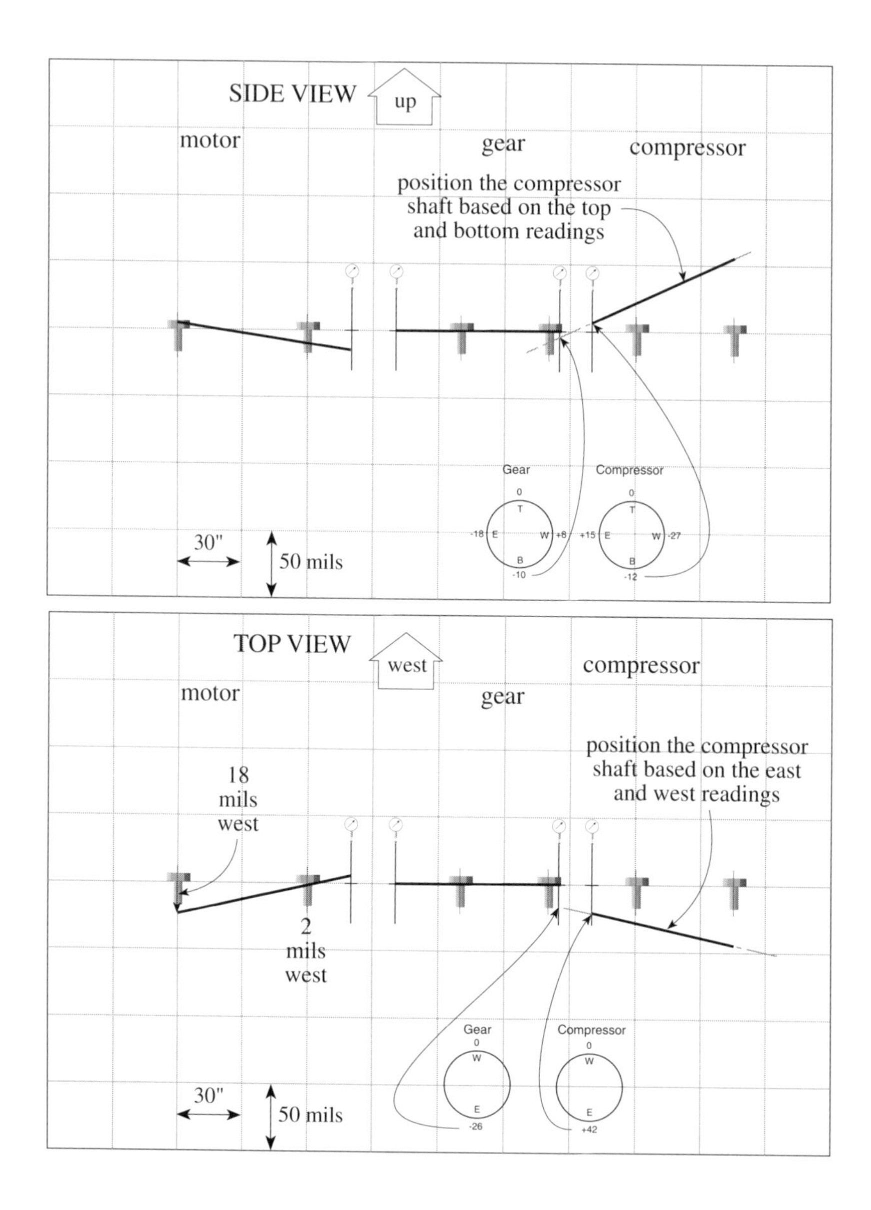

Figure 12-10. Modeling the position of the gear shaft with respect to the compressor shaft using the line to point reverse indicator technique.

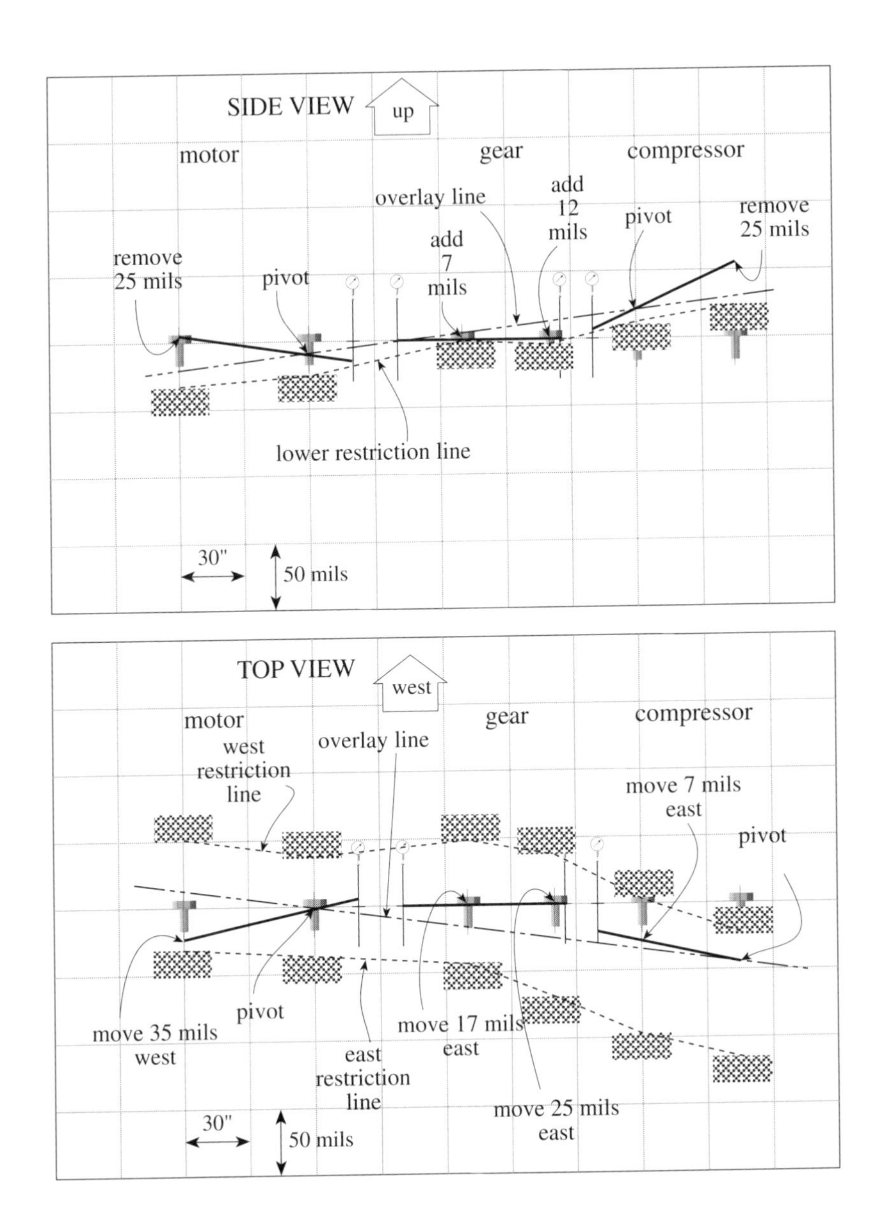

Figure 12-11. Plotting in the vertical and lateral movement envelope and placing an overlay line to pivot at two bolting planes

Mixing Alignment Measurement Methods

In some situations, more than one alignment measurement method could be (or may have to be) employed to measure each set of shafts. Figures 12-12 and 12-13 show how you can use different alignment measurement methods and still model the machinery positions.

A starting steam turbine is flexibly coupled to the intake end of a gas / power turbine assembly. The outside surface of the coupling flange on the gas turbine intake shaft is closely shrouded making it virtually impossible to capture rim readings which is why the face - rim dial indicator technique is used here.

The output shaft of the gas / power turbine is coupled to a low stage compressor by a gear coupling with a long spool piece between the two shaft ends. Here the shaft to coupling spool technique is employed to measure the relative positions of the gas / power turbine, the spool piece, and the low stage compressor shaft.

A laser shaft alignment system was used between the low stage compressor and the high stage compressor. The high stage compressor was randomly called the 'movable' machine even though it would be ridiculous to consider the low stage compressor as the 'stationary' machine.

Despite the fact that three different alignment measurement methods and tools were used to determine the relative positions of each set of shafts, the entire drive train can still be modeled on graph paper. As you can see from the shaft positions, without even plotting the lower and side to side movement restrictions, trying to call one of the machines the 'stationary' machine could lead to a lot of headaches and unnecessary work.

The face-rim dial indicator technique was used between the starter steam turbine and the gas turbine inlet shaft. The dial indicator readings were taken on the gas turbine shaft on a 10" diameter circle.

The shaft to coupling spool dial indicator technique was employed between the gas / power turbine and the low stage compressor.

All of the dial indicator readings have been compensated for bracket sag.

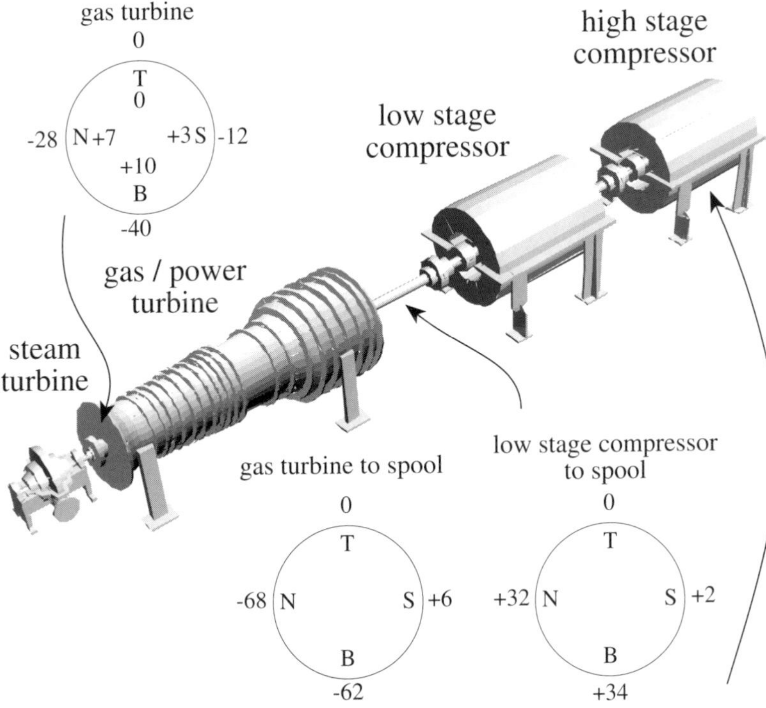

A laser alignment system was used between the low and high stage compressors. The high stage compressor was called the MOVABLE machine. The laser system calculated that the inboard feet of the high stage compressor had to be raised 20 mils and moved 50 mils to the north and that the outboard feet had to be lowered 40 mils and moved 120 mils to the north.

Figure 12-12. Steam turbine - gas / Power turbine - low stage compressor - high stage compressor example problem.

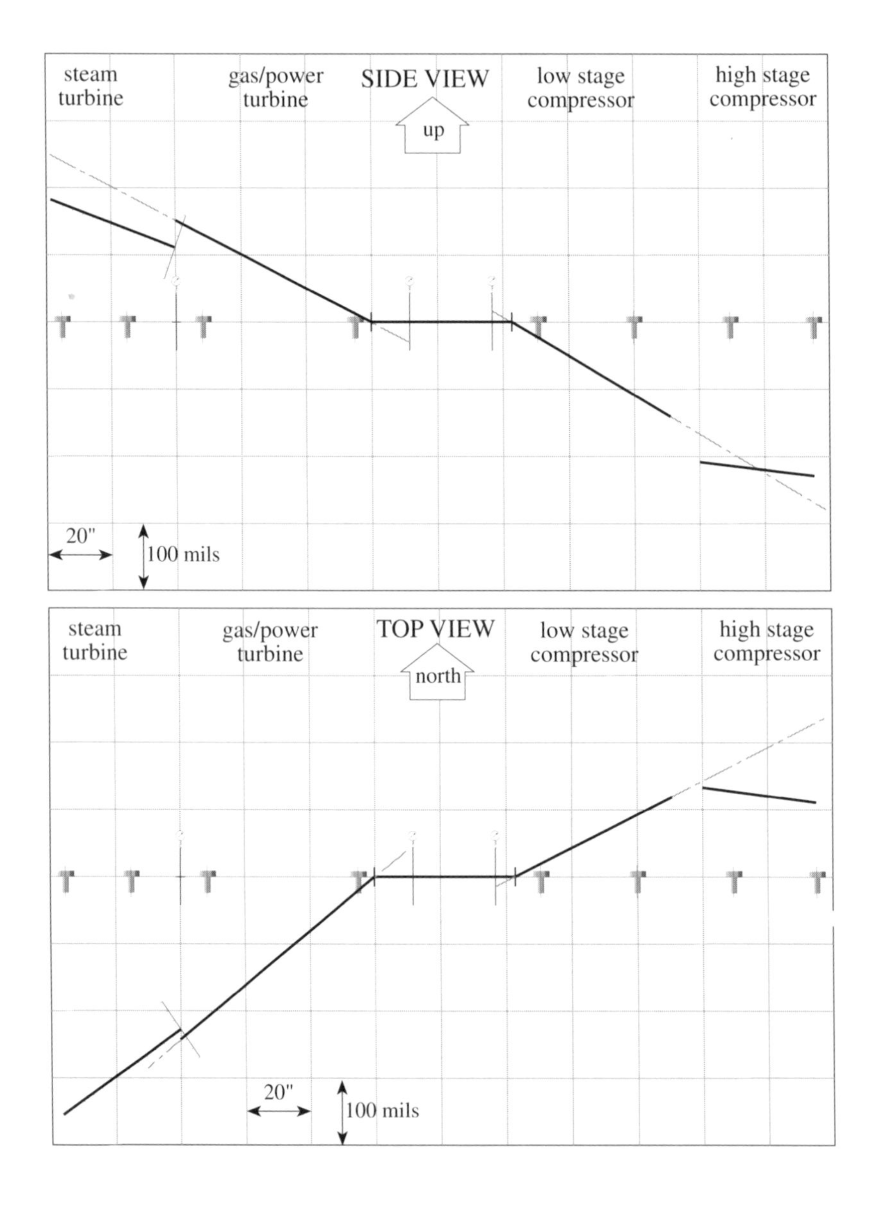

Figure 12-13. Steam turbine - gas / Power turbine - low stage compressor - high stage compressor SIDE and TOP views.

Modeling Right Angle Drive Systems

So far, we have examined rotating machinery drive systems that are horizontally mounted, direct in-line machinery. But not all drive systems are configured that way. Some drive trains

Figure 12-14. Right angle rewinder drive in paper plant.

Figure 12-15. Right angle calendar drive in paper plant.

are arranged in an 'L' shape commonly referred to as a right angle drive. A right angled gear box is flexibly coupled to a driver on the input side and flexibly coupled to some driven machine on its output side. The goal is to align the centerline of rotation of the driver and the input shaft of the gear and the driven unit to the output shaft of the gear.

The trick to this plotting technique is to use the graph in a 'dual scale' mode similar to the method used for face-rim plotting covered in Chapter 7 and to use the 'T' bar overlay to represent the input and output shafts in the right angled gear.

A sample problem shown in figure 12-16, illustrates an electric motor coupled to a speed-reducing right angled gear that drives a paper pulper. A series of reverse indicator readings have been taken between the pulper shaft and the output of the gear. A laser alignment system was used to capture the alignment data between the motor and the input shaft of the gear. The gear was

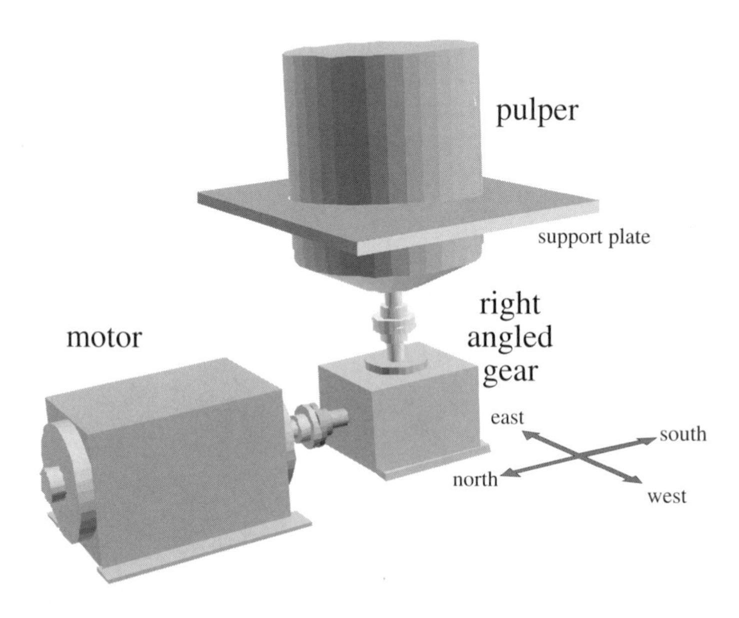

Figure 12-16. Motor - right angle gear - pulper arrangement.

called the 'stationary' machine and the motor was called the 'movable' machine (sounds typical). The readings from both of these techniques are plotted onto the two graphs and the centerlines of rotation of all three units are highlighted for clarity. Keep in mind that there is no 'top' and 'bottom' reading when viewing the pulper shaft and gear output shaft.

In this example, the support plate (that's the square thing surrounding the pulper housing) is rigidly attached to the pulper and consequently will have to be repositioned if we decide to move

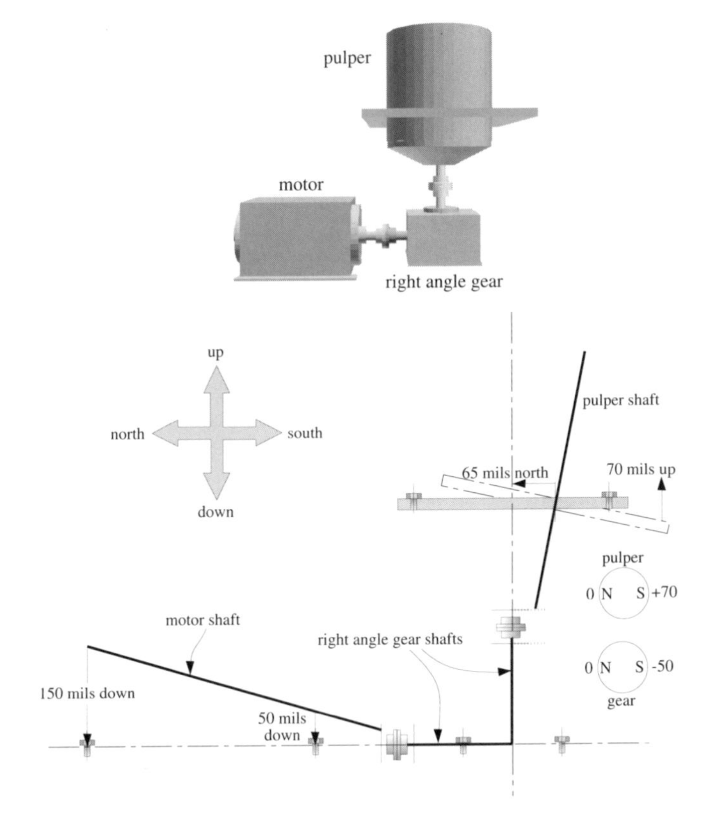

Figure 12-17. Motor - right angle gear - pulper shaft positions in Side View (looking east).

the pulper. Up to this point, alignment moves were usually described in terms of translation in the X-Y-Z planes. To reset any angular misalignment between the pulper and the gear, some people may refer to these 'pitching' moves as rotations (aka Roll-Pitch-Yaw).

The measurements taken with the laser alignment system indicated that the motor would have to be lowered 150 mils at the outboard feet and lowered 50 mils at the inboard feet. This could pose a problem since there aren't any shims under any of the motor feet at the moment.

By assigning the gear to be the 'stationary' unit, the pulper would have to be pitched 70 mils upward by placing shims under both of the south feet of the support plate and then slid to the north 65 mils.

An alternative solution would be to project a right angled overlay onto the graph as shown in figure 12-18 and position the overlay to determine what sort of changes need to be made on the gear feet to arrive at a manageable compromise to prevent from having to move the pulper and lower a motor that has no intention of going down without a fight.

Notice that when the overlay is set in the position shown in figure 12 -18, the motor can be raised upward just at the inboard feet, the pulper doesn't have to be moved at all, and the gear requires just 42 mils of lift at the motor end and 15 mils of lift at the outboard end.

The other view that must be considered is the top or end view. This will assist in determining the correct lateral position (or east / west positions) of the motor, gear, and pulper.

The laser alignment system determined that the motor must be moved 340 mils to the east at the outboard end and 90 mils to the east at the inboard end with the gear assigned as the 'stationary' machine. As you may have guessed already, there probably isn't enough room between the 'shank' of the outboard bolts and the holes drilled in the motor case to allow for almost 3/8" travel

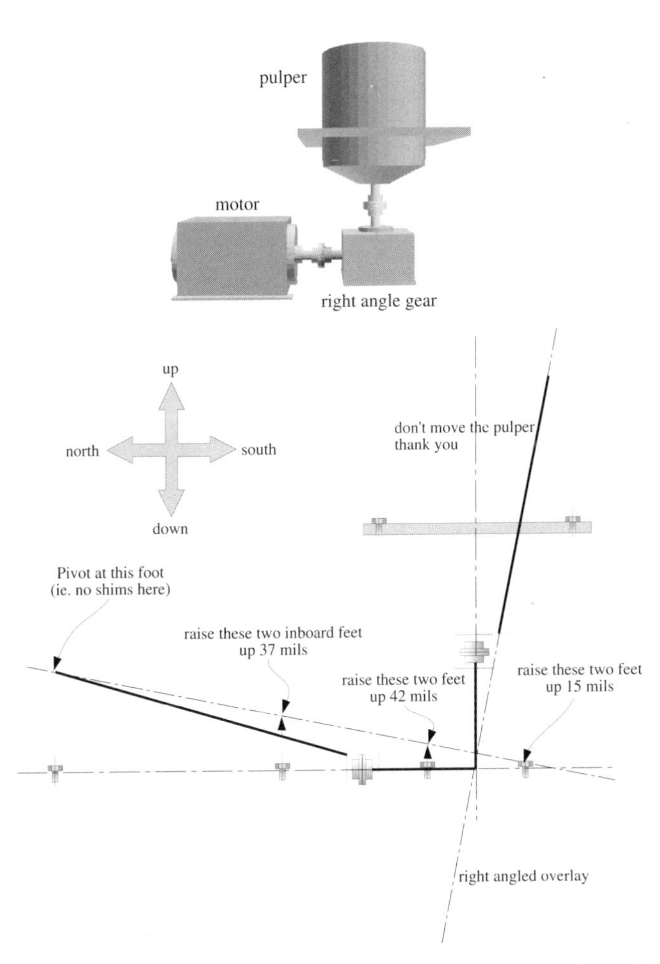

Figure 12-18. Motor - right angle gear - pulper model with right angle overlay applied for movement calculations.

sideways.

Also, if we decided to keep the gear where it is, the pulper would have to be pitched by adding 150 mils under the east bolt set on the support plate and moved 40 mils to the west as illustrated in figure 12-20.

The trick to graphing the machinery in this view is to 'roll' the pulper over so it's in line with the input shaft of the gear and

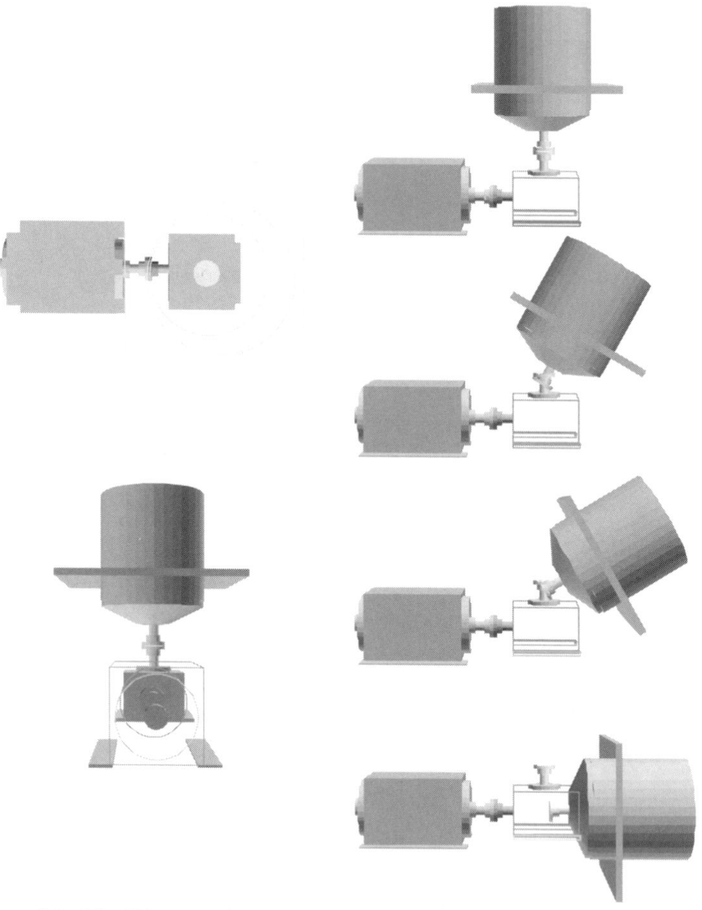

Figure 12-19. To graph the motor - gear - pulper shafts in the top view, the pulper shaft can be 'rolled' over to model its shaft position in the same axis as the motor and gear shafts.

the motor shaft centerline using the point where the readings were taken on the output gear shaft as the axial reference when scaling the pulper dimension onto the new graph as shown in figure 12-20.

Now it's almost as if we are back to aligning three 'in-line' machines like the first example with the motor, gear, and compressor.

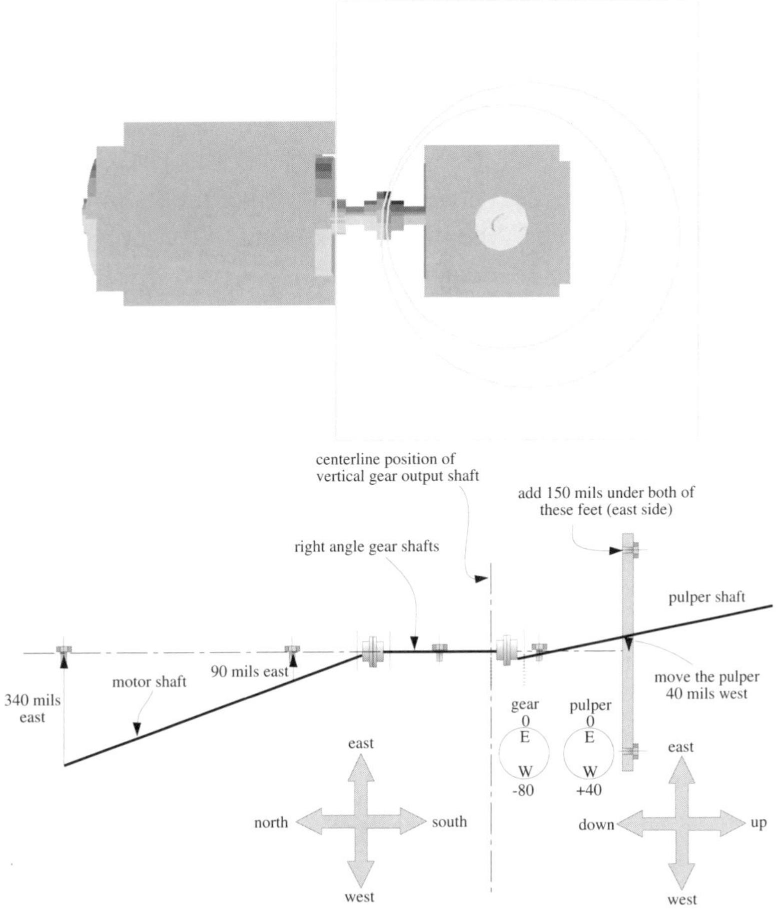

Figure 12-20. Motor - right angle gear - pulper model in the Top View with the pulper shaft 'rolled over' for visualization of its position with respect to the motor and gear shafts.

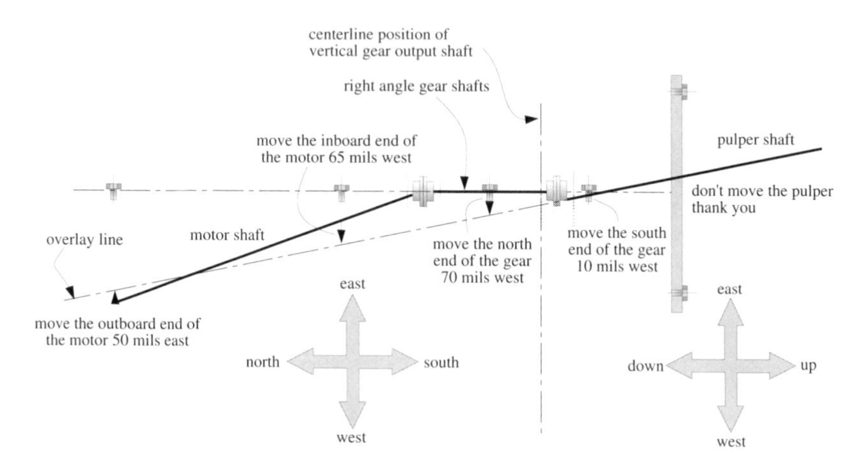

Figure 12-21. Motor - right angle gear - pulper model in the Top View with an overlay line placed on the graph for movement calculations.

By projecting an 'overlay' line onto the graph, one possible move (as shown in figure 12-21) would be to move the outboard end of the motor 50 mils to the east, move the inboard end of the motor 65 mils to the west, move the inboard (north) end of the gear 70 mils to the west, move the outboard (south) end of the gear 10 mils to the west, and keep the pulper exactly where it is.

Comments on aligning multiple element drive trains

Multiple element drive trains can get extremely complex if you don't keep your wits about you when aligning the machinery. Some drive systems can consist of up to 20 or more elements in 'L' shapes, 'U' shapes, 'S' shapes, or other complex arrangements. Regardless of how many pieces of machinery are in the drive train or how the shaft to shaft measurements were taken, the positions of all the machinery can be accurately illustrated on a graphing model, the movement restrictions plotted in, the shaft positions shifted to reflect the off-line to running machinery movement, and usually, by moving more than one element, an effective alignment solution can be performed. The key to success is to follow the multiple element alignment laws covered earlier in this chapter.

Bibliography

Piotrowski, John, "Aligning Multiple Element Drive Trains and Right Angle Drives", P/PM Technology, Vol. 5, Issue 2, March/April 1992.

13
Alignment Considerations for Specific Types of Machinery

Up to this point, fairly broad generalizations have been made about rotating machinery. The shaft alignment measurement methods discussed in Chapter 7 didn't mention any specific types of machinery that the methods should be used on. Intentionally so, since most of these methods can be used on any type of machinery regardless of its specific function. The shafts shown in all of the Chapter 7 figures could have been electric motors, steam turbines, pump, fans, compressors or whatever. It really doesn't matter. The examples of graphing and modeling techniques covered in Chapter 8 had specific machine names on the diagrams, but for all practical purposes the names of the machinery were not relevant, only the graphing concepts. The OL2R methods discussed in Chapter 9 should work on virtually any type of machinery, but certain OL2R methods are better suited for certain situations than some of the others mentioned.

This is not to suggest, however, that the wide variety of rotating machinery behaves the same way or should be aligned the same way. It is important to know a considerable amount about each piece of machinery before, during, and after the alignment process - how it works internally, what operational function it performs, how the process affects its operational performance, and how it interacts with its frame / foundation or external connections such as piping. The Machinery Data Card example in the Appendix illustrates the type of information that should be kept on each piece of rotating machinery.

This chapter will explore some of the specific information relating to alignment considerations for common rotating machinery. The typical OL2R (off-line to running) movement ranges indicated in each machine category contained herein are based on field data measurements made on various types of equipment, but should not be construed as hard and fast rules. This data indicates how the centerline of rotation of the shaft might move from off-line to running conditions (or how the inboard and outboard bearing positions change from OL2R and vice versa). The OL2R movement amounts reflect average ranges of motion, in other words, the high end values could, and in many case have been, in excess of the numbers indicated, sometimes by a factor of 300% or more. Again, most manufacturers of rotating machinery do not conduct OL2R measurements. If you consult the OE (original equipment) manufacturer for OL2R information, ask how the measurements were taken, what the environmental and operational conditions were during the test, and why they feel this data would be indicative of the machinery you have in operation.

There are several case histories that I would have included in this chapter to illustrate the wide variety of off-line to running machinery movement that has been observed, and special alignment procedures that were developed for unique cases, but this would have added another 100+ pages to the book. The information contained below is an attempt to give you a broad perspective of what might happen with these different types of machinery. It is therefore recommended that this information be used strictly as a starting point for you to initially align new rotating machinery systems to allow you the opportunity to safely conduct field studies on your rotating equipment.

Electric Motors

Electric motors (AC induction, synchronous, or DC motors) are perhaps the 'best behaved' types of rotating machinery. Electric motors up to 500 hp are frequently outfitted today with antifriction type bearings so they don't pose any major problems, assuming they are mechanically and electrically sound. Medium to large electric motors are frequently outfitted with sliding type bearings. When the field is applied to motors where the armatures are supported in sliding type bearings, the electro-magnetic field wants to center the rotor with respect to the stator field.

This phenomena is often referred to as 'magnetic center' and needs to be taken into consideration for proper shaft to shaft spacing. Many electric motors have no thrust bearing per se and rely on electromagnetic forces to center the rotor. To find 'mag-

Figure 13-1. Typical electric motor.

netic center', uncouple the motor and run it 'solo'. Once the field is applied, you may notice the shaft 'hunting' back and forth axially for a short period of time and then it will typically settle out at one specific axial position or 'magnetic center'. Very carefully scribe a line on the rotating shaft with a felt tip pen or soap stone near a stationary reference fixture, such as the bearing seal. Keep your fingers away from keys or keyways and don't let any loose clothing, tools, rags or other stationary objects attached to your body or near you hit the shaft. Drop the field (i.e. shut the motor off) and let the rotor stop completely. It is unlikely that the shaft's axial position is directly on magnetic center, so after safety-tagging the breaker, hand rotate the shaft pushing or pulling it axially until the scribe mark you made lines back up with the stationary reference fixture you picked. Now, measure and set the shaft to shaft distance with the machine it is driving.

If misalignment conditions are severe enough on electric motors and the shaft / armature elastically bends a sufficient amount (see figure 1-2), the rotor to stator air gap can get out of tolerance (the accepted tolerance for air gap eccentricity differential is +/- 10% of the total air gap). From a vibration standpoint, eccentric air gap problems will frequently exhibit a spectral peak at twice the line frequency (120 Hz in North America ... see Chapter 14 on vibration and the frequency domain).

Typical OL2R movement range of electric motors (horizontally mounted):

Vertical movement : 1-5 mils upward (5-200 hp) ... 3-30+ mils upward (200+ hp), typically symmetrical (i.e. inboard and outboard ends move up the same amount)

Lateral (sideways) movement : 0-4 mils (usually much less than any vertical movement)

Special notes on electric motors:

Moderate to excessive soft foot conditions have been experienced on virtually every size motor regardless of frame construction design.

Steam Turbines

Steam turbines can range in output from 20 hp to 100,000+ hp with speeds upwards to 25,000+ rpm and, therefore, become some of the more interesting equipment to study, and consequentially some of the more difficult equipment to maintain and operate properly. Steam pressures can range from 200 - 4,000+ psig and temperatures from 400 - 1100+ degrees F. Due to the fact that a high temperature gas is being used to propel blades for shaft rotation, extensive frame / casing design considerations concerning casing / rotor expansion and contraction are taken into account to minimize excessive positional change of the rotor during operation. However, movement of the shaft invariably occurs from off-line to running conditions that can vary considerably from unit to unit. In addition, rotor expansion must be taken into consideration when selecting a flexible coupling to prevent thrust transfer from one rotor to another, causing premature bearing or coupling failure. Again, since there is such a wide variety of equipment in existence, it is always best to consult with your equipment manufacturer for initial installation, design modification, overhaul, or operational problems with these units.

Typical OL2R movement range of steam turbines (horizontally mounted)

- Vertical movement : 5-25 mils upward (5-500 hp) ... 5-40+ mils upward (500+ hp), typically asymmetrical (i.e. inboard and outboard ends do not move up the same amount)
- Lateral (sideways) movement : 0-40+ mils (can be as much or considerably more than the vertical movement)

Special notes on steam turbines :

Moderate to excessive off-line 'soft foot' conditions have been experienced on virtually every size steam turbine regardless of frame construction design. During operation the shape of the casing can change introducing a different 'soft foot' condition than what occurs when off-line. Sometimes, on small to medium-sized steam turbines, one end of the casing is rigidly bolted to the frame and a 'sway bar' or flexible support is mounted at the other end to allow for axial expansion to occur to prevent casing warpage during operation. Sometimes on large steam turbines, the casing is 'keyed' at the casing centerline and the 'hold down' bolts are not tightened to lock the casing against the frame support, but are kept loose to allow for symmetric lateral and axial casing expansion to occur.

The lateral movement that occurs is often directly related to the expansion and contraction of the steam piping connected to the steam turbine casing, and proper design and installation of the piping system is imperative to minimize static (off-line) and dynamic (running) nozzle loads.

Most steam turbines are supported in sliding type bearings and, therefore, exhibit a certain amount of axial clearance between the thrust runner and the active / inactive thrust bearings (often referred to as 'thrust float'). When setting the machinery axial positions off-line, seat the thrust runner against the active thrust bearing before measuring and adjusting the shaft to shaft distance.

Figure 13-2. Gas / power turbine.

Gas / Power Turbines

Industrial gas / power turbine drivers are used in a wide variety of applications ranging from compression of gases and electrical generation to propulsion systems for ships. The Baryton cycle (i.e. a gas turbine) compresses air via a centrifugal or axial flow compressor where the compressed air is mixed with fuel (liquid jet fuel or natural gas) and burned. The hot, high velocity gas is then impinged on a series of several stages of curved blade sets (power turbine) that are used to rotate the driven machinery. Frequently the gas and power turbines, although separate rotors supported in their own bearings, share a common casing and frame. The residual high velocity gas is then vented through ductwork that sometimes houses a heat exchanger for a closed loop system or for use in heating liquids for other purposes.

The gas turbine produces a tremendous amount of forward thrust in reaction to the high velocity gas escaping out of the tail end of the machine. A considerable amount of heat is generated in the cycle and a twisting / torsional counter reaction occurs in the frame during operation. These factors all contribute to some of the most radical off-line to running machinery movement in any type of driver used today.

Typical OL2R movement range of gas / power turbines

Vertical movement :
intake end - 10+ mils downward to 10+ mils upward
exhaust end - 5-80+ mils upward

Lateral (sideways) movement :
intake end - 2-20+ mils
exhaust end - 2-60+ mils

Special notes on gas turbines :

Moderate to excessive off-line 'soft foot' conditions have been experienced on virtually every size gas / power turbine regardless of frame construction design. Movement in the axial direction from OL2R conditions can also be excessive. Forward movement of gas turbines (i.e. toward the intake end) has been observed to translate 180+ mils. Gear or diaphragm type couplings have been employed at the output shaft to drive the equipment. If the coupling is a diaphragm type (or any flexible disk type) and there is movement toward the intake end, damage could occur to the coupling and the thrust forces can be transmitted to the driven machine. The shaft to shaft distance between the power turbine and the driven equipment shaft is usually 40+ inches in an attempt to minimize the effect from large amounts of OL2R movement and to minimize any heat transfer from the exhaust duct work to the driven machine. Although rarely employed, the shaft to coupling spool technique is an excellent choice for measuring the off-line shaft positions.

Internal Combustion Engines

Very few field studies have been conducted on how internal combustion engines move from OL2R conditions. Diesel engines, for example, are frequently used to drive back-up electrical generators, fire pumps, and portable air compressors. The crank shaft is typically set very low in the casing and engine mounts are frequently found at or slightly above the centerline of rotation of the crankshaft. The relatively few studies that have been done have still shown OL2R machinery movement despite the location of the casing support mounting location. Flexible coupling design is somewhat critical since variations in torque occur as each piston delivers rotational force at varying intervals.

Typical OL2R movement range of internal combustion engines

- Vertical movement : 1-5 mils upward (5-200 hp) ... 2-20+ mils upward (200+ hp), typically symmetrical (i.e. inboard and outboard ends move up the same amount)
- Lateral (sideways) movement : 0-4 mils (usually much less than any vertical movement)

Special notes on internal combustion engines :

Moderate to excessive off-line 'soft foot' conditions have been experienced on virtually every size internal combustion engine regardless of frame construction design. Vibration isolators are also frequently used making it somewhat more difficult to ascertain what sort of 'soft foot' corrections are required.

Electric Generators

Electric generators (similar to their cousins - electric motors) are perhaps the 'best behaved' type of rotating machinery as far as OL2R machinery movement is concerned. As with motors, the maximum recommended rotor to stator air gap eccentricity differential is +/- 10% of the total air gap. Where generator armatures are supported in one bearing as found in MG sets, the air gap clearances can also be taken into account when aligning the machinery casings. In figure 13-3, notice that the positions of the shafts are dictated by the positions of the bearing pedestals. The air gap clearances are dictated by the positions of the machine casings which in this case, are independent of the bearing pedestals. Another powerful aspect of the graphing / modeling technique is that the graph can be used not only to align the shafts, but also to center the armature in the stator bore.

Typical OL2R movement range of electric generators (horizontally mounted)

Vertical movement : 1-5 mils upward (5-200 hp) ... 3-30+ mils upward (200+ hp), typically symmetrical (i.e. inboard and outboard ends move up the same amount)

Lateral (sideways) movement : 0-4 mils (usually much less than any vertical movement)

Special notes on electric generators :

Observed 'soft foot' conditions on generators are the same as for motors.

Traditionally, the 16 point method is used to determine the shaft to shaft positions. A slightly more accurate way would be to provide a temporary support for the inboard end of the generator as illustrated below, completely separate the shafts, and capture a set of reverse indicator readings.

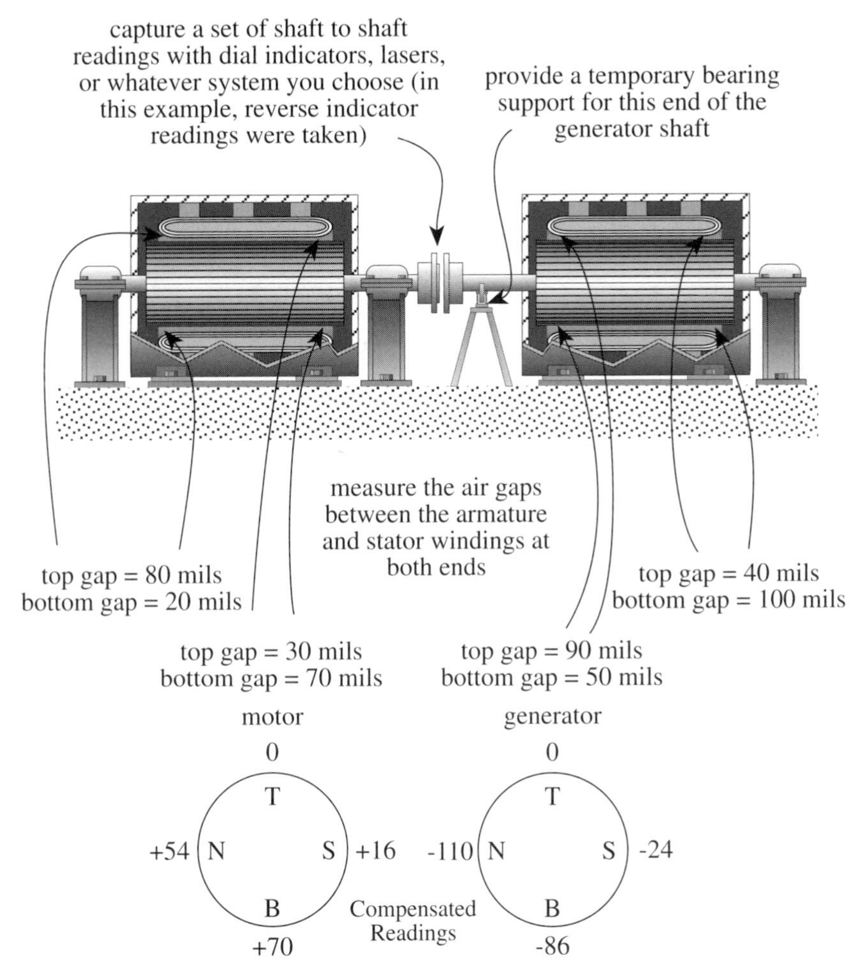

There are two objectives here. One objective is to align the centerlines of rotation of the motor and generator shaft, and the other objective is to center the stator housings to provide an even air gap between the armatures and the stators. The shaft positions are dictated by the positions of the bearing pedestals. The air gap settings are dictated by the positions of the stator housing.

Figure 13-3. Typical 3-bearing motor - generator arrangement.

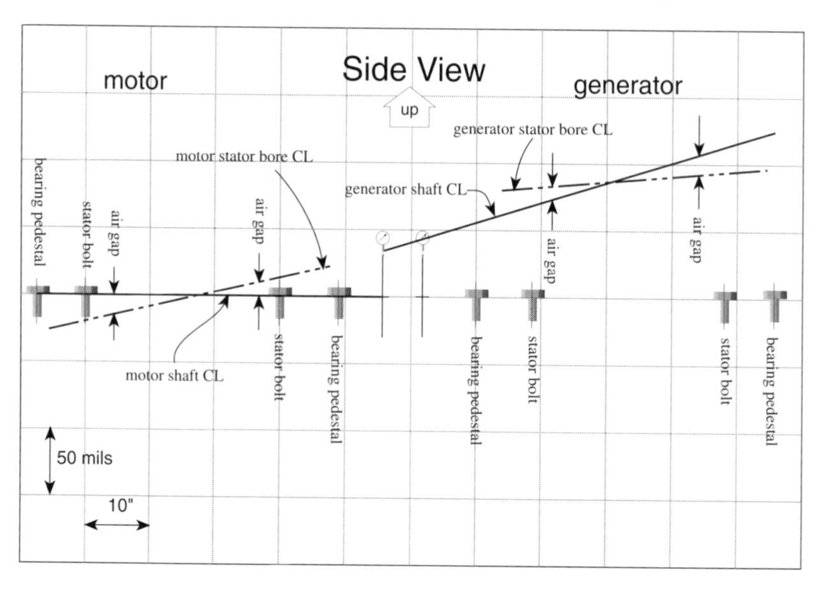

Figure 13-4. 3 bearing motor - generator alignment example.

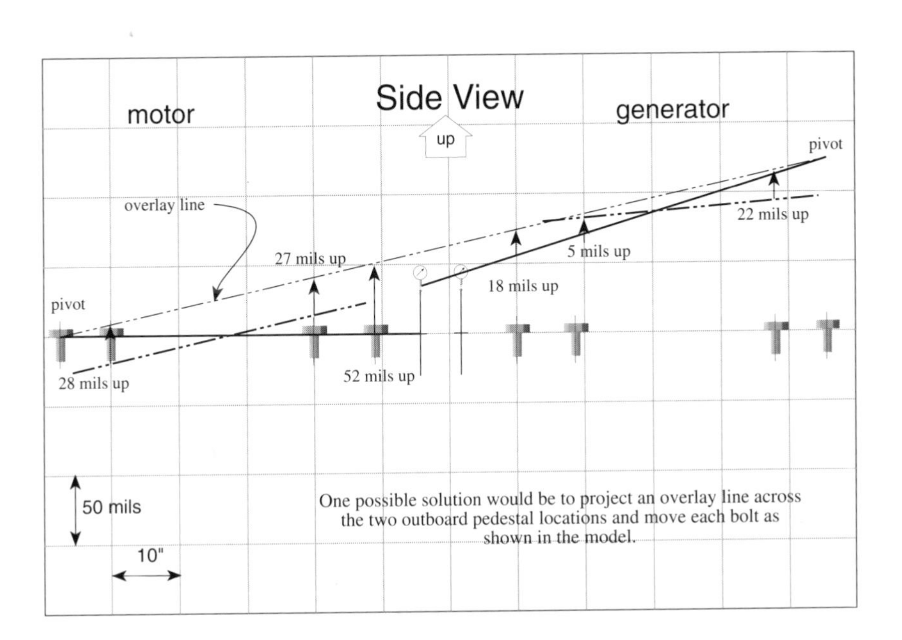

Figure 13-5. 3 bearing motor - generator alignment solution.

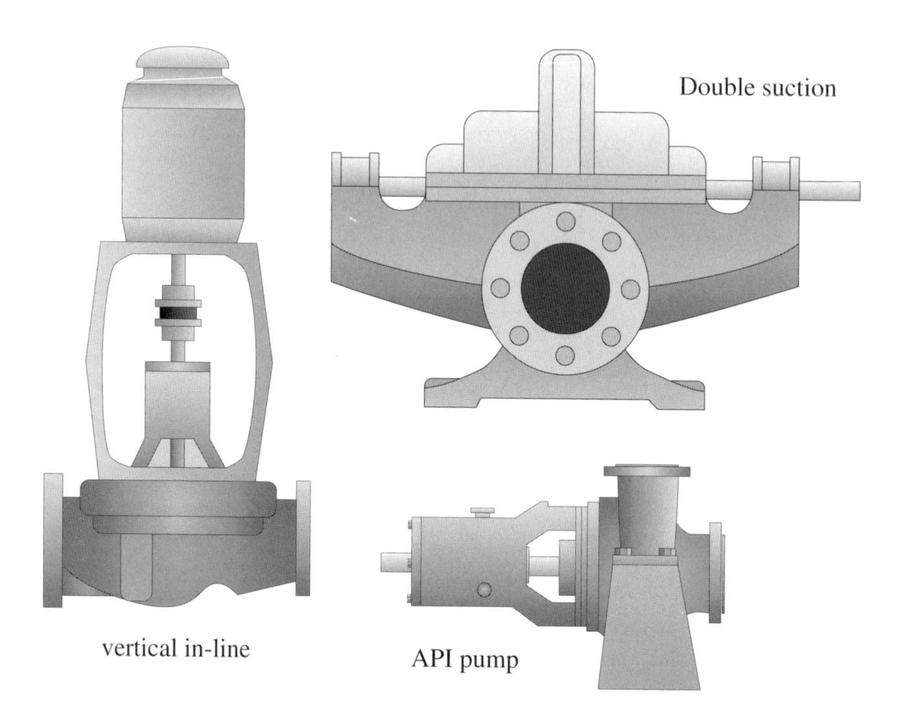

Figure 13-6. Centrifugal pump design variations.

Figure 13-7. Typical double suction centrifugal pump.

Centrifugal Pumps

There are several different designs of centrifugal pumps and it would be difficult to cover every style used in industry. Their purpose is basically to move an incompressible fluid from point A to point B. The temperature of the fluid being conveyed has a great effect on the OL2R conditions of the pump. As discussed in Chapter 3, the piping attached to the pump can have a tremendous influence on obtaining and maintaining accurate alignment, so much so that many people are unwilling to even try to reposition pumps and, henceforth, declare them the 'stationary' machines when aligning them.

There are several different types of vertical pumps such as well water pumps, in-line pumps, and reactor coolant pumps. In most cases, vertical pumps are driven by 'C' flanged motors. These motors are bolted to a cylindrical casting that is attached to the pump casing. In some situations, the pump is supported in its own bearings and the motor is flexibly coupled to the pump. In

Figure 13-8. Multi stage centrifugal pump.

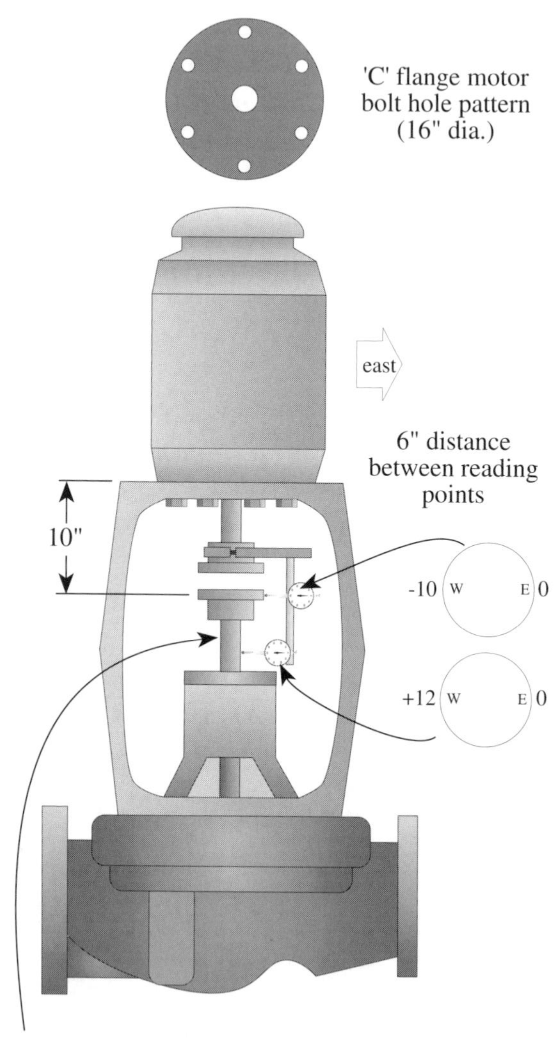

Figure 13-9. Vertical pump with 'C' flanged motor example.

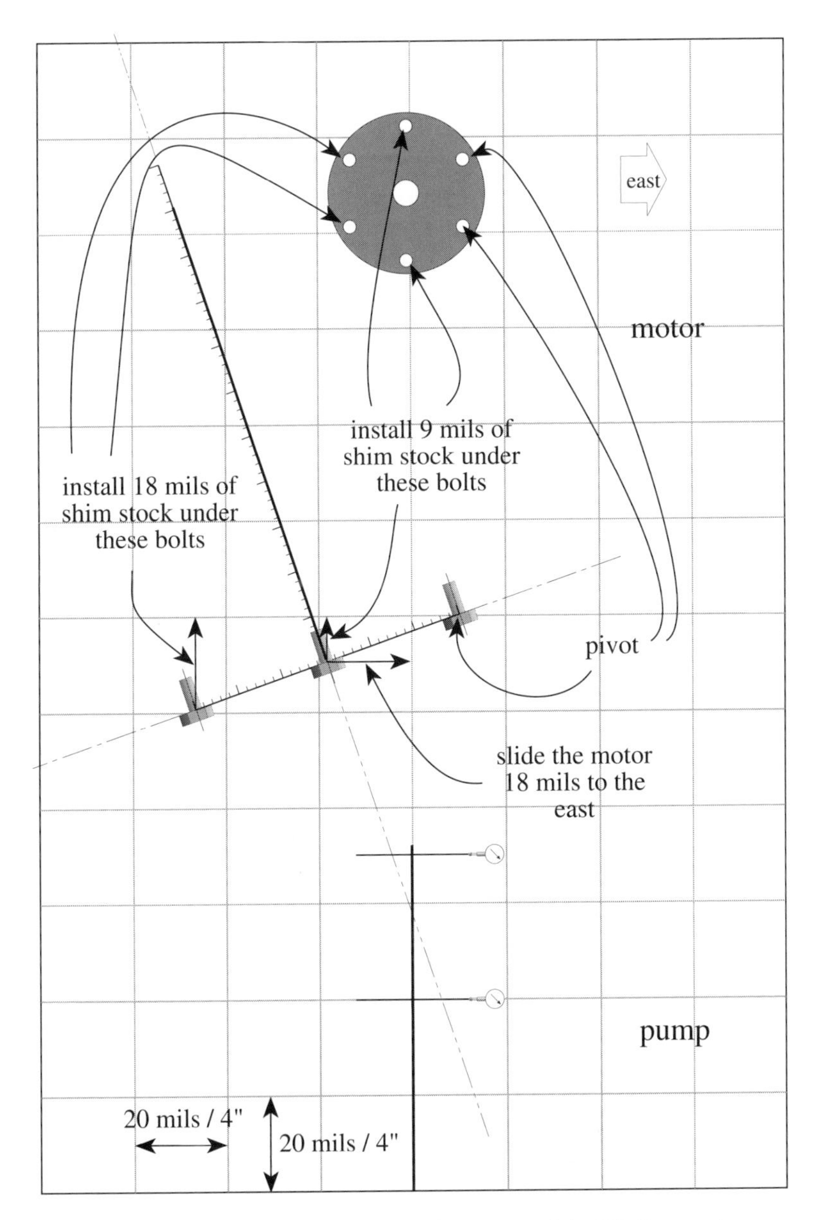

Figure 13-10. Graphing a vertical pump with 'C' flanged motor using the 'T' bar overlay to represent the motor and flange bolt surface.

other situations, the pump is rigidly coupled to the motor shaft and the thrust load is supported by a thrust / radial bearing at the top of the motor. The assumption that many people make is that no alignment is required for these types of machines since the motor, connector casting, and pump casing are machined, rabbeted fits that perfectly align the motor shaft to the pump shaft. Frequently, the machined surfaces are not perfect and there is enough free-play between the rabbeted fits and the bolt holes that alignment is not perfect. Figures 13-9 and 13-10 illustrate how you might align these units when misalignment exists.

Typical OL2R movement range of centrifugal pumps

Vertical movement : 0-80+ mils upward possibly asymmetrical (i.e. inboard and outboard ends do not move up the same amount)

Lateral (sideways) movement : 0-90+ mils (sometimes much greater than any vertical movement and is usually asymmetrical)

Figure 13-11. Vertical 'chair' pump.

Special notes on centrifugal pumps :

Moderate to excessive off-line 'soft foot' conditions have been experienced on virtually every centrifugal pump regardless of frame construction design. On motor driven pump arrangements, 'soft foot' conditions are typically corrected on the motor but rarely on the pump. Maintaining long term alignment of ANSI and API type pumps can be difficult due to the loosely supported inboard (coupling) end of the pump case. Failure of mechanical seals can often be attributed to misalignment conditions. Excessive leakage on mechanically packed pumps can also be attributed to misalignment conditions. Multi-stage pumps can experience internal rubs (i.e. impeller to stator or wear rings) due to rotor distortion caused by moderate to excessive misalignment conditions. On several occasions, boiler feedwater pumps have moved more in the vertical and lateral directions from OL2R conditions than the driving machine (both motors and steam turbines), but are generally set higher than these drivers when off-line (go figure!).

Blowers and Fans

There are several different designs of fans and blowers and again, it would be difficult to cover every aspect of these types of machines. Similar to pumps, their purpose is basically to move large volumes of a compressible fluid at low pressures from point A to point B. Again, the temperature of the gas being conveyed has a great effect on the OL2R conditions of the fan. As discussed in Chapter 3, the ductwork attached to the fan can have a tremendous influence on obtaining and maintaining accurate alignment, so much so that many people are unwilling to even try to reposition fans and blowers and, henceforth, declare them the 'stationary' machine when aligning them. In some situations, where the fan blades are center-mounted on the shaft and the shaft is supported by bearings at each end, the position of the shaft is dictated by the positions of the bearing pedestals which are not directly attached to the fan housing. The fear in altering the positions of the fan bearings is that internal fan blade to shroud clearances could be upset and rubs could occur. Here again, the graphing / modeling technique can be used not only to align the shafts, but also to position the fan housing to properly set fan blade to shroud clearances.

Typical OL2R movement range of blowers and fans

- Vertical movement : 0-80+ mils upward typically asymmetrical (i.e. inboard and outboard ends do not move up the same amount)
- Lateral (sideways) movement : 0-20+ mils (usually much less than the vertical movement)

Figure 13-12. Motor driven centrifugal fan.

Compressors

The variation in compressor design is as diverse as pumps and fans. A sampling of compressor types are illustrated in figure 13-13 and examples of some of these designs are shown in figures 13-14 through 13-16. When gases are compressed, heat is generated and thereby casing expansion typically occurs. Since the compressible fluid is entering the compressor at a much lower temperature than the discharged gas temperature, uneven OL2R movement frequently occurs. Depending on how the suction and discharge piping is attached to the compressor case, lateral (sideways) OL2R movement can occur as the attached piping expands or contracts.

Typical OL2R movement range of compressors
Vertical movement : 0-80+ mils upward typically asymmetrical (i.e. inboard and outboard ends do not move up the same amount)
Lateral (sideways) movement : 0-30+ mils (usually much less than the vertical movement but can be greater than vertical movement in certain applications)

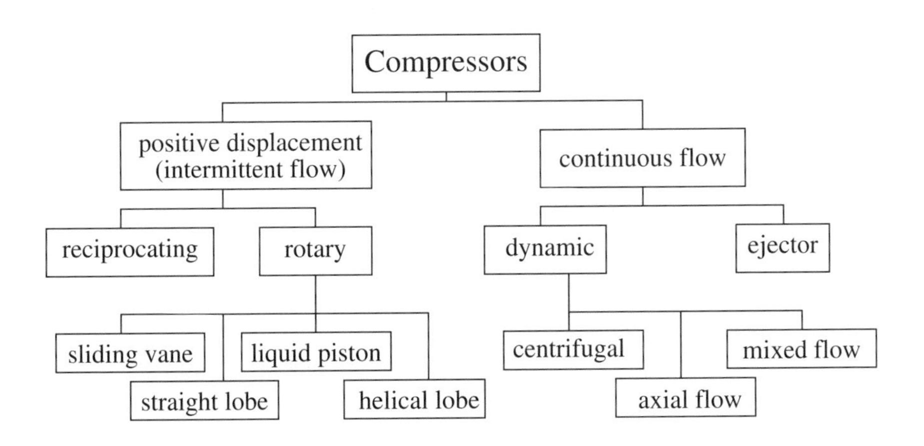

Figure 13-13. Classifications of compressors.

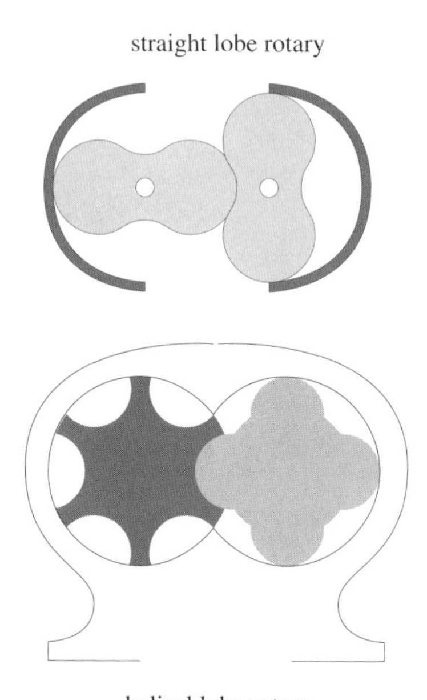

Figure 13-14. Straight & helical lobe compressor designs.

Figure 13-15. Bullgear driven, multi stage, open volute, centrifugal compressor design.

Figure 13-16. Multi stage, closed volute, centrifugal compressor design.

Gearboxes

Gearboxes come in a wide variety of designs and sizes and are used to precisely increase or decrease the shaft speed from one machine to another. A large percentage of gearboxes have a continuous lube oil system where the lower casing half acts as the oil reservoir and an oil pump sprays lubricant onto the meshing gears and floods the bearings where it returns back to the reservoir. Heat is generated in the bearings and meshing gear teeth during operation and the oil temperature is frequently controlled by a heat exchanger. Operating oil temperatures are typically kept around 120-150 degrees Fahrenheit.

Typical OL2R movement range of gearboxes
Vertical movement : 5-20+ mils upward typically slightly asymmetrical (i.e. inboard and outboard ends do not move up the same amount, but very nearly so)
Lateral (sideways) movement : 0-10+ mils (see notes)

Special notes on gearboxes :
Moderate to excessive off-line 'soft foot' conditions have been experienced on virtually every gearbox regardless of frame construction design. Gearboxes are frequently bolted to the frame at more than four points and soft foot correction can be more difficult to correct than machines bolted at three of four points. Uncorrected soft foot can distort the housing causing gear mesh, seal, and bearing problems. Since there is typically a rise in casing temperature from off-line to running conditions, not only will the shafts move upward, but they will also 'spread apart'. If dowel pins are used, casing distortion can occur if all four corners are pinned to the frame. Several OL2R studies have shown that gearboxes 'twist' or 'warp' when operating.

Cooling Tower Fan Drives

Although cooling tower fan drives are not usually thought of as 'glamorous' rotating machinery systems, they are very critical to the operation of the plant and can experience alignment problems as acute as any other type of rotating equipment. In fan drive systems where a right angled gearbox drives a 6-bladed or 8-bladed fiberglass fan assembly where the drive motor is located outside the plenum and the motor is connected to the input shaft of the gear by a long spool piece or 'jackshaft', OL2R movement is usually not measured and in many cases ignored. The saving factor in these designs is that the flexing points in the coupling are separated by a considerable distance, thereby allowing for considerable amounts of centerline to centerline offsets at the flexing points. For example, if there is a 100 inch separation between the flexing points, you could have up to 100 mils of centerline to centerline deviation and still be at 1 mil per inch misalignment (100 mils / 100 inches = 1 mil / inch).

The shaft to coupling spool or the face-face technique shown in Chapter 7 is recommended for aligning these types of drives. Since most cooling towers are located outside, an interesting phenomena can occur when aligning these drive systems during daylight hours with the sun shining. If the drive is kept stationary, the long coupling spool can become unevenly heated from the sun and thermally bow the spool piece. As you begin rotating the shafts to capture a set of readings, the 'hot' side of the spool piece now begins to rotate into the shade and the sun starts to heat a different side of the spool piece. As the hot side cools and the shaded side warms up, the spool piece begins to change its shape causing erroneous readings.

Bibliography

Shepherd, D. G., **Principles of Turbomachinery**, Macmillan Publishing Co., New York, NY, 1956.

Gibbs, Charles W., **Compressed Air and Gas Data - Second Edition**, Ingersoll-Rand Co., Woodcliff Lake, NJ, 1971.

14
Finding Misalignment on Operating Rotating Machinery

The ultimate goal of proper shaft alignment is to extend the operating lifespan of rotating equipment. Machinery shafts want to spin freely with little or no external forces impinging on the rotors, bearings, couplings and seals. When moderate or excessive static or dynamic forces are present, the components begin to slowly degrade, eventually leading to mechanical failure. The performance of the equipment relating to its designed operating conditions such as output horsepower, discharge pressure, flow, speed, etc., may all be just fine, but if excessive amounts of force and vibration are present, it's definitely not going to be running for very long. For some people, frequent overhauls of machinery seem to be taken as a fact of life and many organizations routinely rebuild equipment whether it is needed or not, or worse yet, let their machinery run itself into the ground culminating in a catastrophic failure.

Since the late 1960's, a radically different philosophy has emerged. Preventive maintenance programs are being merged with predictive and pro-active maintenance techniques that have taken the operation of rotating equipment to yet a higher plateau of improved performance by closely monitoring changes in vibration, lubrication, temperature, acoustical emissions, and rotor-bearing loads. Of course, this does not always prevent the instantaneous failures that will occur, but it will stop machinery from slowly beating itself into oblivion.

A successful machinery life extension program requires the commitment by management, engineering, maintenance, and operating personnel to more closely monitor operating parameters, collect and interpret mechanical performance data, and understand how all this relates to the overall productivity of the equipment. Today, it is almost taboo not to invest in some sort of performance monitoring system on new machinery installations and retrofit many existing drive trains with vibration, temperature, and load sensors to prevent many catastrophic failures. For smaller, less expensive rotating machinery, hand held vibration meters and data collectors are used to capture this vital information at regular intervals. This mechanical performance data is trended over periods of time until one or more sensors indicate that a level is beginning to approach, attain, or exceed a predetermined high limit and it is time to correct the problem. But ... what exactly is the problem and what methods can be employed to detect if misalignment is the root cause of the problem?

After years of study, one invariable conclusion can be made: misalignment disguises itself very well on rotating machinery. There are no easy or inexpensive ways to determine if rotating machinery is misaligned while it is running.

We will examine two different analysis methods commonly used in industry to determine if rotating machinery is misaligned - vibration analysis and infrared thermographic equipment. As you shall see, many of the generalizations made by vibration and infrared thermographic equipment vendors concerning detection of misalignment are not always true.

Taking Vibration Data

Vibration data should be taken at five points on each machine element as shown in figure 14-1. If possible, a once per rev signal should be taken and phase angle data be recorded particularly in the axial direction. Be careful when taking phase readings at the coupling end of each unit since the sensor will be pointing in opposite directions in the axial direction which produces a 180 degree phase shift.

If overall vibration amplitude levels are relatively high at any of the sensor locations, a more detailed survey should be taken with a spectrum (FFT) analyzer to aid in determining the vibration

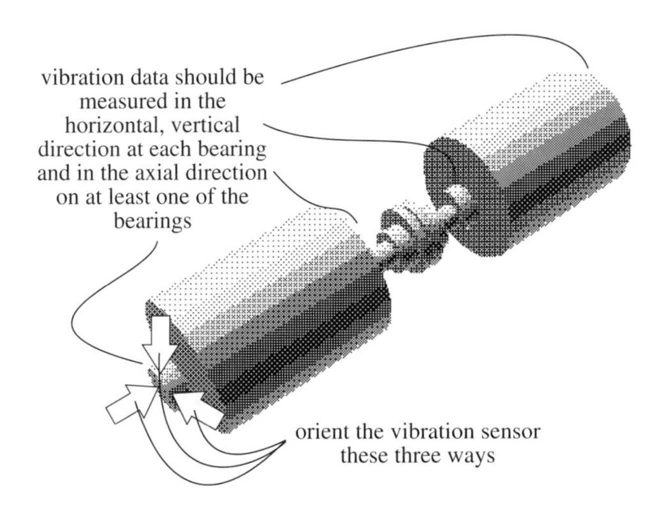

Figure 14-1. Typical vibration monitoring locations and directions.

displacement sensors (aka proximity probes)

- measures distance usually in mils (peak to peak) (1 mil= 0.001")
- useful frequency range 0-80,000 cyles per minute (0-1300 Hz)
- usually 200 mv/mil sensitivity

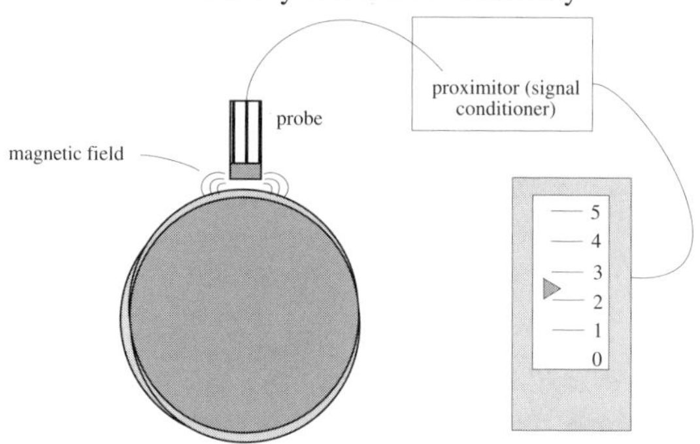

velocity sensors (aka seismometers)

- measures speed or energy usually in in/sec (peak)
- useful frequency range 60-120,000 cyles per minute (10-2000 Hz)
- usually 100 mv/in/sec sensitivity
- good for measuring general machinery condition involving medium frequency vibrations

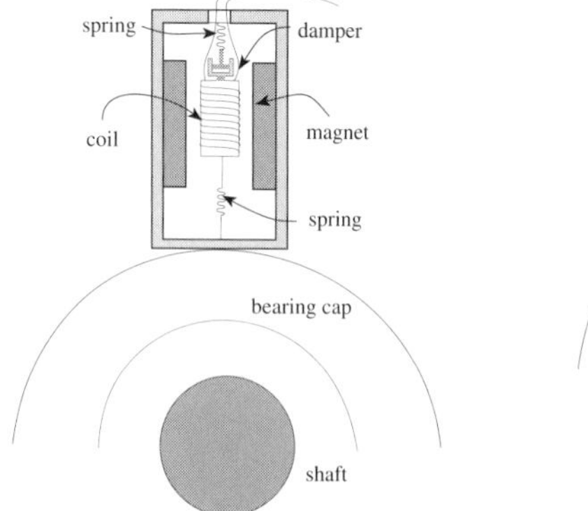

acceleration sensors (aka accelerometers)

- measures rates of acceleration or force usually in g's (peak)
- useful frequency range 1500-3,000,000 cycles per minute (25-50 kHz)
- usually 100 mv/g sensitivity
- good for measuring general machinery condition involving medium and high frequency vibrations

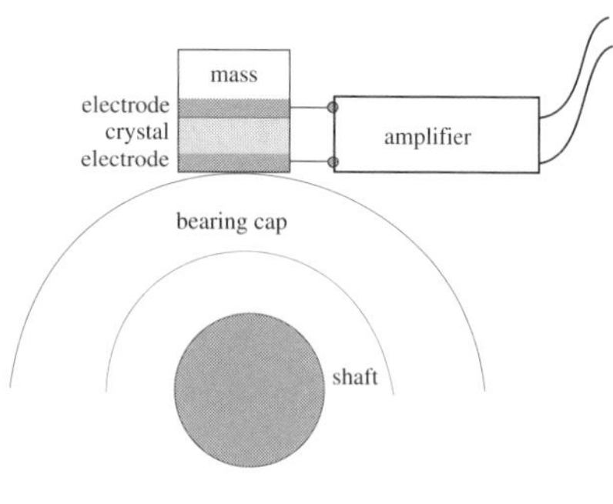

Figure 14-2. The three types of vibration sensors.

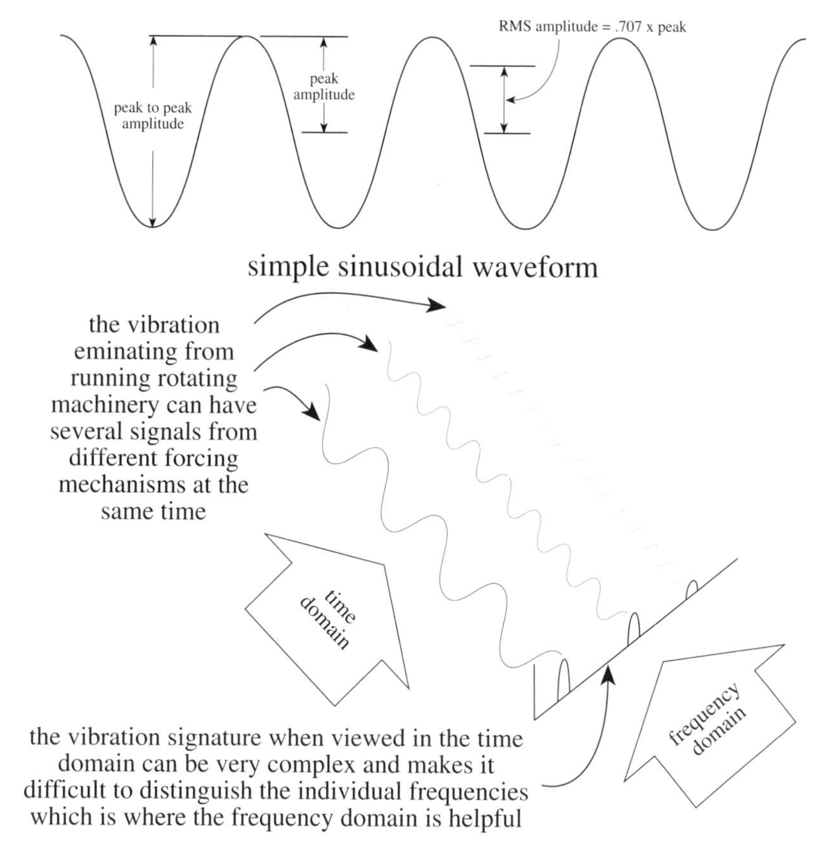

The time domain is amplitude versus time. The frequency domain is amplitude versus frequency. Jean Baptiste Fourier discovered that any complex waveform that is comprised of several distinct frequencies can be mathematically separated into their discrete frequencies. These mathematical equations are called Fourier transforms. With the advent of today's electronics, these mathematical transforms can be done very quickly ... hence the name Fast Fourier Transform was born (aka FFT).

Figure 14-3. The time and frequency domains of vibration.

signature of the machinery. If the source of the problem is misalignment, what will the vibration spectrum look like? Does the signal look the same for all different types of machinery with different kinds of couplings, bearings, and rotors?

Using Vibration Analysis to Detect Misalignment

Discerning what a vibration sensor is telling you happens to be one of the most difficult tasks the machinery diagnostician is faced with. However, with the use of an FFT (fast fourier transform) signal analyzer, vibration 'signatures' can be taken that split the complex overall vibration signal and enables one to look at various frequencies of the sensor's output. Many 'rules of thumb' have emerged that attempt to classify specific machinery problems with specific types of vibration signatures. The experienced vibration analyst quickly learns that these 'rules of thumb' are to be used as guidelines on arriving at the source of the problem. Quite often, more than one problem exists on a piece of rotating machinery, such as a combination of imbalance, misalignment, and damaged bearings that will all appear on the vibration spectrum. One goal of this chapter is to examine the types of vibration signatures misaligned rotating equipment exhibit and the forcing mechanisms involved to generate this signal. Many people who work in the field of vibration analysis feel that shaft misalignment can be detected by the following symptoms:

- high, one or two times running speed frequency components
- high axial vibration levels
- a 180 degree phase shift will occur across the coupling

The above symptoms can happen during shaft misalignment ... but not always. There are three general statements that can be made from a number of controlled tests where rotating machinery was purposely misaligned and also, from many field observations that have been made on equipment that was operating under misalignment conditions.

1. You cannot detect the severity of misalignment using vibration analysis. In other words, there is no relation between the amount of misalignment and the level/amplitude of vibration.
2. The vibration signature of misaligned rotating machinery will be different with different flexible coupling designs. For example, a misaligned gear coupling will not show the same vibration pattern as a misaligned 'rubber tire' type coupling.
3. Misalignment vibration characteristics of machinery rotors supported in sliding type bearings are typically different than the vibration characteristics of machinery rotors supported in anti-friction type bearings.

The Relationship between Vibration Amplitude and Misalignment Severity (or lack thereof)

If two pieces of rotating machinery are coupled together and misaligned by 5 mils per inch and vibration readings are taken on the bearing caps and then the unit is shut down and misaligned to 10 mils per inch (double the initial amount), started back up and vibration readings taken on the bearing caps again, the overall vibration amplitudes measured will not be twice the amount compared to the first set of data collected at misalignment conditions of 5 mils per inch.

Increasing misalignment may actually decrease vibration levels. If a drive train that has been running misaligned for a period of time was shut down and realigned more accurately, the vibration levels may increase after it is restarted. Many drive trains may be slightly to moderately misaligned and vibration analysis will not be able to detect the misalignment condition.

Surprising facts, but true!

To illustrate these points, examine figure 14-5. A 60 hp, 1775 rpm motor and circulating water pump, connected by a metal ribbon type coupling shown in figure 14-4, was purposely misaligned by 21 and 36 mils sideways (laterally) and 55 and 65 mils vertically. Vibration readings were taken at five points on each machine.

Notice with every trend that the vibration only slightly increased, and in some cases, slightly decreased as the equipment was misaligned by 21 and 36 mils. In some cases, when the equipment was at the greatest misalignment condition (65 mils), the vibration decreased from the levels seen at 55 mils.

Figure 14-4. Motor and pump used in Vibration Case Study #1.

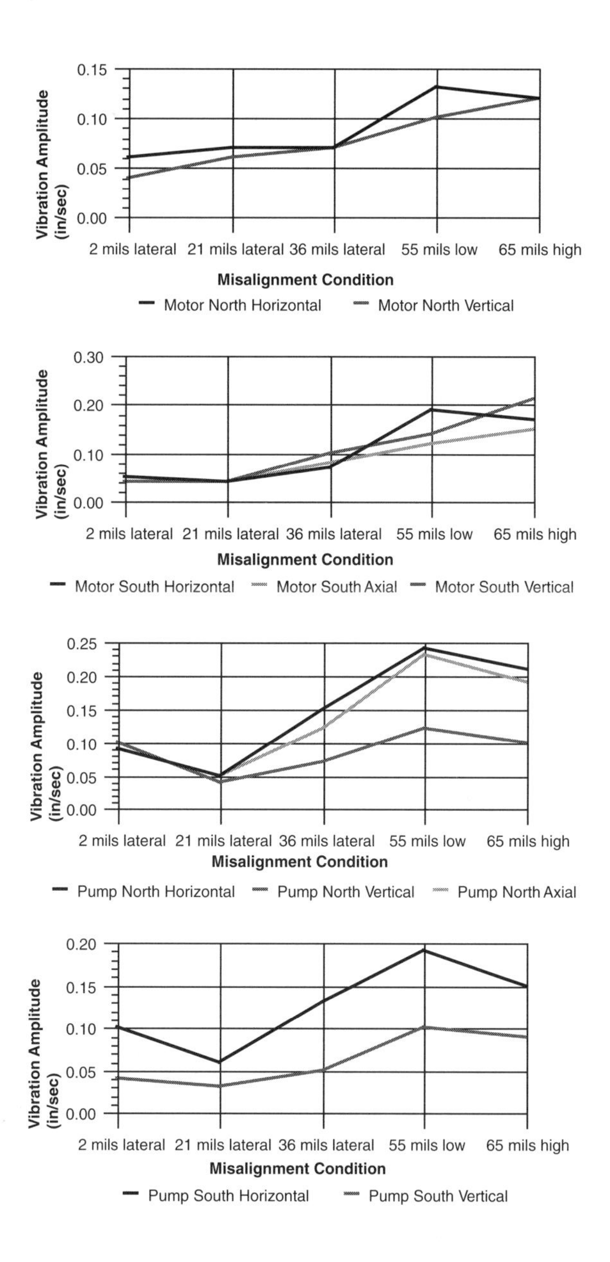

Figure 14-5. Trends of overall vibration vs. increasing misalignment for motor and pump in Vibration Case Study #1.

Why Vibration Levels Decrease with Increasing Misalignment

Rotating machinery shafts can be exposed to two types of forces. Static forces that act in one direction and dynamic forces that change their direction. Static forces are also called pre-loads. Pre-loads on shafts and bearings are caused by many of the the following sources :

- gravitational force
- V-belt or chain tension
- shaft misalignment
- some types of hydraulic or aerodynamic loads

Dynamic loads on shafts and bearings are caused by some of the following sources (not a complete list by any means) :

- out of balance condition (i.e. the center of mass is not coincident with the centerline of rotation)
- eccentric rotor components or bent shafts (another form of unbalance)
- damaged antifriction bearings
- intermittent, period rubs
- gear tooth contact
- pump or compressor impeller blades passing by a stationary object
- electromagnetic forces

Simply stated, vibration is motion. Vibratory motion in machinery is caused by forces that change their direction. For example, a rotor that is out of balance and is not rotating, does not vibrate. As soon as the imbalanced rotor begins to spin, it also begins to vibrate. This occurs because the 'heavy spot' is changing its position causing the (centrifugal) force to change its direction. The rotor / bearing / support system, being elastic, consequentially begins to flex or move as these alternating forces begin to act on the machine.

Another detectable vibration pattern exists in gears and is commonly referred to as gear mesh. Gear mesh can be detected as forces increase or subside as each tooth comes in contact with another. Other types of mechanical or electrical problems that can be detected through vibration analysis can be traced back to the fact that forces are somehow changing their direction.

On the other hand, when two or more shafts are connected together by some flexible or rigid element where the centerlines of each machine are not collinear, the forces transferred from shaft to shaft are acting in one direction only. These forces do not change their direction as an imbalance condition will. If a motor shaft is higher than a pump shaft by 50 mils, the motor shaft is trying to pull the pump shaft upwards to come in line with the motor shaft position. Conversely, the pump shaft is trying to pull the motor shaft downward to come in line with the pump shaft position. The misalignment forces will begin to bend the shafts, not flutter them around like the tail of a fish.

Static forces caused by misalignment act in one direction only, which is quite different than the dynamic forces that generate vibration. Under this pretense, how could misalignment ever cause vibration to occur? If anything, misalignment should diminish the capacity for motion to occur in a rotor / bearing / support system.

How Varying Degrees of Misalignment Affect Rotating Machinery - Case Study #1

The pump and motor arrangement as shown in figure 14-4 was used to study what effect different kinds of shaft alignment have on the vibration of this equipment. The motor used for the test was a 60 hp., 1775 rpm, cast aluminum construction. The rotor was supported by antifriction bearings and the rotation was clockwise when standing at the outboard end of motor looking at the pump. The pump used for the test was a center-mounted double suction impeller and, again, the rotor was supported by antifriction bearings. The size of the pump was 8x10-12 rated at 2000 gpm with 85 ft. of head. The flexible coupling was a metallic ribbon type as shown in figure 14-6.

OL2R movement of the pump and motor arrangement was calculated by taking surface temperature measurements at various points on the pump and motor. The OL2R movement on the motor was very slight with the inboard end raising 5 mils and the outboard end raising only 2 mils. Since no change in temperature occurred on the pump from 'off-line' to running conditions, there was no change in position of the pump shaft centerline.

It is important to note that the surface temperatures at various points on the motor, particularly the inboard end bell housing of the motor, increased as the amount of misalignment increased. The maximum temperature observed occurred at one of the extreme misalignment conditions, attaining 145 degrees Fahrenheit. Each test run lasted approximately 30 minutes so the motor did not have a chance to stabilize thermally.

Figure 14-6. Flexible coupling type used on motor and pump in Vibration Case Study #1.

A total of seven test runs were made under various alignment conditions. Vibration data was taken at five points on each unit using a hand held vibration meter with an accelerometer sensor to record overall readings on the bearing housings. A seismometer with a magnetic base was also attached at each sensor location and the signal was fed into an analyzer and an X-Y plotter to record the vibration signature. Since the motor housing was of aluminum construction, one quarter inch thick carbon steel plates were epoxied to the motor in the horizontal and vertical planes at both bearings and another plate was attached on the inboard end bell to capture the axial vibration levels. The vibration monitoring equipment is shown in figure 14-7.

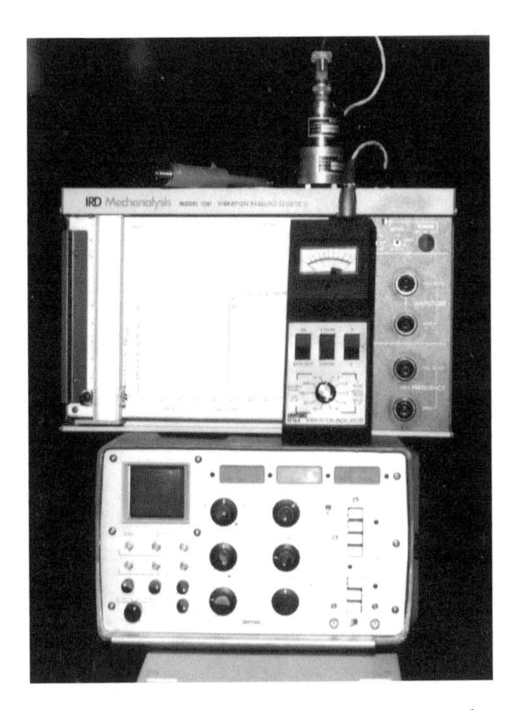

Figure 14-7. Vibration analysis equipment used on motor and pump in Vibration Case Study #1.

Shaft Alignment Positions During the Test Runs

The reverse - indicator shaft alignment method was used to determine the relative shaft positions for all test runs. The nomenclature used to identify each run indicates the maximum amount of motor shaft deviation with respect to the pump shaft at the point of grid contact in the coupling. In Run # 2 M2W, for instance, the maximum amount of shaft deviation occurred in the lateral (i.e. side to side) position of the shafts. The motor coupling hub contact point is 0.002" to the west of the contact point on the pump coupling hub and is so designated as M2W or Motor - 2 mils - west.

Run # 1 (Motor uncoupled and running solo). The first run was conducted with the motor uncoupled to determine whether the motor had an unbalanced condition, damaged bearings, or some other problem that could affect the vibration response when coupled to the pump.

Run # 2 M2W (Motor - 2 mils - west). The pump and motor were initially aligned well within acceptable alignment tolerances.

Run # 3 M21W (Motor - 21 mils - west). The motor was positioned 0.021" to the west with no shims being added or removed from the pump or the motor from the second run.

Run # 4 M36W (Motor - 36 mils - west). The attempt was made in this run to slide the motor further to the west. The motor, however, became bolt bound and was unable to be moved any further sideways.

Run # 5 M65H (Motor - 65 mils - high). The motor is now positioned well from side to side, but is 0.065" higher than the pump shaft centerline.

Run #6 M55L (Motor - 55 mils - low). In this test run, the motor shaft centerline is set considerably below the pump shaft centerline while still maintaining good side to side alignment.

Run #7 M6W (Motor - 6 mils - west). The pump and motor were again aligned well within acceptable alignment tolerances (similar to Run # 2 M2W) to determine if the vibration response at the bearings would repeat.

Vibration Data for Case Study #1

The overall vibration amplitude levels measured on the bearing housings with the hand held vibration meter are shown in figure 14-5. The vibration spectrum plots are shown in figures 14-8 through 14-17 and were arranged to observe the vibration signature changes during each test run at each of the sensor locations on the motor and the pump.

Observations and Comments on Case Study #1

-The largest change in overall vibration levels occurred at the inboard bearings of both the pump and the motor.

-The axial vibration level on the motor attained its highest value when its shaft was higher (above) than the pump shaft. Likewise, the axial vibration level on the pump attained its highest value when its shaft was higher than the motor shaft.

-The highest horizontal amplitude readings on both the pump and the motor occurred when the motor shaft was low with respect to the pump shaft centerline (M55L), even though the amount of misalignment was not as severe as in the M65H case.

-The pump bearings increased in amplitude when the motor shaft was higher than the pump shaft (from Run 5 to Run 6) whereas the motor bearings decreased or stayed the same.

-The outboard bearing of the pump experienced a greater increase in overall vibration levels compared to the out-

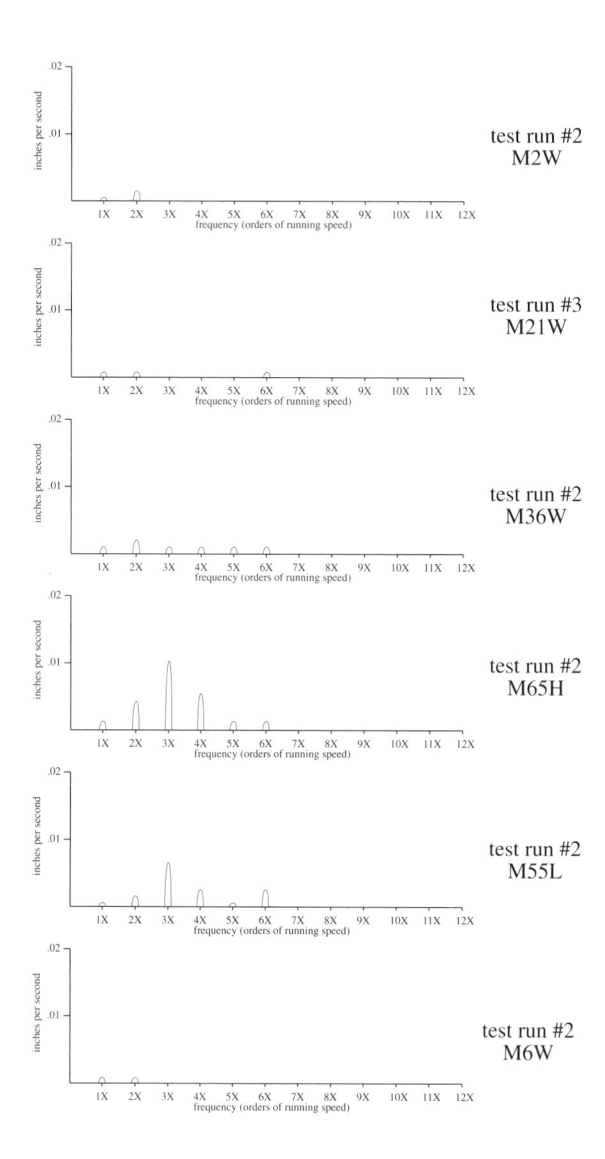

Figure 14-8. Outboard motor bearing, horizontal direction vibration spectrums for Case Study #1.

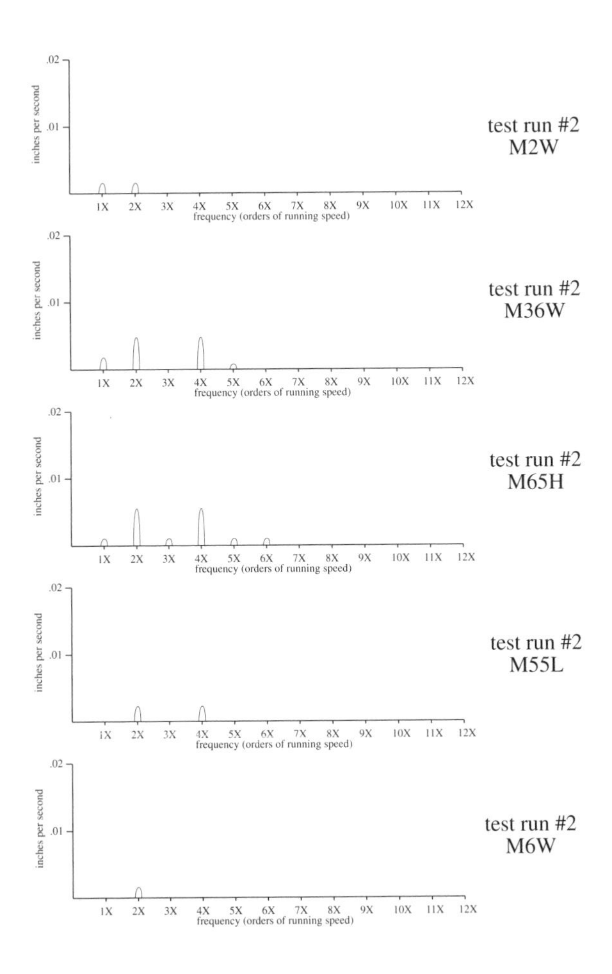

Figure 14-9. Outboard motor bearing, vertical direction vibration spectrums for Case Study #1.

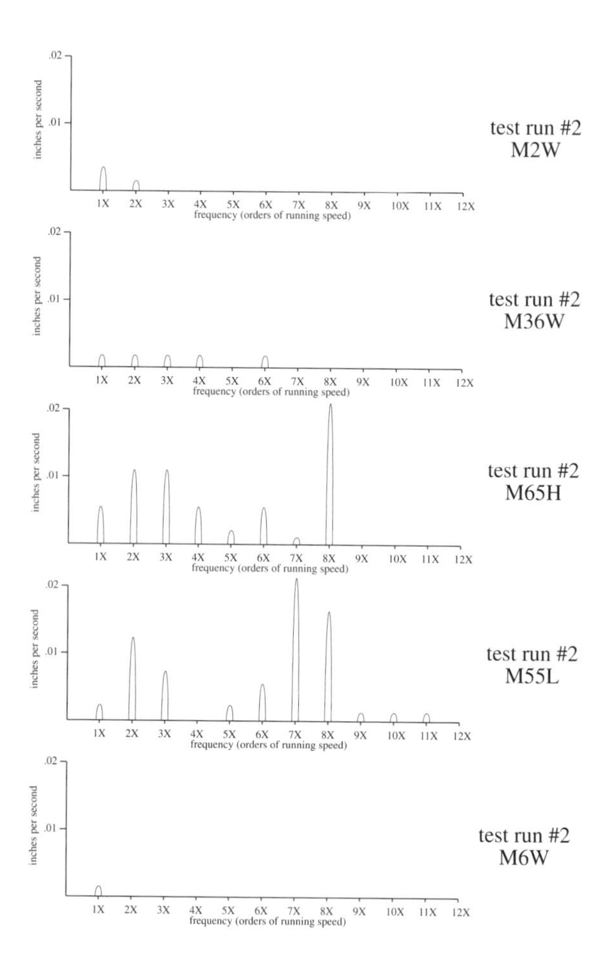

Figure 14-10. Inboard motor bearing, horizontal direction vibration spectrums for Case Study #1.

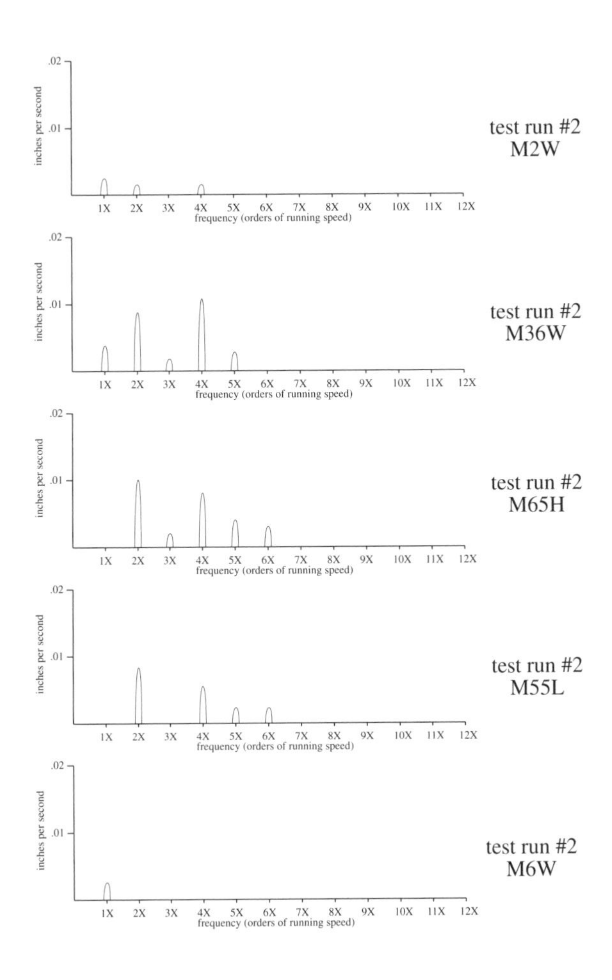

Figure 14-11. Inboard motor bearing, vertical direction vibration spectrums for Case Study #1.

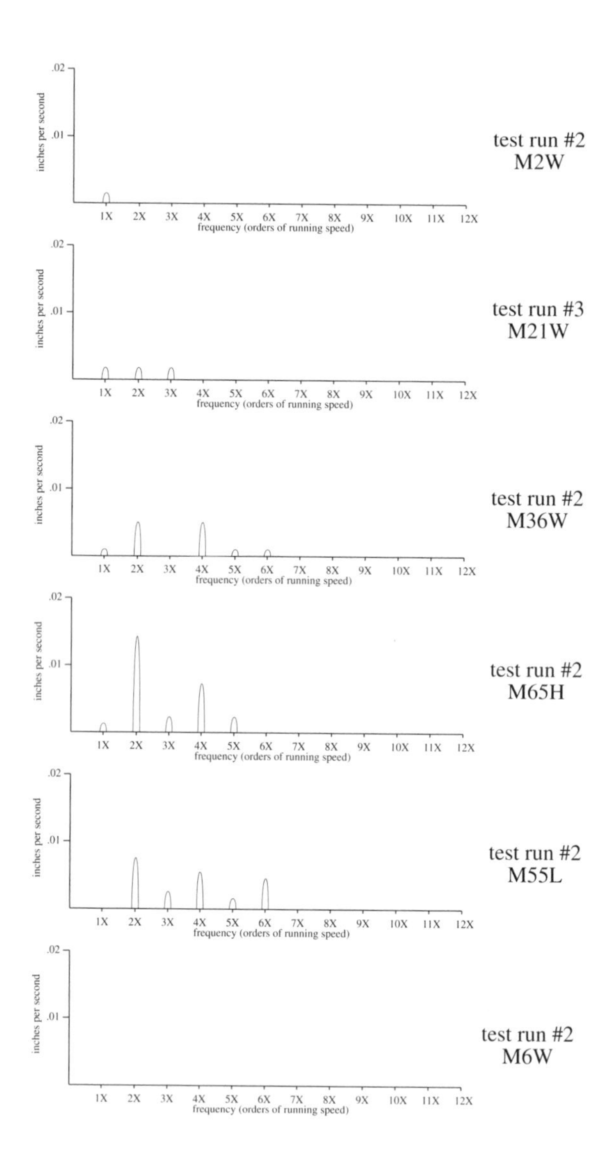

Figure 14-12. Inboard motor bearing, axial direction vibration spectrums for Case Study #1.

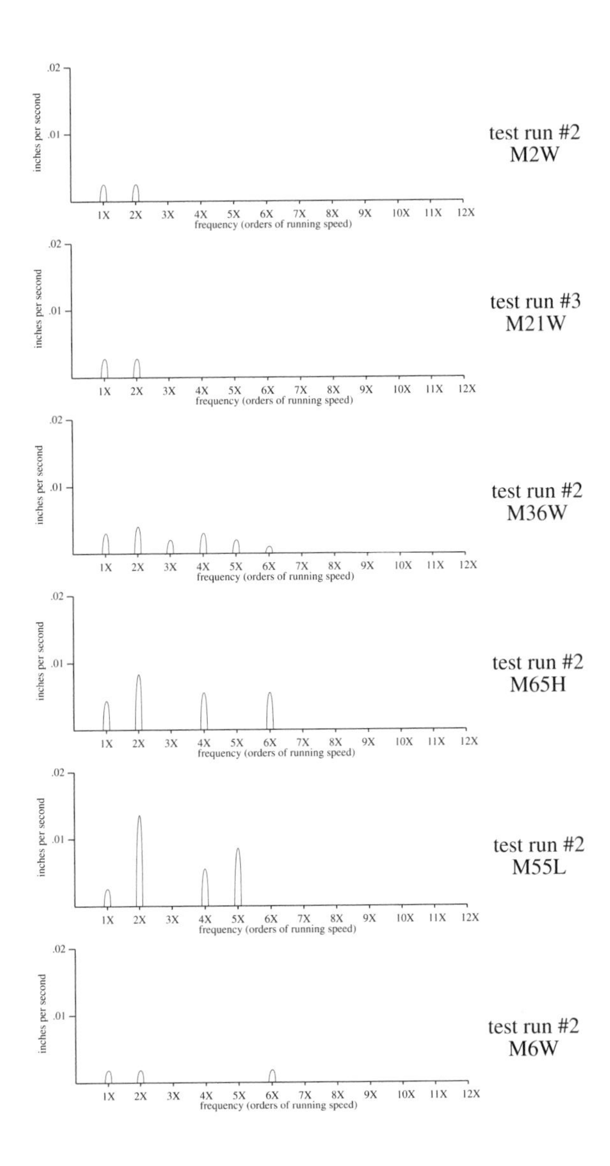

Figure 14-13. Inboard pump bearing, axial direction vibration spectrums for Case Study #1.

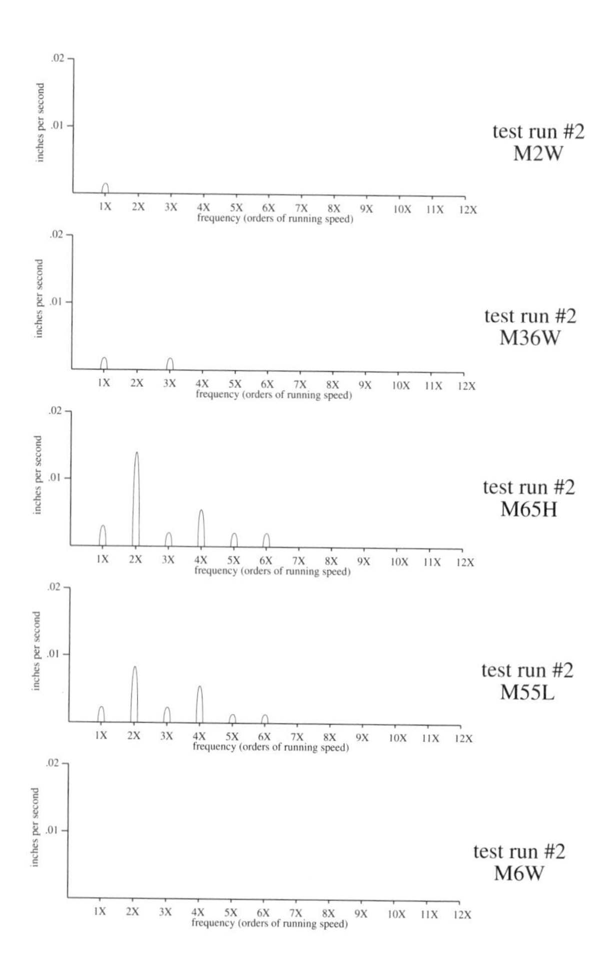

Figure 14-14. Inboard pump bearing, horizontal direction vibration spectrums for Case Study #1.

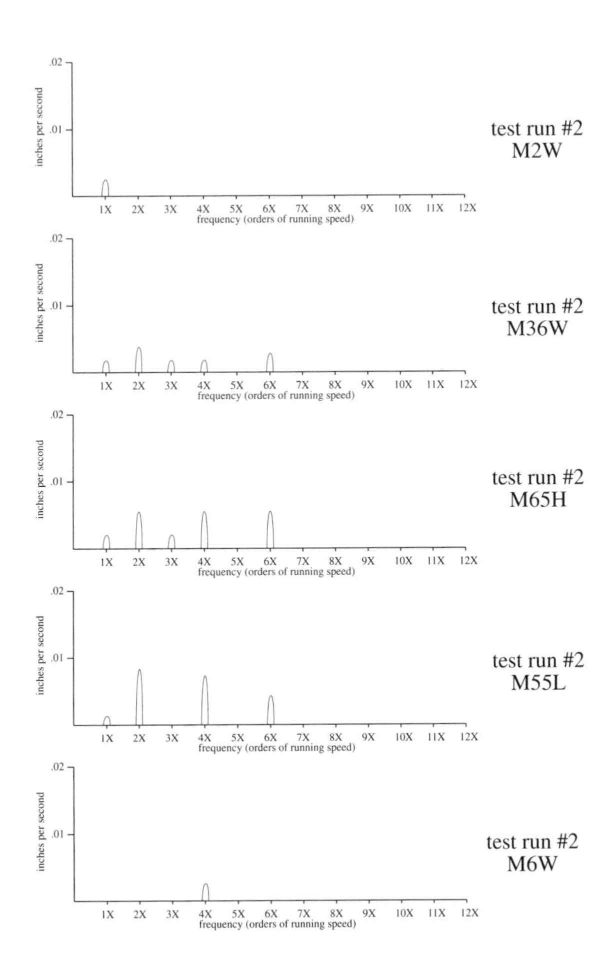

Figure 14-15. Inboard pump bearing, vertical direction vibration spectrums for Case Study #1.

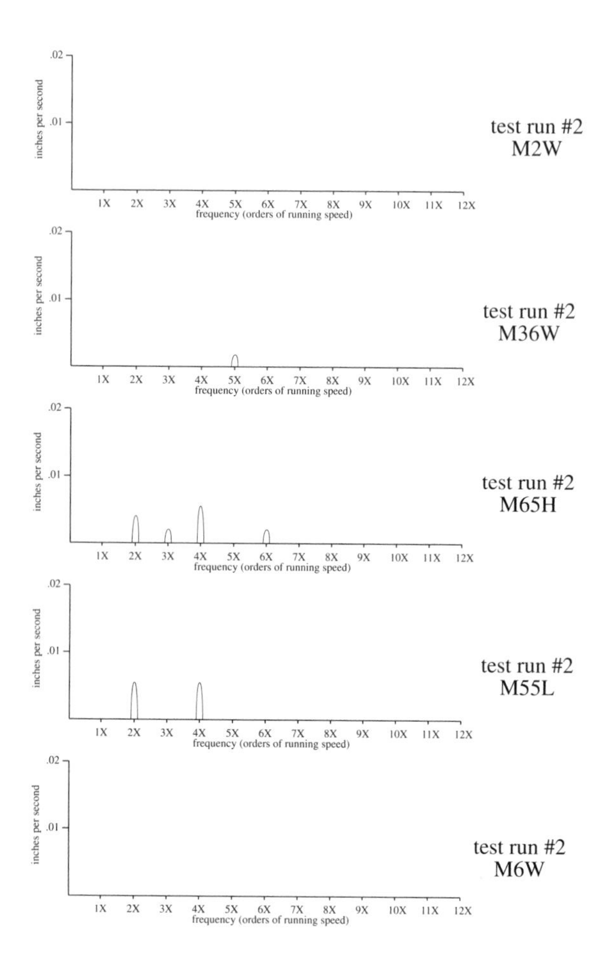

Figure 14-16. Outboard pump bearing, horizontal direction vibration spectrums for Case Study #1.

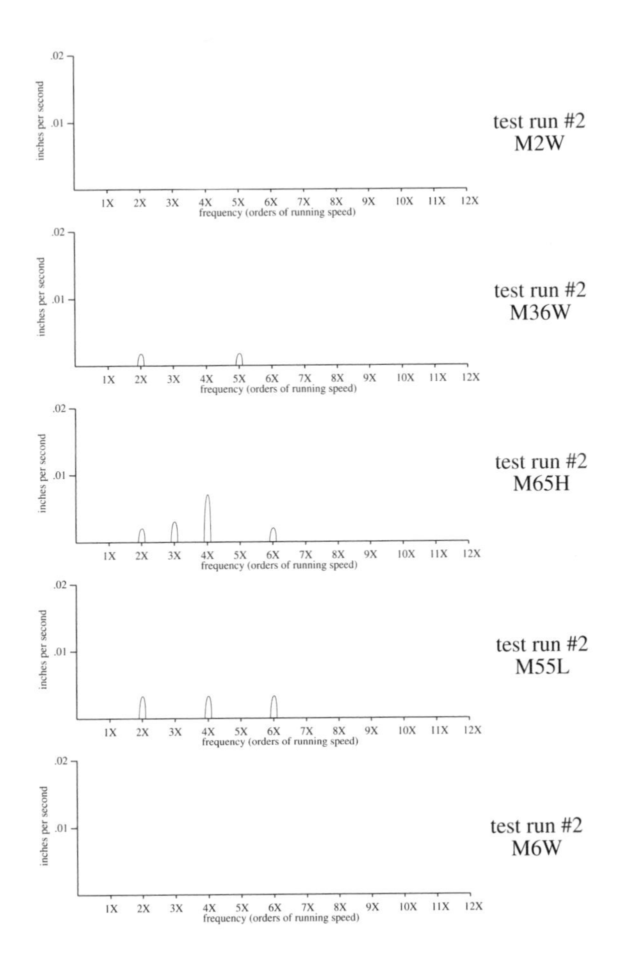

Figure 14-17. Outboard pump bearing, vertical direction vibration spectrums for Case Study #1.

board bearing of the motor.

- -The pump experienced a decrease in overall vibration levels on both bearings when the motor was misaligned laterally (notice Runs 2 and 3).
- -The horizontal and axial amplitude levels of the inboard pump bearing increased as the amount of the misalignment increased. The vertical levels however changed only slightly during all the different shaft configurations.
- -The overall vibration levels on the pump and motor taken during Run no.7 were, for the most part, the same as the levels taken during Run# 2 where the alignment conditions were nearly identical verifying that the vibration was due to misalignment and not other factors.
- -The once per rev amplitude levels were not affected by any misalignment conditions and, in fact, decreased in many instances when the misalignment increased. Consequently, the attempts at taking phase angle data when tracking the once per rev signal was inconclusive as the phase angle would continually drift.
- -The highest peaks occurred at the maximum misalignment conditions (M65H and M55L) on the inboard sensor locations on the motor.
- -The 3X peaks on the motor were higher in amplitude in the horizontal directions than in the vertical direction.
- -The 2X and 4X peaks on the motor were more dominant in the vertical and axial directions than in the horizontal direction.
- -The 2X, 4X, and 6X peaks prevailed in the pump bearings.
- -Higher multiples of running speed occurred on the pump from 40 to 100 kcpm, particularly during the vertical misalignment runs.

How Varying Degrees of Misalignment Affect Rotating Machinery - Case Study #2

Another misalignment test was conducted on an alignment and balance demonstrator unit shown in figure 14-18. The overall vibration data was collected with an accelerometer. Motor current was taken on one of the leads going into the motor with a clamp-on ammeter. In addition, two proximity probes were mounted across the bearing blocks of the center shaft unit to monitor any shaft deflection that might occur during misalignment conditions. All of the data collected is summarized in figure 14-20.

As you can see, virtually every vibration level decreased as the misalignment increased. Although the motor amperage increased slightly after misaligning the center shaft laterally 31 mils, the amperage only slightly changed from 31 mils to 62 mils lateral misalignment.

The only data parameter that increased as the misalignment increased was the distances between the rotor balance disks and the proximity probe tips. The only way this could have happened is if the shaft was elastically bending to accommodate the misalignment condition.

Figure 14-18. Alignment and balance demonstrator used for misalignment case study #2.

Demonstrator Alignment at Start

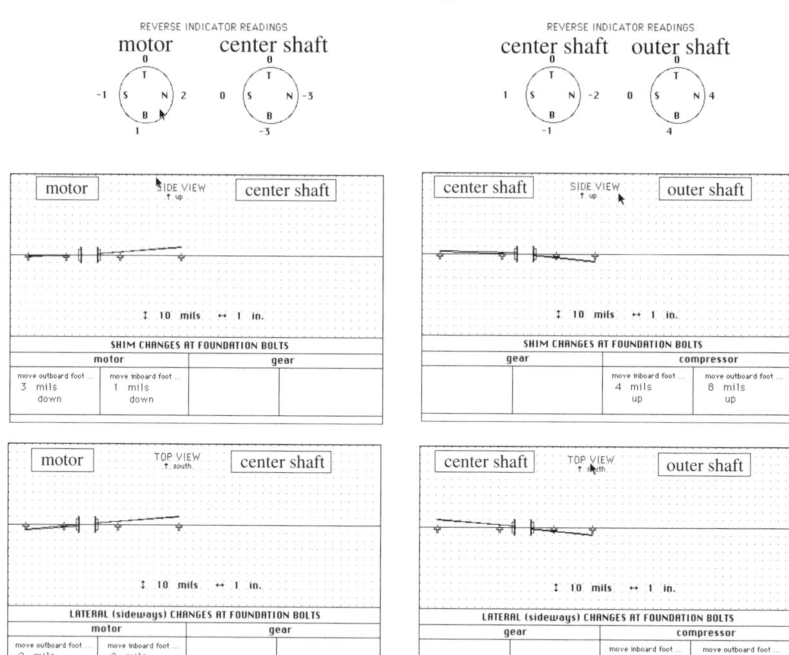

Misalignment Tests

Purposely misalign the center rotor laterally different amounts.

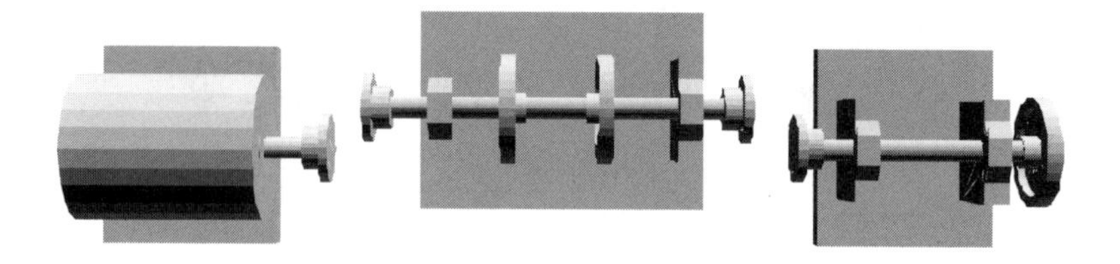

Figure 14-19. Initial shaft alignment and test summay for vibration case study #2.

Misalignment Vibration, Rotor Distortion, and Power Consumption Study

A motor, center shaft, and outer shaft demonstrator model was used for the misalignment study. The drive train was initially aligned within 1/2 mil per inch misalignment, started, and vibration, motor amperage, and proximity probe gap data was collected. The unit was stopped and the center shaft was purposely misaligned by 31 mils sideways, locked in that position, started, and data was collected. Again, the unit was stopped and the center shaft was purposely misaligned by 62 mils sideways, locked in that position, started, and data was collected. Below is the recorded data.

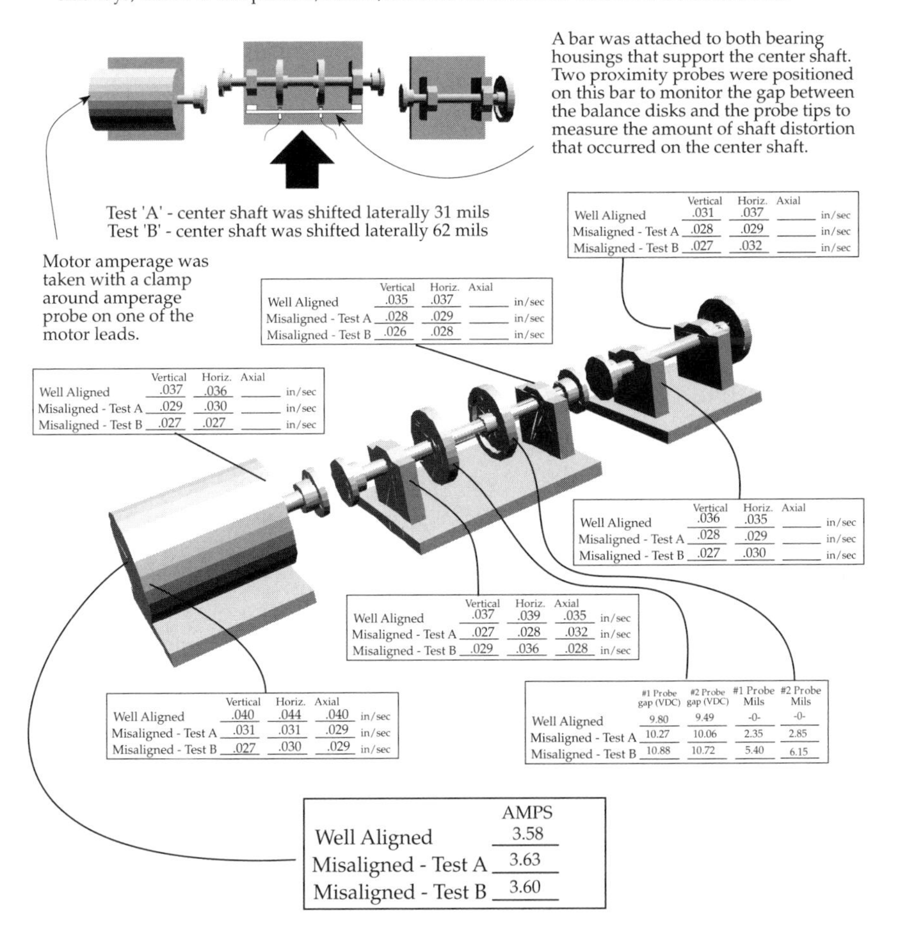

Figure 14-20. Vibration, motor amperage, and shaft defletion data for vibration case study #2.

Vibration Signatures of Common Flexible Couplings

As previously mentioned, vibration can and does occur on rotating machinery that is being subjected to misalignment conditions. Case Studies #1 and #2 have shown samples of the vibration levels that occur and vibration spectral patterns that emerge (at least in Case #1, with a metal ribbon type coupling).

You probably noticed that the vibration spectral patterns changed in Case Study # 1 when the motor and pump were subjected to different kinds of misalignment conditions. But what would have happened if there was a different type of flexible coupling being used between the motor and the pump? Would the patterns have been the same?

Over the past ten years, several individuals have conducted controlled tests on rotating machinery where the drive train was subjected to misalignment conditions and vibration data was collected and observed. In addition, data was collected on machinery where there was a known misalignment condition present. Figures 14-21 through 14-25 show samples of this data.

As you review these figures you will notice that the vibration patterns on equipment with the same coupling design is not exactly the same and in some cases radically different or nonexistent.

Flexible Disk Type Coupling
Various vibration responses to misalignment

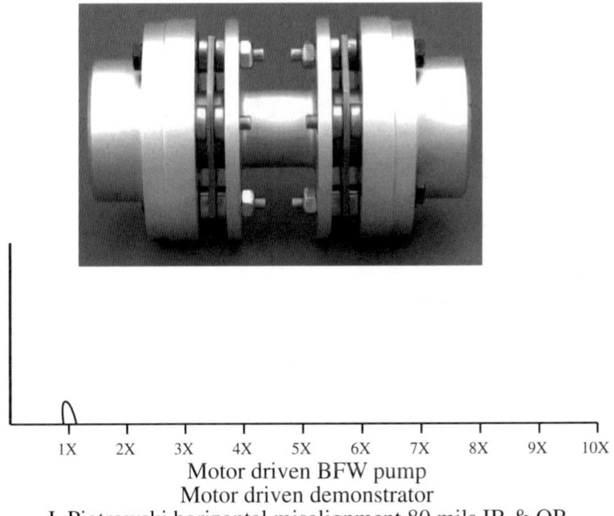

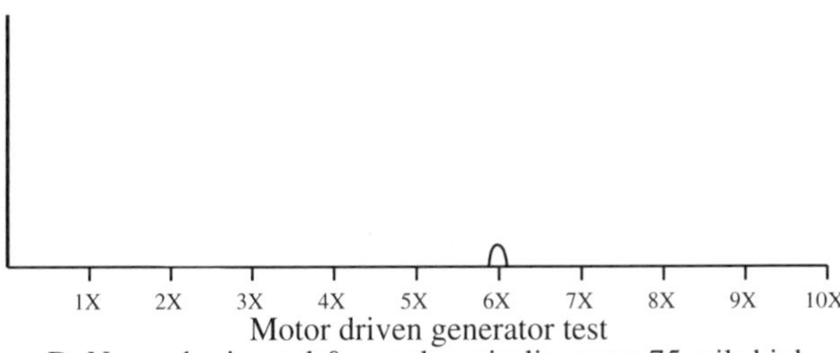

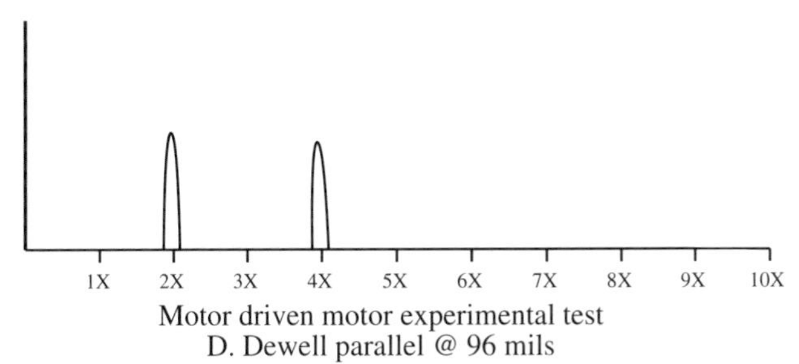

Figure 14-21. Observed vibration patterns on flex disk type couplings.

Gear Coupling
Various vibration responses to misalignment

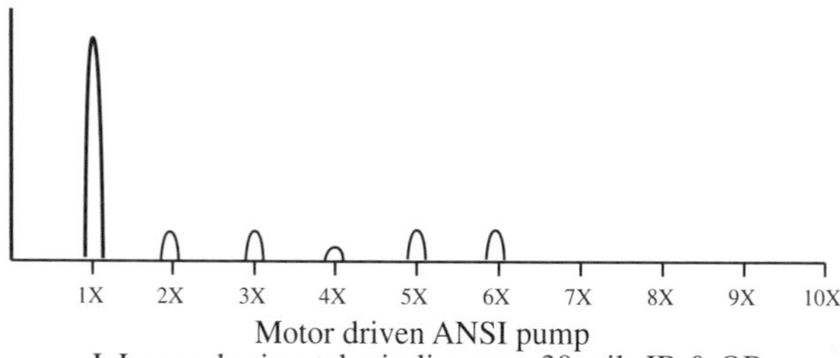

Motor driven ANSI pump
J. Lorenc horizontal misalignment 30 mils IB & OB

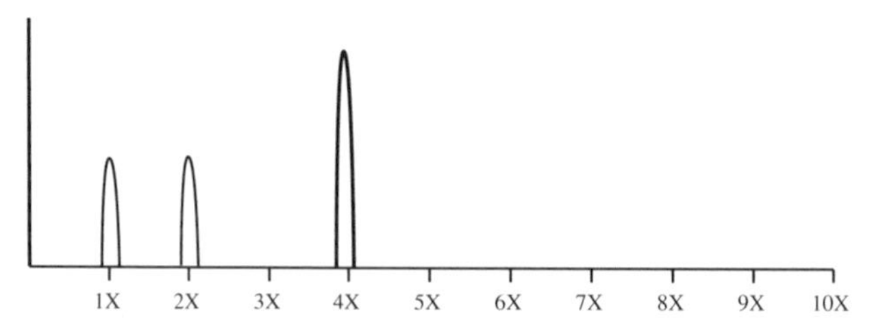

Gas/Power Turbine driven compressor
J. Piotrowski horizontal misalignment 65 mils IB & OB

Figure 14-22. Observed vibration patterns on gear type couplings.

Jaw Coupling
Various vibration responses to misalignment

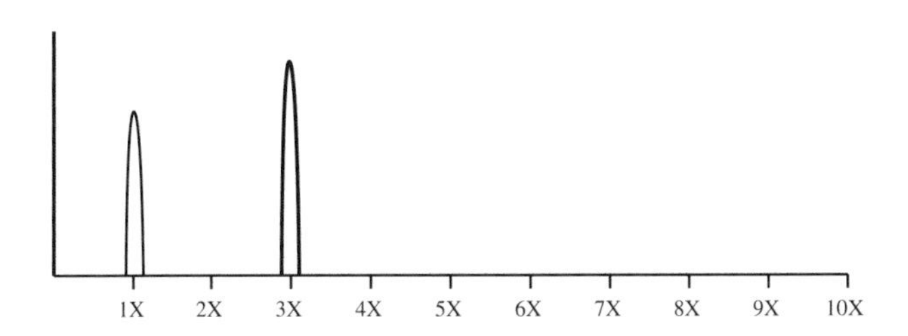

Motor driven ANSI pump
J. Lorenc horizontal misalignment 90 mils IB & OB

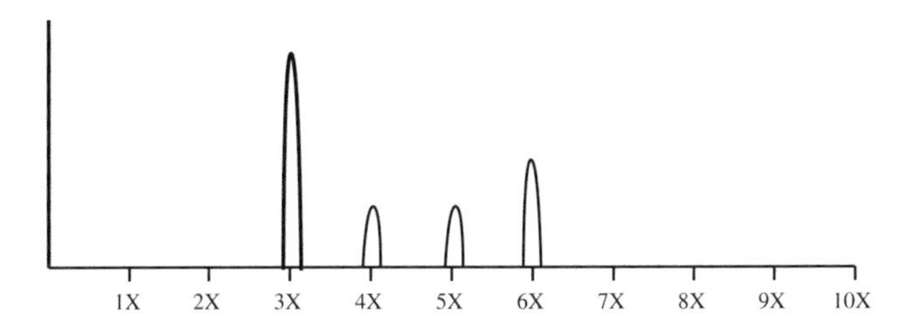

Motor driven generator test
D. Nower horizontal & angular misalignment @ 15 mils per inch

Figure 14-23. Observed vibration patterns on jaw type couplings.

Metal Ribbon Coupling
Various vibration responses to horizontal misalignment

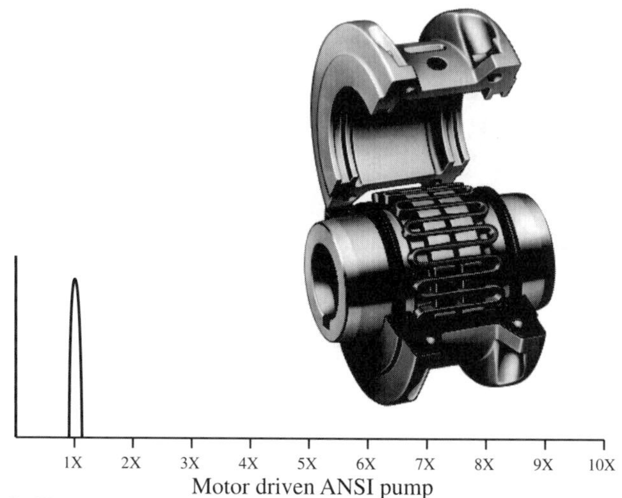

Motor driven ANSI pump
S. Chancey vertical misalignment 50 mils @ IB & 75 mils @ OB
J. Lorenc horizontal misalignment 90 mils IB & OB

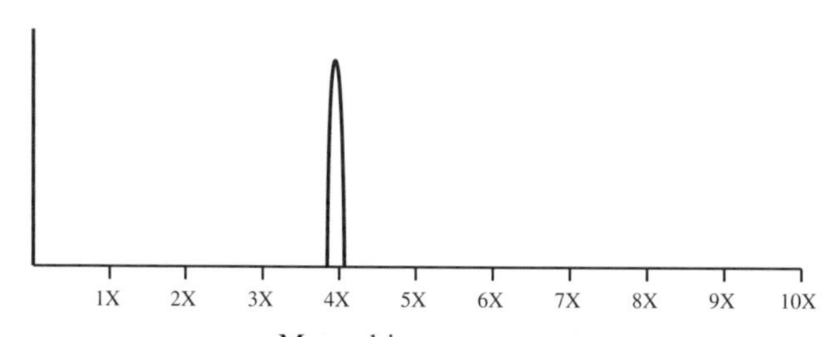

Motor driven generator test
D. Nower horizontal misalignment 50 mils IB & OB

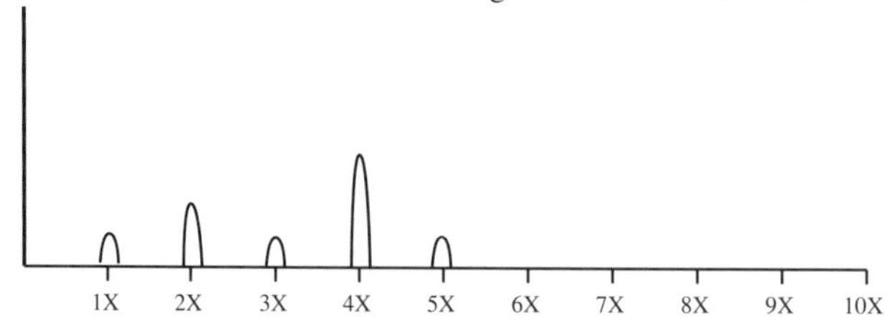

Motor driven centrifugal pump
J. Piotrowski horizontal misalignment 36 mils IB & OB

Figure 14-24. Observed vibration patterns on grid type couplings.

'Rubber Tire' Type Coupling
Various vibration responses to misalignment

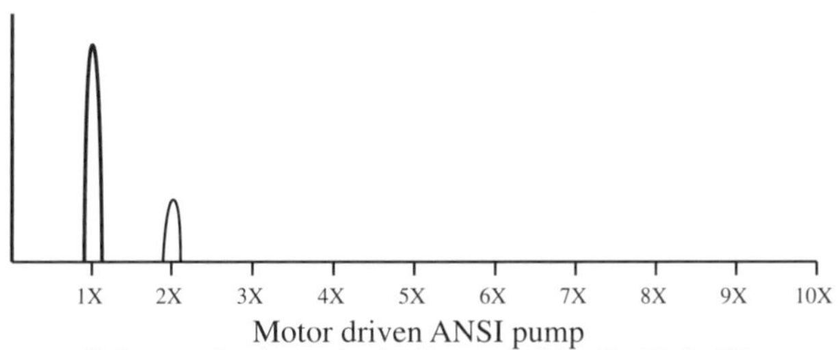

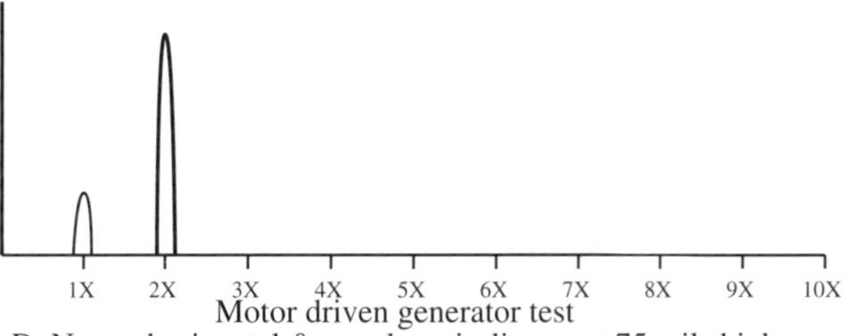

Figure 14-25. Observed vibration patterns on tire type couplings.

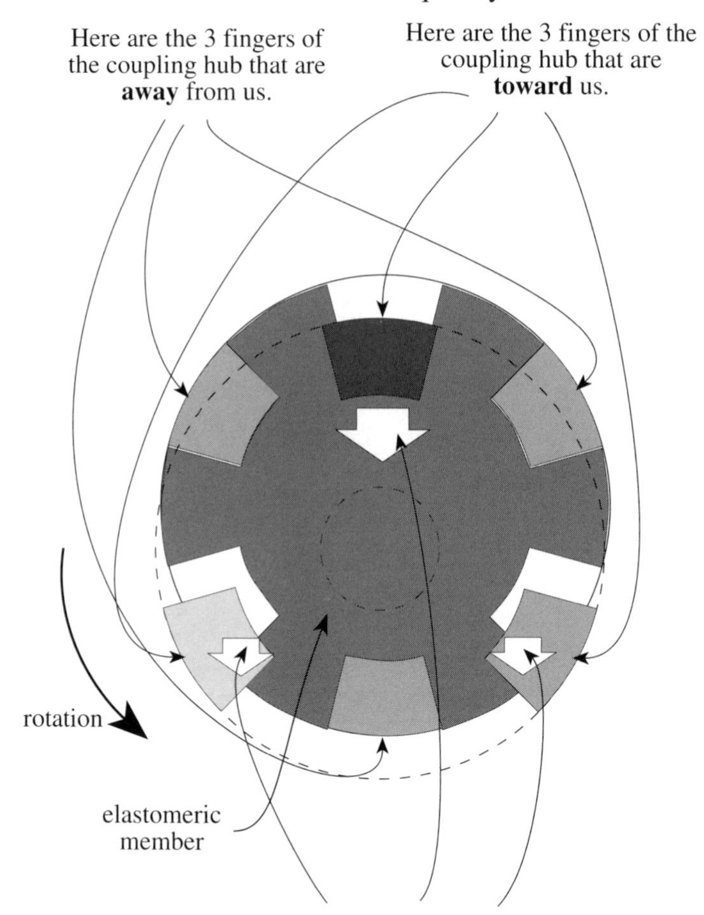

When the three 'jaws' (on the coupling hub that is toward us) rotate to the position shown above, there is a downward force being exerted due to compression of the elastomeric member. This force is transmitted through the elastomeric member to the other coupling hub (and shaft) that is away from us. This downward force subsides as the hubs begin to rotate and then, reoccurs again after 120 degrees of rotation (1/3 of a revolution) therby causing a downward force to be exerted 3 times during one revolution of the shaft.

Figure 14-26. Force analysis of jaw type coupling.

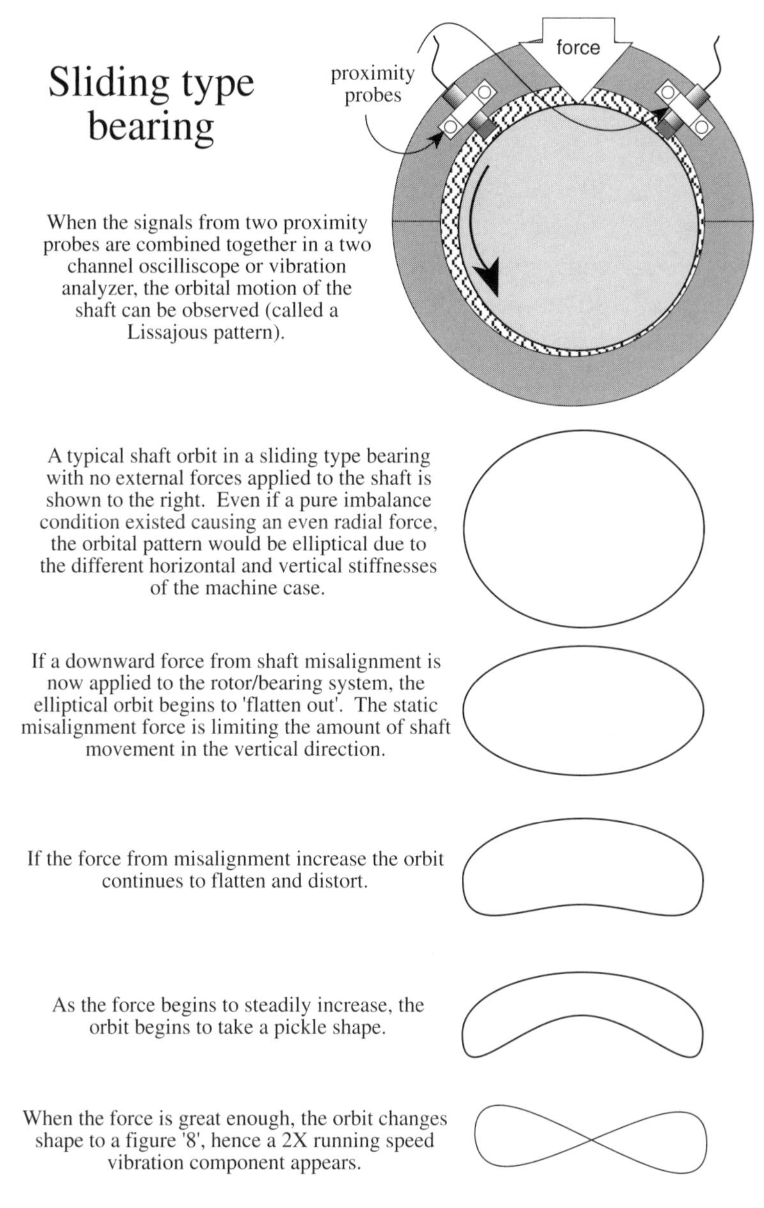

Figure 14-27. Observed vibration orbital patterns on rotors supported in sliding type bearings.

Using Infrared Thermography to Detect Misalignment Conditions - Case Study # 3

Predictive maintenance technicians from a bottling company collected the data in figures 14-28 through 14-33. The data was compiled by Infraspection Institute in Shelburne, VT.

The test was conducted by coupling a 10 hp motor to a 7200 watt electric generator. A specific flexible coupling was installed between the motor and the generator, the unit was then accurately aligned, and then started up. Vibration, ultrasound, and thermal imaging data was then collected after 10 minutes run time. The unit was then shut down, misaligned by 10 mils, started back up and the data collected again. This was repeated several times with an additional 10 mils of misalignment introduced.

After a series of misalignment runs were made, the unit was then outfitted with a different coupling design and the process was repeated.

Notice that as the misalignment increases, so too did the temperature of the coupling or of the flexing element. The increase in temperature is somewhat linear as illustrated in the temperature graphs with each coupling tested. Disappointingly however, the vibration and ultrasound data were never published with the infrared data.

In addition, there must be a word of caution here since it is very tempting to make generalizations from this data.

Not every flexible or rigid coupling will increase in temperature when subjected to misalignment conditions. The flexible couplings used in this test were mechanically flexible couplings (the chain and metal ribbon types) or elastomeric types.

In mechanically flexible couplings, the heat is being generated as the metal grid slides back and forth across the tooth slots in the coupling hubs or as the chain rollers slide across the sprocket teeth as the coupling elements attempt to accept the misalignment condition. In the elastomeric couplings, the elastomer is being

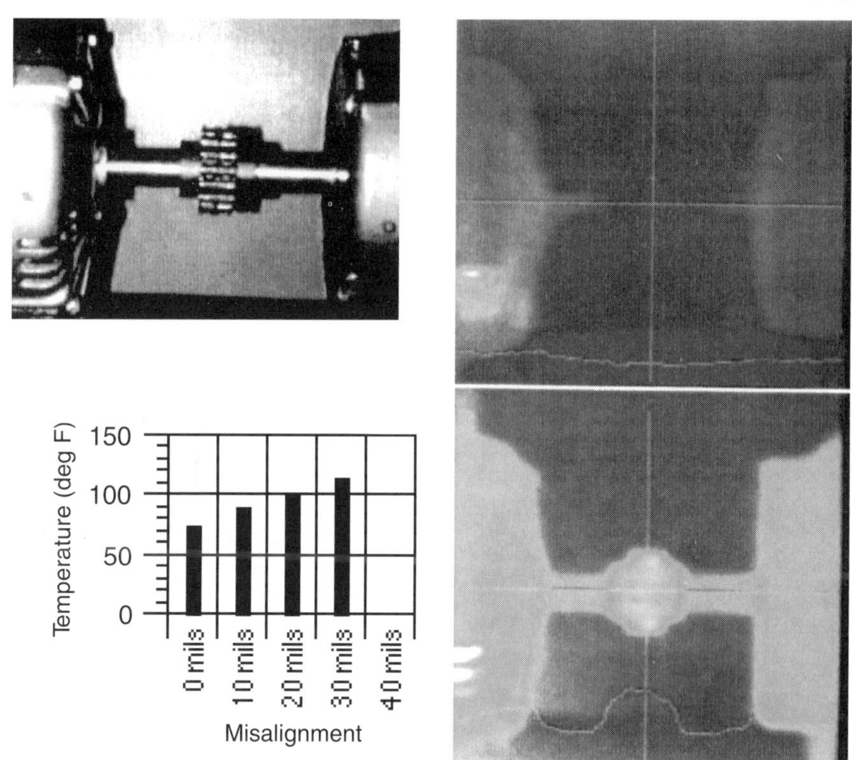

Figure 14-28. Observed temperature patterns on misaligned chain type coupling. Upper right photo shows infrared image of coupling running under good alignment conditions. Lower right photo shows coupling running under 'worst case' misalignment condition indicated by rightmost bar on temperature graph. Photos and data courtesy of Infraspection Institute, Shelburne, VT.

heated through some sliding friction but primarily by shear and compression forces as these coupling elements attempt to accept their misalignment conditions.

What would have happened if a flexible disk or diaphragm

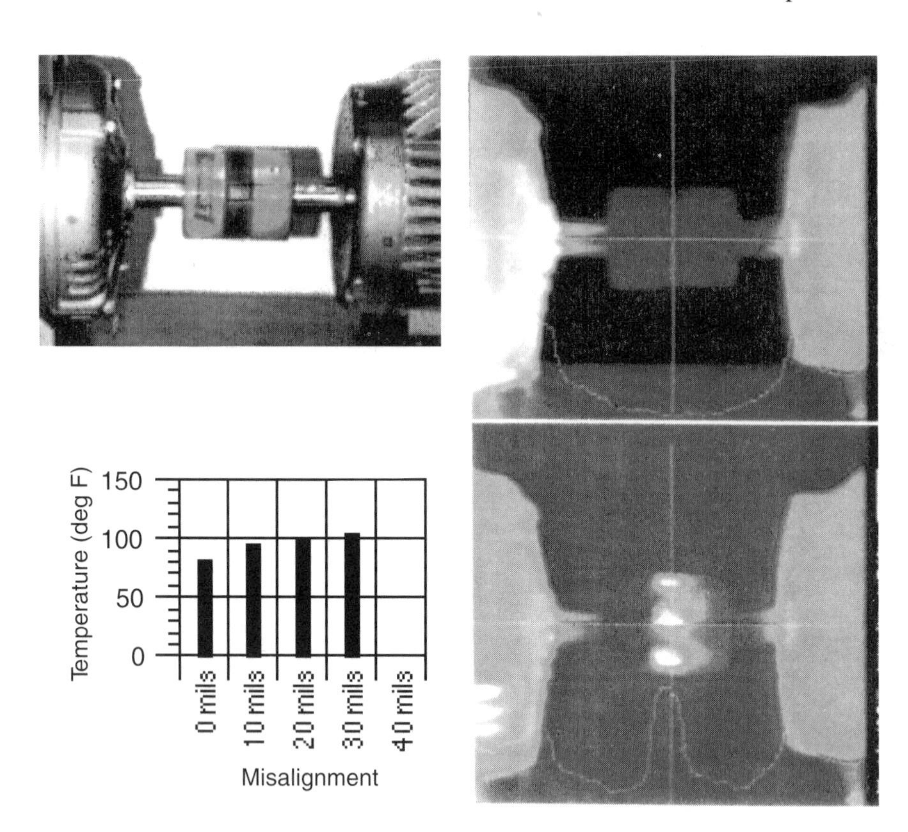

Figure 14-29. Observed temperature patterns on misaligned jaw type coupling. Upper right photo shows infrared image of coupling running under good alignment conditions. Lower right photo shows coupling running under 'worst case' misalignment condition indicated by the rightmost bar on temperature graph. Photos and data courtesy of Infraspection Institute, Shelburne, VT.

type coupling was tested also? Flexible disk or diaphragm couplings accept misalignment conditions by elastically bending the two disk packs or diaphragms and virtually no heat will be generated by the flexure of metal disks as these types of couplings attempt to accommodate any misalignment conditions.

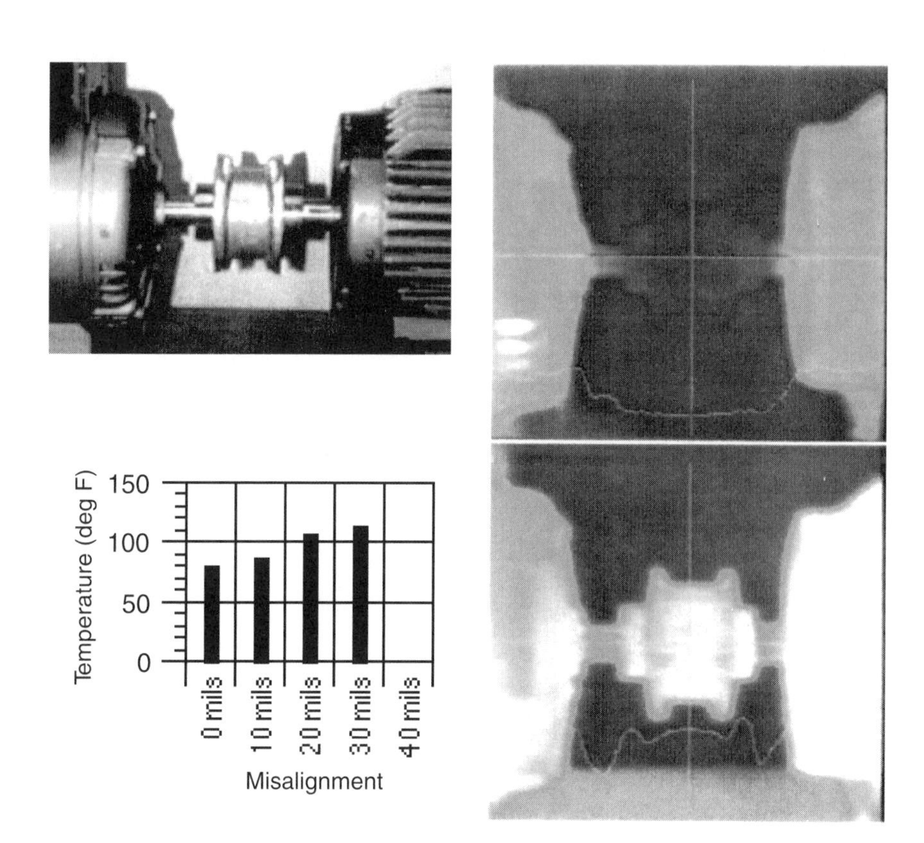

Figure 14-30. Observed temperature patterns on misaligned metal ribbon type coupling. Upper right photo shows infrared image of coupling running under good alignment conditions. Lower right photo shows coupling running under 'worst case' misalignment condition indicated by the rightmost bar on temperature graph. Photos and data courtesy of Infraspection Institute, Shelburne, VT.

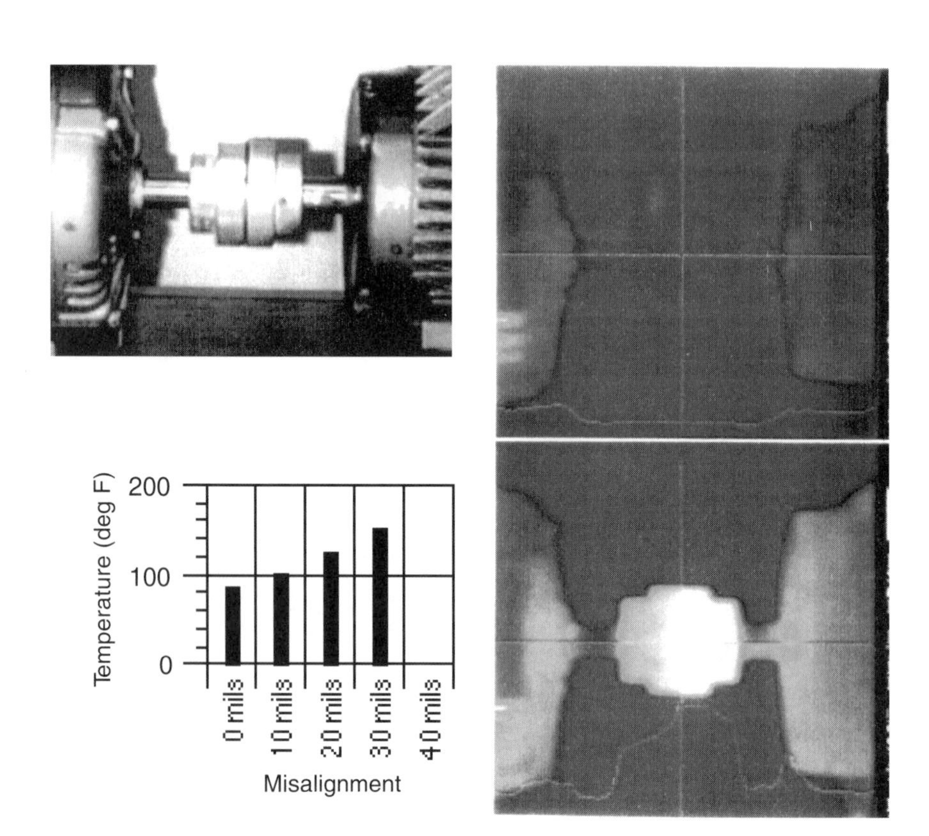

Figure 14-31. Observed temperature patterns on misaligned elastomeric insert type coupling. Upper right photo shows infrared image of coupling running under good alignment conditions. Lower right photo shows coupling running under 'worst case' misalignment condition indicated by the rightmost bar on temperature graph. Photos and data courtesy of Infraspection Institute, Shelburne, VT.

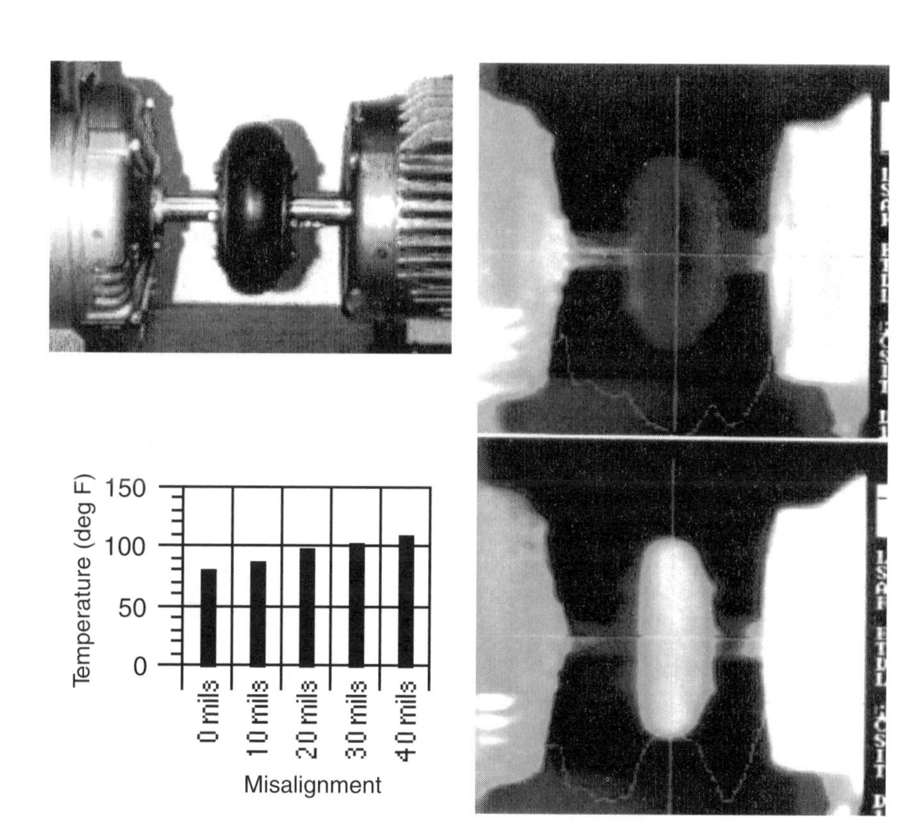

Figure 14-32. Observed temperature patterns on misaligned rubber tire type coupling. Upper right photo shows infrared image of coupling running under good alignment conditions. Lower right photo shows coupling running under 'worst case' misalignment condition indicated by the rightmost bar on temperature graph. Photos and data courtesy of Infraspection Institute, Shelburne, VT.

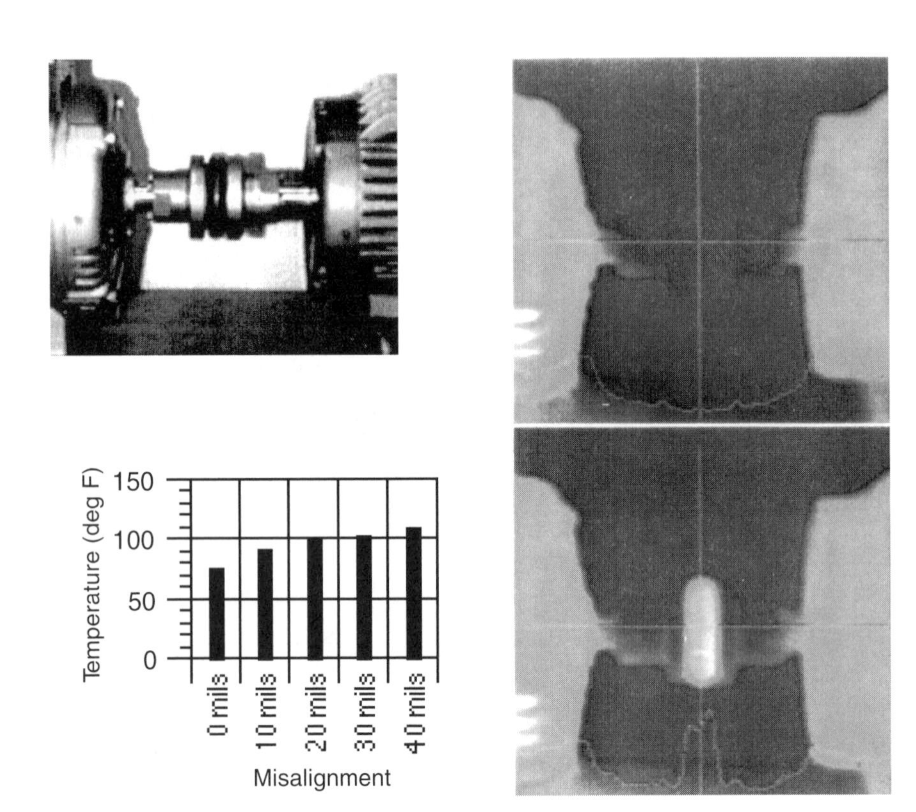

Figure 14-33. Observed temperature patterns on misaligned rubber insert type coupling. Upper right photo shows infrared image of coupling running under good alignment conditions. Lower right photo shows coupling running under 'worst case' misalignment condition indicated by the rightmost bar on temperature graph. Photos and data courtesy of Infraspection Institute, Shelburne, VT.

Bibliography

Sohre, John S., "Turbomachinery Analysis and Protection", Proceedings - First Turbomachinery Symposium, Gas Turbine Labs, Texas A&M University, College Station, Texas, 1972.

Audio Visual Customer Training - Instruction Manual, IRD Mechanalysis Inc., Publication no. 414E, Columbus, Ohio, 1975.

Eshleman, Ronald L., "Torsional Vibration of Machine Systems", Proceedings - Sixth Turbomachinery Symposium, Gas Turbine Labs, Texas A&M University, College Station, Texas, Dec. 1977

"Alignment Loading of Gear Type Couplings", Application Notes no. (009)L0048, Bently Nevada Corp., Minden, Nevada, March, 1978.

Jackson, Charles, **The Practical Vibration Primer**, Gulf Publishing Co., Houston, Texas, 1979.

Maxwell, J.H., "Vibration Analysis Pinpoints Coupling Problems", Hydrocarbon Processing, Jan. 1980, pgs. 95-98.

Eshleman, Ronald L.,"Effects of Misalignment on Machinery Vibrations", Proceedings - Balancing / Alignment of Rotating Machinery, Galveston, Texas, Feb. 23-26, 1982, Vibration Institute, Clarendon Hills, Ill.

Baxter, Nelson L., "Vibration and Balance Problems in Fossil Plants : Industry Case Histories", Electric Power Research Institute, Palo Alto, CA, publication no. CS-2725, Research Project no. 1266-27, Nov. 1982.

Mannasmith, James, Piotrowski, John D., "Machinery Alignment Methods and Applications", Vibration Institute meeting, Cincinnati Chapter, Sept. 8, 1983.

Bertin, C.D., Buehler, Mark W.. "Typical Vibration Signatures - Case Studies", Turbomachinery International, Oct. 1983, pgs. 15-21.

Eshleman, Ronald L., "The Role of Couplings in the Vibration of Machine Systems", Vibration Institute meeting, Cincinnati Chapter, Nov. 3, 1983.

Dewell, D. L., Mitchell, L. D., “Detection of a Misaligned Disk Coupling Using Spectrum Analysis”, Journal of Vibration, Acoustics, Stress, and Reliability in Design, January 1984, Volume 106, Pgs. 9-16.

Piotrowski, John D.,”How Varying Degrees of Misalignment Affect Rotating Machinery - A Case Study”, Proceedings - Machinery Vibration Monitoring and Analysis Meeting , New Orleans, La., June 26-28, 1984, Vibration Institute, Clarendon Hills, Ill.

Piotrowski, John D., “Aligning Cooling Tower Drive Systems”, Proceedings - Machinery Vibration Monitoring and Analysis, Ninth Annual Meeting, May 20-24, 1985, New Orleans, La., Vibration Institute, Clarendon Hills, Ill.

Lorenc, Jerome A., "Changes in Pump Vibration Levels Caused by the Misalignment of Different Style Couplings", Proceedings 8th International Pump Users Symposium, Houston, TX, March 1991.

Murphy, Dan, "Cooling Tower Vibration Analysis", Maintenance Technology, July 1991, Vol. 4, No. 7, pgs. 28-33.

Schultheis, Steven M., "Diagnostic Techniques Using ADRE 3 for Evaluation of Radial Rubs in Rotating Machinery", Orbit, October 1991, Pgs. 6-11.

Jackson, C. J., "Back to Fundamentals - Part 5 - Preloads (Alignment)", presented at 1992 Vibration Institute annual meeting.

Nower, Daniel L., "Preliminary Report on Characterizing Shaft Misalignment Effects Using Dynamic Measurements", Presented at the Vibration Institute Meeting, Wilmington, VA, June 1992.

Bond, Terry, “Application Update - Deltaflex Coupling - Vibration Analysis : Motor to Centrifugal Pump”, Lovejoy, Inc., October, 1992, personal correspondence.

Mitchell, John S., **Introduction to Machinery Analysis and Monitoring**, 1993, Pennwell Publishing Co., Tulsa, OK, ISBN 0-87814-401-3.

Chancey, Steve, Piotrowski, J. D., "Vibration Study on a Motor Driven Centrifugal Pump - The Effects of Varying Amounts of Unbalance, Misalignment, and Fluid Flow", field study, 1993.

Campbell, A. J., "Static and Dynamic Alignment of Turbomachinery", Orbit, June 1993, pgs. 24-29.

A Cost Effective, Pro-Active Method to Find, Prioritize, and Correct Coupling Misalignment Using Infrared Thermography and Laser Alignment Technologies, 1994, distributed by the Infraspection Institute, Shelburne, VT, data taken and collected by Frank Pray and Bruce Bortnem, Miller Brewing Co., Irwindale, CA.

Nower, Daniel, "Misalignment : Challenging the Rules", Reliability, May/June 1994, Vol. 1, Issue 2.

Xu, Ming, Marangoni, R. D., "Vibration Analysis of a Motor - Flexible Coupling - Rotor System Subject to Misalignment and Unbalance, Part I : Theoretical Model and Analysis", Journal of Sound and Vibration (1994), 176(5), 663-679.

Xu, Ming, Marangoni, R. D., Vibration Analysis of a Motor - Flexible Coupling - Rotor System Subject to Misalignment and Unbalance, Part II : Experimental Validation", Journal of Sound and Vibration (1994), 176(5), 681-691.

Appendix

Machinery Data Card

Equipment Name [ ] Location [ ]

Equipment Photo / Sketch

DRIVER UNIT Information

○ induction motor ○ synchronous motor ○ steam turbine ○ gas turbine ○ diesel ○ other ________

Mfg.	Model No.	Serial No.
HP	Speed RPM	Total Unit Weight/Rotor weight / lbs.
Trip Speed RPM	'Critical' Speed(s) RPM	Service Factor

Bearing Information Type of bearings : ○ antifriction ○ sliding ○ magnetic

Inboard (coupling) Brg.	Mfg.	Mdl. / Brg. No.	clearance
Outboard Brg.	Mfg.	Mdl. / Brg. No.	clearance
Thrust Brg.	Mfg.	Mdl. / Brg. No.	axial float

Off-line to Running Machinery Movement

	vertical movement	lateral movement	
outboard end	______ mils ○ up ○ down	______ mils N S E W	foot bolt size [ ] in.
inboard end	______ mils ○ up ○ down	______ mils N S E W	wrench size [ ] in.

DRIVEN UNIT Information

○ pump ○ gear ○ compressor ○ generator ○ fan ○ other ________

Mfg.	Model No.	Serial No.
HP	Speed RPM	Total Unit Weight/Rotor weight / lbs.
Trip Speed RPM	'Critical' Speed(s) RPM	Service Factor

Bearing Information Type of bearings : ○ antifriction ○ sliding ○ magnetic

Inboard (coupling) Brg.	Mfg.	Mdl. / Brg. No.	clearance
Outboard Brg.	Mfg.	Mdl. / Brg. No.	clearance
Thrust Brg.	Mfg.	Mdl. / Brg. No.	axial float

Off-line to Running Machinery Movement

	vertical movement	lateral movement	
outboard end	______ mils ○ up ○ down	______ mils N S E W	foot bolt size [ ] in.
inboard end	______ mils ○ up ○ down	______ mils N S E W	wrench size [ ] in.

COUPLING Information

○ elastomeric ○ gear ○ flex disc ○ diaphragm ○ metal ribbon ○ chain ○ other ________

Mfg.	Model No.	Serial No.

Shaft to shaft spacing in. +/- in.	Bolt Torque ft-lbs	Wrench size in.
Driver shaft diameter in. +/- in.	Interference fit mils	Shaft taper in/ft
Driven shaft diameter in. +/- in.	Interference fit mils	Shaft taper in/ft

Lubrication Info ○ grease ○ oil ○ continuous feed oil

Viscosity or type	Amount of lube oz/pints
Recommended Mfg.	Secondary Mfg.

Shaft Alignment Information

○ Reverse Indicator ○ Face-Peripheral ○ Double Radial
○ Face-Face ○ Shaft to Coupling spool
○ Laser ○ LVDT ○ Prox probe ○ Other ________

'Desired' Dial Indicator Readings
(no bracket sag)

0 T B   0 T B

Indicate where any 'soft foot' corrections were made, the shape and thickness of the shims or custom 'wedges', and indicate how the corrections are oriented under each of the machinery feet.

Top View of Machinery

Desired Off - Line Vertical Shaft Positions

Desired Off - Line Lateral Shaft Positions

Maintenance History

date of installation ________

Date	Work Performed

Preliminary Alignment Checks

Company : ______________________________
Equipment Name : ______________________________
Location : ______________________________
Equipment Identification # : ______________________

Static Piping Stress Checks

Date : ____________

dial indicators were located on the ...
O motor O pump

With piping attached and after loosening the pump foot bolts, the dial indicators read the following values ...

_______ mils vertical
_______ mils lateral / horizontal

Shaft & Coupling Hub Runout Checks

motor shaft pump shaft

Date : ______

key key

'Soft Foot' Checks & Corrections

Loosen or remove all foot bolts. Clean underside of machinery feet and points of contact on baseplate. Rough align both units. With the foot bolts in place but not tightened down, measure four points around each bolt hole with a set of feeler gauges and record the readings.

Indicate where any 'soft foot' corrections were made, the shape and thickness of the shims or custom 'wedges', and indicate how the corrections are oriented under each of the machinery feet.

Piping Stress Test by : ______________
Runout Checks by : ______________
'Soft Foot' Checks by : ______________

Notes :

Installation and Shaft Alignment Report

Company : ______________________
Equipment Name : ______________________
Location : ______________________
Equipment Identification # : ______________

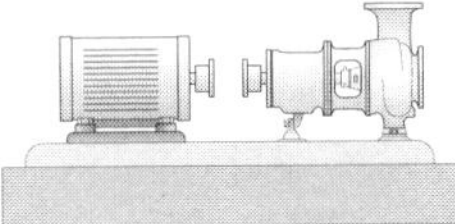

Static Piping Stress Checks

Date : __________

dial indicators were located on the ...
O motor O pump

With piping attached and after loosening the pump foot bolts, the dial indicators read the following values ...

_______ mils vertical
_______ mils lateral / horizontal

Shaft & Coupling Hub Runout Checks

motor shaft pump shaft
Date : ______

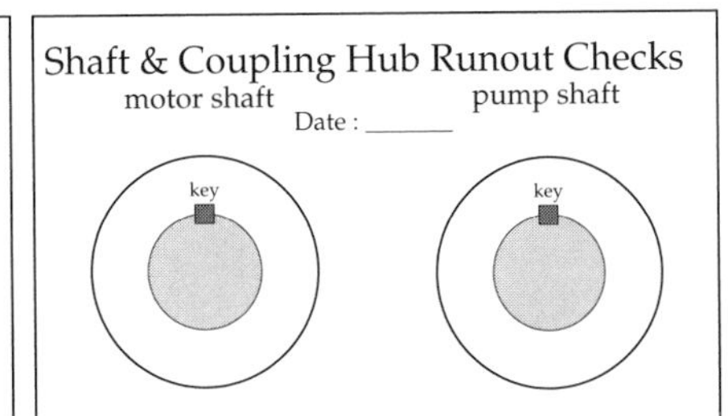

'Soft Foot' Checks & Corrections

Indicate where any 'soft foot' corrections were made, the shape and thickness of the shims or custom 'wedges', and indicate how the corrections are oriented under each of the machinery feet.

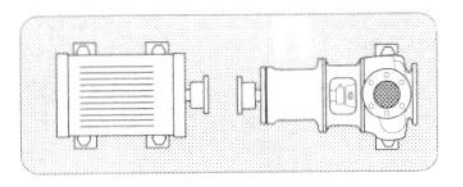

Piping Stress Test by : ______________
Runout Checks by : ______________
'Soft Foot' Checks by : ______________
Final Alignment Readings by : __________

Shaft Alignment Information

Indicate how the shaft positions were measured with the machinery off-line.

O Reverse Indicator O Face-Peripheral
O Shaft to Coupling Spool O Face-face
O Double Radial O Laser O Other ________
amount of bracket sag _______ mils

'Desired' / 'Shoot-for'
Dial Indicator Readings

0 T B 0 T B

Final Readings

0 T B 0 T B

Recommended Torque Values for Nuts & Bolts

SAE Grade 2 Unified National Coarse (UNC)

(55 kpsi proof strength - 69 kpsi tensile strength for sizes .250' to .500")
(52 kpsi proof strength - 64 kpsi tensile strength for sizes .500' to .750")
(28 kpsi proof strength - 55 kpsi tensile strength for sizes .750' to 1.500")

Bolt Size (in.)	Threads per inch	Torque (dry)	Torque (lubricated)* (75% of dry torque)	Clamp Load (lbf +/- 25%)
0.1640	32	19 in-lb	14 in-lb	546
0.1900	24	28 in-lb	21 in-lb	734
0.2500	20	67 in-lb	50 in-lb	1313
0.3125	18	120 in-lb	90 in-lb	2175
0.3750	16	21 ft-lb	16 ft-lb	3188
0.4375	14	32 ft-lb	24 ft-lb	4388
0.5000	13	52 ft-lb	39 ft-lb	5850
0.5625	12	69 ft-lb	52 ft-lb	7500
0.6250	11	100 ft-lb	75 ft-lb	9300
0.7500	10	190 ft-lb	142 ft-lb	13800
0.8750	9	293 ft-lb	220 ft-lb	11400
1.0000	8	427 ft-lb	320 ft-lb	15000
1.1250	7	8678 ft-lb	650 ft-lb	18900
1.2500	7	1155 ft-lb	865 ft-lb	24000
1.3750	6	1467 ft-lb	1100 ft-lb	28575
1.5000	6	1667 ft-lb	1250 ft-lb	34800

* also applies to plated bolts
Based on IFI 5th Edition Technical Data.
torque coefficient, K=0.20 nonplated steel fasteners
torque coefficient, K=0.15 plated steel fasteners
These figures are given for estimated torque as a function of friction, not bolt tension.
Since there are a number of variables that directly or indirectly affect friction (surface finish, type of coating or lubricant, speed of tightening, human error) the clamp loads may be +/- 25% of the listed values.

Recommended Torque Values for Nuts & Bolts

SAE Grade 2 Unified National Fine (UNF)

(55 kpsi proof strength - 69 kpsi tensile strength for sizes .250' to .500")
(52 kpsi proof strength - 64 kpsi tensile strength for sizes .500' to .750")
(28 kpsi proof strength - 55 kpsi tensile strength for sizes .750' to 1.500")

Bolt Size (in.)	Threads per inch	Torque (dry)	Torque (lubricated) (75% of dry torque)	Clamp Load (lbf +/- 25%)
0.1640	36	23 in-lb	17 in-lb	622
0.1900	32	35 in-lb	26 in-lb	852
0.2500	28	75 in-lb	56 in-lb	1500
0.3125	24	156 in-lb	108 in-lb	2400
0.3750	24	23 ft-lb	18 ft-lb	3600
0.4375	20	37 ft-lb	28 ft-lb	4913
0.5000	20	55 ft-lb	41 ft-lb	6600
0.5625	18	80 ft-lb	60 ft-lb	8400
0.6250	18	110 ft-lb	83 ft-lb	10575
0.7500	16	192 ft-lb	144 ft-lb	15375
0.8750	14	184 ft-lb	138 ft-lb	12600
1.0000	12	274 ft-lb	205 ft-lb	16425
1.0000	14	280 ft-lb	210 ft-lb	16800
1.1250	12	397 ft-lb	297 ft-lb	21150
1.2500	12	553 ft-lb	415 ft-lb	26550
1.3750	12	746 ft-lb	559 ft-lb	32550
1.5000	12	979 ft-lb	734 ft-lb	39150

* also applies to plated bolts
Based on IFI 5th Edition Technical Data.
torque coefficient, K=0.20 nonplated steel fasteners
torque coefficient, K=0.15 plated steel fasteners
These figures are given for estimated torque as a function of friction, not bolt tension.
Since there are a number of variables that directly or indirectly affect friction (surface finish, type of coating or lubricant, speed of tightening, human error) the clamp loads may be +/- 25% of the listed values.

Recommended Torque Values for Nuts & Bolts

SAE Grade 5 Unified National Coarse (UNC)

(85 kpsi proof strength - 120 kpsi tensile strength for sizes .250' to .750")
(78 kpsi proof strength - 115 kpsi tensile strength for sizes .750' to 1.000")
(74 kpsi proof strength - 105.5 kpsi tensile strength for sizes 1.000' to 1.500")

Bolt Size (in.)	Threads per inch	Torque (dry)	Torque (lubricated)* (75% of dry torque)	Clamp Load (lbf +/- 25%)
0.1640	32	52 in-lb	40 in-lb	890
0.1900	24	73 in-lb	55 in-lb	1114
0.2500	20	96 in-lb	76 in-lb	2025
0.3125	18	204 in-lb	156 in-lb	3338
0.3750	16	31 ft-lb	23 ft-lb	4950
0.4375	14	50 ft-lb	37 ft-lb	6788
0.5000	13	76 ft-lb	57 ft-lb	9075
0.5625	12	109 ft-lb	82 ft-lb	11625
0.6250	11	150 ft-lb	112 ft-lb	14400
0.7500	10	266 ft-lb	200 ft-lb	21300
0.8750	9	430 ft-lb	322 ft-lb	29475
1.0000	8	644 ft-lb	483 ft-lb	38625
1.1250	7	794 ft-lb	596 ft-lb	42375
1.2500	7	1120 ft-lb	840 ft-lb	53775
1.3750	6	1470 ft-lb	1100 ft-lb	64125
1.5000	6	1950 ft-lb	1462 ft-lb	78000

* also applies to plated bolts
Based on IFI 5th Edition Technical Data.
torque coefficient, K=0.20 nonplated steel fasteners
torque coefficient, K=0.15 plated steel fasteners
These figures are given for estimated torque as a function of friction, not bolt tension.
Since there are a number of variables that directly or indirectly affect friction (surface finish, type of coating or lubricant, speed of tightening, human error) the clamp loads may be +/- 25% of the listed values.

Recommended Torque Values for Nuts & Bolts

SAE Grade 5 Unified National Fine (UNF)

(85 kpsi proof strength - 120 kpsi tensile strength for sizes .250' to .750")
(78 kpsi proof strength - 115 kpsi tensile strength for sizes .750' to 1.000")
(74 kpsi proof strength - 105.5 kpsi tensile strength for sizes 1.000' to 1.500")

Bolt Size (in.)	Threads per inch	Torque (dry)	Torque (lubricated) (75% of dry torque)	Clamp Load (lbf +/- 25%)
0.1640	36	53 in-lb	40 in-lb	1023
0.1900	32	66 in-lb	49 in-lb	1279
0.2500	28	120 in-lb	87 in-lb	2325
0.3125	24	228 in-lb	168 in-lb	3675
0.3750	24	35 ft-lb	26 ft-lb	5588
0.4375	20	55 ft-lb	41 ft-lb	7575
0.5000	20	85 ft-lb	64 ft-lb	10200
0.5625	18	122 ft-lb	91 ft-lb	12975
0.6250	18	170 ft-lb	128 ft-lb	16350
0.7500	16	297 ft-lb	223 ft-lb	23775
0.8750	14	474 ft-lb	355 ft-lb	32475
1.0000	12	705 ft-lb	529 ft-lb	42300
1.0000	14	721 ft-lb	541 ft-lb	32275
1.1250	12	890 ft-lb	668 ft-lb	47475
1.2500	12	1241 ft-lb	930 ft-lb	59550
1.3750	12	1672 ft-lb	1254 ft-lb	72975
1.5000	12	2194 ft-lb	1645 ft-lb	87750

* also applies to plated bolts
Based on IFI 5th Edition Technical Data.
torque coefficient, K=0.20 nonplated steel fasteners
torque coefficient, K=0.15 plated steel fasteners
These figures are given for estimated torque as a function of friction, not bolt tension.
Since there are a number of variables that directly or indirectly affect friction (surface finish, type of coating or lubricant, speed of tightening, human error) the clamp loads may be +/- 25% of the listed values.

Recommended Torque Values for Nuts & Bolts

SAE Grade 8 Unified National Coarse (UNC)

(120 kpsi proof strength - 150 kpsi tensile strength for sizes .250' to 1.500")

Bolt Size (in.)	Threads per inch	Torque (dry)	Torque (lubricated) (75% of dry torque)	Clamp Load (lbf +/- 25%)
0.1380	32	22 in-lb	16 in-lb	815
0.1640	32	41 in-lb	31 in-lb	1254
0.1900	24	59 in-lb	44 in-lb	1508
0.2500	20	143 in-lb	107 in-lb	2850
0.3125	18	24 ft-lb	18 ft-lb	4725
0.3750	16	43 ft-lb	32 ft-lb	6975
0.4375	14	69 ft-lb	52 ft-lb	9600
0.5000	13	106 ft-lb	79 ft-lb	12750
0.5625	12	153 ft-lb	115 ft-lb	16350
0.6250	11	211 ft-lb	158 ft-lb	20325
0.7500	10	376 ft-lb	282 ft-lb	30075
0.8750	9	606 ft-lb	454 ft-lb	41550
1.0000	8	908 ft-lb	681 ft-lb	54525
1.1250	7	1288 ft-lb	966 ft-lb	68700
1.2500	7	1817 ft-lb	1362 ft-lb	87225
1.3750	6	2381 ft-lb	1786 ft-lb	103950
1.5000	6	3161 ft-lb	2371 ft-lb	126450

* also applies to plated bolts
Based on IFI 5th Edition Technical Data.
torque coefficient, K=0.20 nonplated steel fasteners
torque coefficient, K=0.15 plated steel fasteners
These figures are given for estimated torque as a function of friction, not bolt tension.
Since there are a number of variables that directly or indirectly affect friction (surface finish, type of coating or lubricant, speed of tightening, human error) the clamp loads may be +/- 25% of the listed values.

Recommended Torque Values for Nuts & Bolts

SAE Grade 8 Unified National Fine (UNF)

(120 kpsi proof strength - 150 kpsi tensile strength for sizes .250' to 1.500")

Bolt Size (in.)	Threads per inch	Torque (dry)	Torque (lubricated) (75% of dry torque)	Clamp Load (lbf +/- 25%)
0.1380	40	25 in-lb	18 in-lb	910
0.1640	36	43 in-lb	32 in-lb	1321
0.1900	32	68 in-lb	51 in-lb	1792
0.2500	28	163 in-lb	122 in-lb	3263
0.3125	24	27 ft-lb	20 ft-lb	5113
0.3750	24	49 ft-lb	37 ft-lb	7875
0.4375	20	77 ft-lb	58 ft-lb	10650
0.5000	20	119 ft-lb	89 ft-lb	14400
0.5625	18	171 ft-lb	128 ft-lb	18300
0.6250	18	239 ft-lb	179 ft-lb	23025
0.7500	16	419 ft-lb	314 ft-lb	33600
0.8750	14	668 ft-lb	501 ft-lb	45825
1.0000	12	994 ft-lb	745 ft-lb	59700
1.0000	14	1019 ft-lb	764 ft-lb	61125
1.1250	12	1444 ft-lb	1083 ft-lb	77025
1.2500	12	2011 ft-lb	1508 ft-lb	96600
1.3750	12	2711 ft-lb	2033 ft-lb	118350
1.5000	12	3557 ft-lb	2667 ft-lb	142275

* also applies to plated bolts
Based on IFI 5th Edition Technical Data.
torque coefficient, K=0.20 nonplated steel fasteners
torque coefficient, K=0.15 plated steel fasteners
These figures are given for estimated torque as a function of friction, not bolt tension.
Since there are a number of variables that directly or indirectly affect friction (surface finish, type of coating or lubricant, speed of tightening, human error) the clamp loads may be +/- 25% of the listed values.

Shaft Alignment and related patents filed in the United States

US Patent #	Held by :	Month	Year	Description
39608	Williams	August	1863	shaft centerer
283627	John Logan	August	1883	improvement to gauges (dial indicator)
431054	Henderson	July	1890	pipe clamping device
458055	George Hunt	December	1890	gage for aligning engines
487427	Poole	December	1892	level hanger for shafting
521306	Humphrey Campbell	June	1893	Shaft setting device
541754	Isgrig	June	1895	line shaft alignment
575857	Sly	January	1897	lantern bracket
651024	Thomas	June	1900	aligning & leveling instrument for shafting
679591	James Barns	March	1901	centering device for lining up engines
685288	Miller	October	1901	indicator for lathes
685290	John C. Miller	March	1901	aligning work piece to CL of lathe spindle
685455	Kinkead	October	1901	instrument for hanging & lining up shafting
807085	Newton	December	1905	shaft aliner
868074	Ernest Clark	April	1906	shaft liner and leveler
958736	Edgar Ferris	August	1908	shaft aligning device
1221507	Buesse	April	1917	pipe clamp base
1231479	Blumer	June	1917	roll locating device
1295936	Spellman	March	1919	measuring instrumnet
1339384	Douglas	May	1920	gage
1351663	Koch	August	1920	measurement gage comparator
1477257	Lewis Fritz	July	1922	shaft alignment gauge
1505313	Alvah J. Colwell	June	1923	crank shaft throw parallel gauge
1516288	Frank Godfrey	April	1924	bevel protractor
1559230	William Eccles	January	1921	aligning meter for gears & shafts
1591485	Albert Guillet	August	1925	lining block & measure for leveling & lining machinery
1616084	Albert Guillet	April	1926	leveling & lining spinning frames
1799739	John Elering et. al.	November	1926	precision measuring device
1907959	Albert Guillet	April	1928	lining & leveling means for machinery
2395393	Arche Brilliantine	June	1944	electric alignment micrometer
2402567	Milner	October	1942	gauge for airfoils
2451720	Davis	October	1948	centering device
2461143	Clifford	February	1949	gaging device / shaft bracket
2499753	Hubbard	May	1946	chain tightening & securing device
2516854	Joseph Christian	January	1946	gauging apparatus for aligning shafts
2634939	Robert Voss	July	1949	shaft aligning mechanism
2638676	Callahan	May	1953	shaft alignment device
2656607	Harding	October	1953	aligning device (line shaft)
2726058	Luther Foltz	March	1954	shaft align bracket
2833051	Cunningham	May	1958	shaft align bracket
2929922	Townes & Schawlow		1958	maser
3176403	Meyer	April	1965	measuring tube (no sag)
3244392	Sheets	April	1966	shaft align bracket
3525158	Torlay	August	1970	shaft align bracket
3604121	Harold Hull	June	1968	roll alignment
3631604	Schenavar	January	1972	shaft align bracket
3664029	Glucoft, Westerfield	May	1972	shaft align bracket
3733706	Arthur Blohm	July	1970	machine aligning device
3783522	V. Ray Dodd	April	1972	method & apparatus for shaft alignment

3789507	Malcolm Murray	February	1972	machine element alignment system
3849857	Malcolm Murray	July	1973	machine element alignment positioner
4033042	Donald E. Bently	October	1974	shaft alignment apparatus & method
4053845	Gordon Gould	October	1977	laser
4060719	Daltnowski	July	1976	goemetric calculator
4102052	Heinz Bloch	December	1976	deflection indicator for couplings
4161068		July	1979	shaft align bracket
4161068	Thomas McMaster	November	1977	shaft align bracket
4161436	Gordon Gould	July	1979	laser
4215482	Richard Szewczyk	April	1978	workpiece centering device 4 lathe
4216587	Willice Stone	January	1979	shaft align bracket
4231161	Flavio Belfiore	April	1979	pulley alignment tool
4234924	LaVance et. al.	September	1978	baseleine measure for electronic positioning
4244111	James Heard	April	1979	shaft align bracket
4367594	Murray	January	1983	shaft align brackets
4428126	Burke Banks	December	1981	continous monitoring shaft positions
4439925	Brian Lock	March	1982	concentricity measuring
4447962	Joseph Grosberg	May	1982	adjustable bore target/gauge
4451992	Stephen Malak	September	1982	shaft positioning device/method
4463438	Zatezalo	July	1984	shaft alignment calculator
4502233	Gary Boitz et. al.	July	1983	shaft align apparatus & method
4516328	Massey	May	1985	shaft align brackets
4518855	Malak	May	1985	shaft alignment bracket
4586264	Zatazelo	November	1986	alignment methods
4623979	John Zatezalo et. al.	January	1984	shaft alignment calculator
4698491	Lysen	May	1984	laser alignment system
4704583	Gordon Gould	November	1987	laser
4746021	Gordon Gould	May	1988	laser
4790507	Brian Morrissey	May	1987	Tool for precise movement of machinery
4928401	Murray	May	1990	vernier strobe system
4964224	Lawrence Jackson	July	1989	shaft align bracket
4991965	Dieter Busch	May	1988	laser monitoring machinery movement
5026998	Roland Holzl (Germany)	April	1990	alignment methods
5056237	Paul R. Saunders	July	1990	electronic shaft alignment device
5185937	K. Piety & D.Nower			Alignment bracket
5263261	K. Piety & D.Nower	November	1993	shaft alignment device

Converting from one vibration unit to another

acceleration = 0.0000000142 x frequency^2 x displacement [0.0000000142 F^2 D]
acceleration = 0.00027 x velocity x frequency [0.00027 VF]
velocity = 0.00005236 x frequency x displacement [0.00005236 FD]
velocity = 3870 x (acceleration/frequency) [3870 (g/F)]
displacement = 19100 x (velocity / frequency) [19100 (V/F)]
displacement = 70400000 x (acceleration / (frequency)^2) [70400000 $(g/F)^2$]

note : frequency is expressed in cycles per minute CPM

Vibration Characteristic	**Common Units of Measurement**
Frequency	cycles per minute (CPM)
Displacement	mils peak to peak
Velocity	in/sec peak
Acceleration	G peak
Phase	degrees

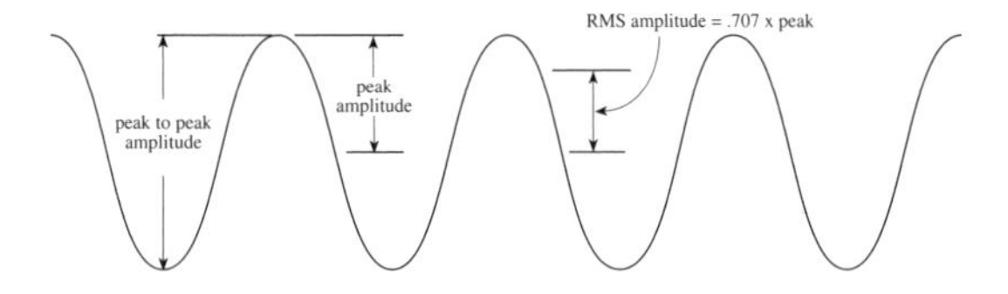

Index